The Faculties

OXFORD **PHILOSOPHICAL** CONCEPTS

OXFORD PHILOSOPHICAL CONCEPTS

Christia Mercer, Columbia University
Series Editor

PUBLISHED IN THE OXFORD PHILOSOPHICAL CONCEPTS SERIES

Efficient Causation
Edited by Tad Schmaltz

The Faculties
Edited by Dominik Perler

FORTHCOMING IN THE
OXFORD PHILOSOPHICAL CONCEPTS SERIES

Health
Edited by Peter Adamson

Memory
Edited by Dmitri Nikulin

Evil
Edited by Andrew Chignell

Self-Knowledge
Edited by Ursula Renz

Dignity
Edited by Remy Debes

Sympathy
Edited by Eric Schliesser

Animals
Edited by G. Fay Edwards and Peter Adamson

Pleasure
Edited by Lisa Shapiro

Space
Edited by Andrew Janiak

Consciousness
Edited by Alison Simmons

Eternity
Edited by Yitzhak Melamed

Moral Motivation
Edited by Iakovos Vasiliou

OXFORD PHILOSOPHICAL CONCEPTS

The Faculties

A HISTORY

Edited by Dominik Perler

OXFORD
UNIVERSITY PRESS

Oxford University Press is a department of the
University of Oxford. It furthers the University's objective
of excellence in research, scholarship, and education
by publishing worldwide.

Oxford   New York
Auckland   Cape Town   Dar es Salaam   Hong Kong   Karachi
Kuala Lumpur   Madrid   Melbourne   Mexico City   Nairobi
New Delhi   Shanghai   Taipei   Toronto

With offices in
Argentina   Austria   Brazil   Chile   Czech Republic   France   Greece
Guatemala   Hungary   Italy   Japan   Poland   Portugal   Singapore
South Korea   Switzerland   Thailand   Turkey   Ukraine   Vietnam

Oxford is a registered trade mark of Oxford University Press
in the UK and certain other countries.

Published in the United States of America by
Oxford University Press
198 Madison Avenue, New York, NY 10016

© Oxford University Press 2015

All rights reserved. No part of this publication may be reproduced,
stored in a retrieval system, or transmitted, in any form or by any means,
without the prior permission in writing of Oxford University Press,
or as expressly permitted by law, by license, or under terms agreed with
the appropriate reproduction rights organization. Inquiries concerning
reproduction outside the scope of the above should be sent to the
Rights Department, Oxford University Press, at the address above.

You must not circulate this work in any other form
and you must impose this same condition on any acquirer.

Library of Congress Cataloging-in-Publication Data
The faculties : a history / edited by Dominik Perler.
pages   cm.—(Oxford philosophical concepts series)
Includes bibliographical references.
ISBN 978-0-19-993525-3 (hardcover : alk. paper)
ISBN 978-0-19-993527-7 (pbk. : alk. paper)
1. Ability—Philosophy—History.   2. Personality—Philosophy—History.
3. Intellect—Philosophy—History.   I. Perler, Dominik.
BF431.F256   2015
128'.3—dc23   2014041237

1 3 5 7 9 8 6 4 2
Printed in the United States of America
on acid-free paper

Contents

CONTRIBUTORS vii

SERIES EDITOR'S FOREWORD xi

Introduction 3
DOMINIK PERLER

1 Faculties in Ancient Philosophy 19
KLAUS CORCILIUS

Reflection Faculties and Self-Debate 59
HELENE P. FOLEY

2 Faculties in Arabic Philosophy 66
TANELI KUKKONEN

3 Faculties in Medieval Philosophy 97
DOMINIK PERLER

Reflection Faculties and Imagination 140
VERENA OLEJNICZAK LOBSIEN

4 Faculties in Early Modern Philosophy 150
STEPHAN SCHMID

5 Faculties in Kant and German Idealism 198
JOHANNES HAAG

Reflection Faculties and Phrenology 247
REBEKKA HUFENDIEK AND MARKUS WILD

6 Faculties and Modularity 254
REBEKKA HUFENDIEK AND MARKUS WILD

Reflection Faculties and Neuroenhancement 299
SASKIA K. NAGEL

BIBLIOGRAPHY 309

INDEX OF NAMES 331

INDEX OF CONCEPTS 337

Contributors

KLAUS CORCILIUS is Associate Professor of Philosophy at the University of California at Berkeley. He is the author of *Streben und Bewegen: Aristoteles' Theorie der animalischen Ortsbewegung* (2008) and of several articles on Aristotle's theory of desire, animal motion, and human action. Recent publications include a volume coedited with Dominik Perler, *Partitioning the Soul: Debates from Plato to Leibniz* (2014).

HELENE P. FOLEY is Professor of Classics, Barnard College, Columbia University. She is the author of books and articles on Greek epic and drama, on women and gender in antiquity, and on modern performance and adaptation of Greek drama. Author of *Ritual Irony: Poetry and Sacrifice in Euripides* (1985), *The Homeric Hymn to Demeter* (1994), *Female Acts in Greek Tragedy* (2001), *Reimagining Greek Tragedy on the American Stage* (2012), and *Euripides: Hecuba* (2015); coauthor of *Women in the Classical World: Image and Text* (1994). She edited *Reflections of Women in Antiquity* (1981) and coedited *Visualizing the Tragic: Drama, Myth and Ritual in Greek Art and Literature* (2007) and *Antigone on the Contemporary World Stage* (2011).

JOHANNES HAAG is Professor of Theoretical Philosophy at the University of Potsdam. He has published on early modern philosophy as well as contemporary theories of perception and the foundations of intentionality. He is the author of *Der Blick nach innen. Wahrnehmung und Introspektion* (2001) and *Erfahrung und Gegenstand. Das Verhältnis von Sinnlichkeit und Verstand* (2007), and coeditor of *Ideen. Repräsentationalismus in der Frühen Neuzeit* (2010) and *Übergänge—diskursiv oder intuitiv?* (2013).

REBEKKA HUFENDIEK is a postdoctoral researcher at the University of Basel. Her main research interests are in the area of the philosophy of mind, especially embodied cognition, naturalism, and emotion theories. In her dissertation, she investigated embodied emotions. She coedited a volume on embodied cognition, *Philosophie der Verkörperung. Grundlagentexte zu einer aktuellen Debatte* (with Joerg Fingerhut and Markus Wild, 2013).

TANELI KUKKONEN has held appointments at the Universities of Victoria, Jyväskylä, and Otago and New York University Abu Dhabi. He specializes in classical Arabic philosophy and the Aristotelian tradition, especially in the fields of cosmology and philosophy of mind. He is the author of *Ibn Tufayl* (2014) as well as over thirty essays on topics in the Aristotelian and Platonic philosophical traditions.

VERENA OLEJNICZAK LOBSIEN is Professor of English Literature at the Humboldt-Universität zu Berlin. Her major research interests are early modern English literature and culture with a focus on transformations of antiquity and their aesthetic potential. She is the author of *Subjektivität als Dialog* (1994), *Skeptische Phantasie* (1999), *Transparency and Dissimulation: Configurations of Neoplatonism in Early Modern English Literature* (2010), *Jenseitsästhetik: Literarische Räume letzter Dinge* (2012), and, with Eckhard Lobsien, coauthor of *Die unsichtbare Imagination* (2003).

SASKIA K. NAGEL is Assistant Professor at the University of Twente. Her background is in cognitive science and in philosophy. She is interested in the anthropological, ethical, and social dimensions of scientific and technological progress and in the role of technologies for human self-understanding. Her recent work focuses on questions about self-determination throughout the lifespan. She seeks to understand public attitudes toward scientific developments. She is author of *Ethics and the Neurosciences* (2010) and has published various articles on ethical questions related to neuroscientific progress, in particular on neuroenhancement and questions of autonomy.

DOMINIK PERLER is Professor of Philosophy at the Humboldt-Universität zu Berlin. His research focuses on medieval and early modern philosophy, mostly in the areas of metaphysics, epistemology, and philosophy of mind. He is the author of *Theorien der Intentionalität im Mittelalter* (2002), *Zweifel und Gewissheit. Skeptische Debatten im Mittelalter* (2006), *Transformationen der Gefühle*.

Philosophische Emotionstheorien 1270–1670 (2011), coauthor of *Occasionalismus. Theorien der Kausalität im arabisch-islamischen und im europäischen Denken* (2000), editor of *Ancient and Medieval Theories of Intentionality* (2001), and coeditor of *Selbstbezug und Selbstwissen. Texte zu einer mittelalterlichen Debatte* (2014) and *Partitioning the Soul. Debates from Plato to Leibniz* (2014).

STEPHAN SCHMID is Wissenschaftlicher Mitarbeiter at the Humboldt-Universität zu Berlin, where he works on early modern and medieval philosophy as well as on contemporary analytic philosophy, focusing mainly on metaphysics (modality, causality, ontology) and philosophy of mind (intentionality). He is the author of *Finalursachen in der frühen Neuzeit* (2011) and coeditor of *Final Causes and Teleologial Explanations* (2011) and *Dispositionen. Texte aus der zeitgenössischen Debatte* (2014) and has published various articles on Aquinas, Suárez, and Spinoza.

MARKUS WILD is Professor of Theoretical Philosophy at the University of Basel. His research focuses mainly on early modern philosophy, philosophy of mind, and naturalism. He has worked on Montaigne, Descartes, Hume, animal minds, mental representations, consciousness, and teleosemantics. He is the author of *Die anthropologische Differenz* (2006) and coeditor of *Animal Mind and Animal Ethics* (with Klaus Petrus, 2013).

Series Editor's Foreword

Oxford Philosophical Concepts (OPC) offers an innovative approach to philosophy's past and its relation to other disciplines. As a series, it is unique in exploring the transformations of philosophy's central concepts from their ancient sources to their modern use.

OPC has several goals: to make it easier for historians of philosophy to contextualize key concepts in the history of philosophy, to render that history accessible to a wide audience, and to enliven contemporary philosophy by displaying the rich and varied sources of concepts still in use today. The means to these goals are simple enough: eminent historians of philosophy come together to rethink a central concept in philosophy's past. The point of this rethinking is not to offer a broad overview, but to identify problems the concept was originally supposed to solve and investigate how approaches to those problems shifted over time, sometimes radically. Each OPC volume is *a history* of its concept in that it tells a story about changing solutions to specific philosophical concerns.

Recent scholarship has made evident the benefits of reexamining the standard narratives about the history of western philosophy. OPC's editors look beyond the canon and explore their concepts over a wide philosophical landscape. Each volume traces a concept from its inception as a solution to specific problems through its historical transformations to its modern use, all the while acknowledging its historical context.

Many editors have found it appropriate to include long-ignored writings drawn from the Islamic and Judaic traditions and the philosophical contributions of women. Volumes also explore ideas drawn from Buddhist, Chinese, Indian, and other philosophical cultures when doing so adds an especially helpful new perspective. By combining scholarly innovation with focused and astute analysis, OPC encourages a deeper understanding of our philosophical past and present.

One of the most innovative features of *Oxford Philosophical Concepts* is its recognition that philosophy bears a rich relation to art, music, literature, religion, science, and other cultural practices. The series speaks to the need for informed interdisciplinary exchanges. Its editors assume that the most difficult and profound philosophical ideas can be made comprehensible to a large audience and that materials that are not strictly philosophical often bear a significant relevance to philosophy. To this end, each OPC volume includes "Reflections." These are short, stand-alone essays written by specialists in art, music, literature, theology, science, or cultural studies that *reflect on* the concept from other disciplinary perspectives. The goal of these essays is to enliven, enrich, and exemplify the volume's concept and reconsider the boundary between philosophical and extraphilosophical materials. OPC's Reflections display the benefits of using philosophical concepts and distinctions in areas that are not strictly philosophical, and encourage philosophers to move beyond the borders of their discipline as presently conceived.

The volumes of OPC arrive at an auspicious moment. Many philosophers are keen to invigorate the discipline. OPC aims to provoke philosophical imaginations by uncovering the brilliant twists and unforeseen turns of philosophy's past.

Christia Mercer
Gustave M. Berne Professor of Philosophy
Columbia University in the City of New York
January 2015

The Faculties

Introduction

Dominik Perler

1. Faculties and Their Explanatory Function

In daily life it seems quite natural to explain the activities of nonhuman as well as human animals by referring to their special capacities or abilities. In some situations we also tend to talk about their faculties. Thus, we say that dogs can see and smell things in their environment because they have perceptual faculties. Or we claim that human beings can grasp thoughts and make decisions because they have rational faculties. Sometimes we even try to distinguish different types of animals by focusing on their distinctive faculties. We may say, for instance, that cats are utterly different from dogs because they have a unique visual faculty that enables them to see things from a long distance—something dogs could never do. And we often affirm that human beings are a special type of animal because they have unique rational faculties that enable them to do sophisticated things such as learning languages and composing music. The better we describe its distinctive faculties, the better we can characterize a certain type of animal. This is especially true for human beings, a rather complex type of animal. It hardly suffices to make the trivial statement that they are rational animals because they are endowed with rational faculties. We need to spell out what these faculties consist in, for instance by describing in detail those that are responsible for concept formation, speech, and perhaps even music composition. The better we know what these faculties are, what kind of activities they make possible, and how they interact, the better we can give an account of the so-called rational nature of human beings.

Given the crucial explanatory function of faculties, it is not surprising that they have a long history. Since antiquity philosophers have been studying faculties, trying to understand what they are, how many there are in a human being or in a nonhuman animal, how they enable it to have a wide range of activities, and how they all work together. The general idea is that exploring the special nature of a given animal amounts to analyzing its special faculties. This is most evident in Aristotle and his followers up to the early modern period. In his *De anima*, the key text for the development of faculty theories, Aristotle famously stated that we cannot give an account of a living being unless we appeal to its soul. He then explained what it amounts to have a soul by referring to a set of faculties, and he distinguished different sets for different types of living beings.[1] On his view, very basic faculties—those responsible for nourishment and growth—make a very basic form of life possible, namely that of plants. More complex faculties—those responsible for local movement and perception—make a more complex form possible, namely that of animals like cats and dogs. And still more complex faculties—those responsible for thinking—make the most complex form possible, namely that of human beings. The important point is that Aristotle established a hierarchy of faculties and thereby introduced a principle of classification: the enumeration and ordering of different types of faculties makes it possible to distinguish different types of living beings. To be sure, this is first and fore-

---

[1] See *De anima* II.2 (413b11–15), and the analysis in chapter 1, section 4 here.

most a biological principle. When employing this principle, Aristotle was as much interested in trees, worms, and dogs as in human beings, and he attempted to establish a biological order for all of them. To do so, he looked at basic organic faculties as well as at higher cognitive faculties: what enables plants to grow is as important as what enables human beings to think. It would therefore be inappropriate to understand Aristotle's theory of faculties as aiming exclusively at cognitive or even rational faculties. It was designed to explain faculties on all levels in all kinds of living beings. But it paved the way for a theory of rational faculties. For as soon as human beings are characterized as having rational faculties on top of the other faculties, a number of questions inevitably arise. What are the rational faculties? How are they related to the lower faculties? What kind of activities do they produce? And why do they make human beings so special? A long tradition of Greek, Arabic, and Latin authors focused on these questions, thus making faculties the cornerstone of their theories of human nature.

However, it was not just human nature that attracted the interest of many Aristotelians. The nature of nonhuman animals was equally puzzling and provoked extensive discussion. This is not surprising, for it is hardly satisfactory to simply draw a line between human animals that are endowed with rational faculties and nonhuman animals that lack them. Clearly, complex animals like cats and dogs are not only capable of perceiving things, they are also capable of engaging in many more cognitive activities: they can distinguish different types of things, remember them, use them for specific purposes, make plans, and so on. Hence, they must have some "quasi-rational faculties" that make all these activities possible. Arabic philosophers were fully aware of this fact and therefore focused on the faculties that are responsible for memory, purposeful behavior, and many more sophisticated activities.[2] It is only an examination of these faculties, they realized, that makes it possible to distinguish higher animals like dogs and cats from

---

[2] Avicenna's distinguishing five internal senses—and hence five nonrational yet cognitive faculties—had a strong impact on later discussions; see chapter 2, section 3.

lower ones like worms. And when we enumerate and describe these faculties, we come to understand that it is not rationality alone that is important to human nature. We share many faculties with nonhuman animals—faculties that are cognitive but not rational.

Yet it was not only the Aristotelian tradition that provided detailed analysis of faculties. In the seventeenth century, when Aristotelianism came increasingly under attack, the traditional metaphysical account of faculties as the core of the soul was widely rejected, but the leading idea that we cannot describe and classify different types of living beings unless we look at their faculties was not abandoned. In particular, the thesis that human beings are special animals because they have special higher faculties was still maintained. Thus, René Descartes clearly denied that there are any lower (i.e. vegetative and sensory) faculties, as the scholastic Aristotelians had assumed. On his view, what makes nourishment in plants or perception in animals possible is nothing but an arrangement of their material parts—there is no need to appeal to faculties. Nevertheless, Descartes still subscribed to the thesis that human beings have two special faculties, namely intellect and will, and he still defended the idea that we cannot give an account of the "rational nature" of human beings if we fail to analyze these faculties.[3] It is therefore quite understandable that he dealt with similar problems as his predecessors. How do the rational faculties work, how do they interact, and what kind of activities do they make possible?

In the eighteenth century, Immanuel Kant also posed questions along this line. While showing no interest in a metaphysical account of faculties that would locate them in a soul or even in an immaterial substance, he still maintained the thesis that it is of crucial importance to examine what kind of faculties human beings have and how they all fit together. This is indispensable, Kant assumed, because we cannot explain what special activities—in particular what epistemic activities—we can engage in unless we look at the special faculties that make them possible.

---

[3] See Descartes, *Meditations* IV (AT VII, 56, CSM II, 39), and the analysis in chapter 4, section 2.

Otherwise we would simply accept the existence of these activities as a brute fact. Moreover, we cannot understand what kinds of activities we are *not* able to bring about unless we also reflect upon what faculties we lack. That is why we need to analyze faculties "from within" as well as "from without."[4] That is, on the one hand we should look at our faculties by starting with a detailed analysis of the activities we detect in ourselves; on the other hand we should also look at our faculties by exploring our limits and by asking what would be required for activities that go beyond these limits. In any case, it is only an exploration of faculties that enables us to have an understanding of the specific domain of our own activities.

Later authors tended to be more skeptical about the possibility of saying something illuminating about faculties we lack. But many philosophers were and still are convinced that we should investigate those we do have. Rationalists up to the present do not hesitate to speak about "our faculties" and aim at giving a detailed account of their structure and interaction. Noam Chomsky conceives of them as "mental organs" that must not be missing in a theory of the human mind,[5] and Jerry Fodor, who vigorously reintroduced faculties into contemporary philosophy of mind and psychology, calls them "modules" of the mind.[6] We cannot give a satisfactory explanation of the human mind, he holds, unless we spell out what these modules are, what specific features they have, and how they fit together. Or, in short: analyzing the mind amounts to exploring its modular structure.

This revival of faculty theories shows that the concept of faculty is not exclusively linked to a specific theoretical program, say to the Aristotelian theory of the soul. Nor is it bound to a specific historical period, say to the premodern era when Aristotelianism was dominating. It is rather a concept that is used in different contexts, albeit in different ways, and that serves to tackle a simple but intriguing

---

4 On this double perspective, see chapter 5, section 1.
5 Noam Chomsky, "Rules and Representations," *Behavioral and Brain Sciences* 3 (1980): 2–3.
6 Jerry Fodor, *The Modularity of Mind: An Essay on Faculty Psychology* (Cambridge, MA: MIT Press, 1983), and the analysis in chapter 6, section 3.

problem: on the one hand we take a human mind (or an entire living being) to be a single, unified thing that can engage in a number of activities; on the other hand we see that these activities are often quite diverse and that each of them has its own causal mechanism. How can there be diverse activities in a single, unified thing? Clearly, a human mind (or an entire living being) is not just an aggregate of distinct things so that each type of activity could be assigned to a different thing. Being unified, it is *one* thing and hence *one* agent. So how can one thing act in many ways? A natural response is: this is possible because the unified thing has an internal complexity. It has a number of powers, capacities, or faculties that enable it to produce a wide range of activities. So, when we speak about an animal that can breathe, move its legs, and perceive objects in its environment, we talk about one and the same thing that can do all these things in virtue of its faculties. Or when we speak about the human mind that can form concepts, utter sentences, and compose music, we also talk about one and the same thing that can do all these things in virtue of its rational faculties. An appeal to faculties enables us to resolve the "one and many problem": one thing can bring about diverse activities in virtue of its faculties.[7]

Of course, this answer is far from being complete. It does not explain what faculties really are. Are they just powers? And if so, what exactly are powers? Nor does the answer elucidate what the "in virtue of" relation is supposed to be. How can a thing produce an activity in virtue of a faculty? Does it simply use the faculty? And does it use it by activating it, or does something external activate it? Moreover, it is not clear how there can be a coordination or even a hierarchical order of all the faculties so that one and the same thing can bring about a well-ordered series of activities. How can a human being use her rational faculty on the basis of her perceptual faculty and thereby engage in or even produce a thought

---

7 Note that this problem is different from the "one *in* many problem" that is at stake in debates about universals. The problem here is how an *individual* thing can bring about different *individual* activities. Whether or not there are universal faculties instantiated in many individuals is a different problem that will not be examined here.

that is based on, say, a visual experience? No doubt, all these questions (and many more) need to be examined. But the fact that faculties are invoked in order to resolve the "one and many problem" shows that they play an important explanatory role. They make it understandable that a human mind (or an entire living being) is not just a black box that miraculously produces different activities—there is a *reason* for this richness. It is therefore not surprising that faculties play a key role in cognitive theories as well as in biological theories, ranging from Aristotle to contemporary evolutionary anthropology. And when we find different conceptual frameworks, we see different attempts to spell out what faculties really are and in what sense they are responsible for the perplexing diversity of activities in a human mind or in an entire living being.

2. Philosophical Puzzles and Problems

Clear and elegant as the appeal to faculties seems to be, it is not as innocent as it looks at first glance. There is a simple but fundamental question one might ask. Do we really explain anything when we refer to faculties? The ancient philosopher and physician Galen already asked this question and gave a negative answer. We do not provide any satisfactory explanation of a human mind or of a living being, he claimed, for we do not make clear what makes a certain activity possible. For instance, when talking about a visual faculty we do not explain what makes acts of seeing possible. To do that, we would have to spell out a number of optical and physiological processes and we would have to specify the cause—or perhaps even a network of causes—of all these processes. When talking about a visual faculty we simply try to conceal that we have no idea what the relevant cause is. Or as Galen put it, "so long as we are ignorant of the true essence of the cause which is operating, we call it a *faculty*."[8] The word "faculty" is therefore

---

[8] Claudius of Pergamon Galen, *On the Natural Faculties*, with an English translation by Arthur J. Brock (Cambridge, MA: Harvard University Press, 1916), I.iv, also quoted and discussed in chapter 1, section 5.

nothing but a placeholder: it refers to something we do not know and cannot name. It is even a deceptive placeholder, for, instead of searching for the real cause and describing it on the basis of empirical investigations, we use this word as if it were referring to some inner thing—as if there were an inner "seer" that is responsible for acts of seeing. We thereby reify the faculty, that is, we turn it into a real thing that is said to act inside us. And when we then talk about a plurality of faculties we treat a human being or some other animal as if it were populated with many inner things that produce many activities.

This critique was repeated by many early modern authors and became the standard objection against faculty theories.[9] These theories, it was said, are not only explanatorily vacuous because they do not indicate the real cause of activities; they are also ill-conceived because they introduce inner entities that are supposed to do some causal work. It is therefore hardly surprising that faculty theories were often presented as metaphysically misguided or even utterly false theories. In the twentieth century, Gilbert Ryle famously claimed that anyone referring to intellect and will as mental faculties producing inner activities commits a "category mistake": a mere disposition is mistakenly taken to be an inner thing that does some work.[10] If the entire human mind is then characterized as a bundle of inner things, we end up with a "ghost in the machine" that mysteriously acts inside us. According to Ryle, we cannot escape from this misleading picture unless we avoid all talk of faculties doing or producing things and start with a rigorous dispositional analysis of our activities.

In light of this critique it becomes clear that the concept of faculty is a rather controversial concept. It can hardly be regarded as a natural concept that anyone who wants to explain the multiplicity of activities in a human mind or in an entire living being appeals to. When being used, it gives rise to a number of serious questions and needs to be clar-

---

9 See, for instance, Locke, *An Essay concerning Human Understanding* II, ch. 21, sec. 20; Hobbes, *Leviathan* IV, ch. 46; and the discussion in chapter 3, section 1.
10 See Gilbert Ryle, *The Concept of Mind* (Chicago: University of Chicago Press, 1949), 18–23.

ified and justified within a larger theoretical framework. At least five bundles of problems ought to be taken into account.

The categorization problem. The most basic problem is a metaphysical one. To which category of things or entities do faculties belong? No doubt, one could immediately respond that they are powers, capacities, or abilities. And in fact, this response was chosen by many advocates of faculties, especially by Greek, Arabic, and Latin authors who mostly used the word "power" (*dynamis*, *quwwa*, or *potentia*) when talking about faculties. It is obvious, however, that this simple answer leads to the further question of what kind of powers faculties are. Are they on a par with powers like the burning power of fire and the heating power of the sun? Or do they constitute a special type of power, perhaps one that has its own conditions of actualization? Is it even necessary to distinguish different kinds of powers when speaking about different types of faculties? For instance, should the nutritive faculty as a "lower" power be distinguished from the perceptual faculty as a "higher" power, and should special conditions of actualization be indicated for each kind of power? Should we even distinguish natural powers that are necessarily actualized when triggered from nonnatural powers (e.g. the will) that are freely used and therefore not necessarily actualized? There is an even more fundamental problem. It is not clear what identifying faculties with powers amounts to. What is a power? Is it a special feature of a thing? If so, is it an essential or an accidental feature? And how does it relate to other features? Or should we avoid talking about a feature and rather speak about the structure of a thing? If so, what does referring to a power structure amount to?

The individuation problem. Suppose that we have agreed on a metaphysical classification of faculties. How can we then individuate faculties? How can we pick out a faculty and distinguish it from other ones? What would be the relevant criteria? Here, again, there is a quick answer at hand. We can identify different types of activities, which are clearly manifestations of faculties, and then establish a corresponding list of faculties. Thus, we can say that there must be a visual faculty that makes seeing possible, a conceptual faculty that makes concept formation possible, and so on. However, this method leads to a proliferation of faculties: there will be as many faculties as activities. For instance, we need to postulate a faculty for composing music that differs from the faculty for thinking, or even a faculty for composing classical music that differs from the faculty for composing jazz music. How can we limit the number of faculties and assign a single faculty to a range of activities? Or must limiting the number of faculties elude us? Is, then, the thought that there is a clearly defined set of faculties for a certain type of animal an illusion? Or can we individuate some faculties—perhaps even the basic faculties—that are constitutive of a human or a nonhuman animal? It seems necessary to define such faculties, for otherwise we could not classify animals and distinguish different types.

The ordering problem. Let us be optimistic and assume that we can indeed individuate a restricted number of faculties. Can we then establish an order for all of them? Can we distinguish "lower" from "higher" faculties, as Aristotelians and Cartesians assumed, and reach a hierarchical order? Of course we can, one might immediately respond, because it is clear that, say, the nutritive faculty is lower than the perceptual faculty. But why is it lower? Simply because of the activity it produces? And what are the relevant criteria for ranking a given activity as lower than another one? Moreover, it is unclear how the so-called higher cognitive faculties should be treated. Can they also be hierarchically ordered? Are we entitled to say that the visual faculty is inferior to the faculty that is responsible for judging and reasoning? Or does it not make sense to speak about a vertical order, and should we rather look for a horizontal order, that is, for a network of functionally definable faculties that constitute together a complex cognitive system? Perhaps it is even necessary to have two kinds of order, a vertical one for various levels of organic faculties and a horizontal one for the interconnectedness of the cognitive faculties.[11] In any case, both the criteria for establishing an order and the order itself need to be spelled out.

The localization problem. Once we have established a list of faculties and even found some order among them, it is quite natural to ask where they can be found. It is again tempting to give a simple answer: we can look at the place in the body where a certain activity occurs and then locate the corresponding faculty in that place. Thus, we can say that the visual faculty is located in the eyes because it is precisely in this organ that acts of seeing occur. It is evident, however, that this answer can hardly be satisfactory since it is far from clear that an activity can be located in a single organ. Does seeing really occur in the eyes and not in the brain? Or are both parts of the body involved, and is the faculty of seeing therefore a multilocal faculty? Perhaps one might respond that seeing occurs in the brain only and that the activity in the eyes is nothing but the necessary causal antecedent for seeing. Can we then localize the activity of seeing and the visual faculty in a special part of the brain? And can we subdivide the brain into many areas so that each faculty can be assigned to a special area, as some nineteenth-century physiologists thought?[12] Or is it misleading to compartmentalize the brain because it has such a high degree of plasticity that one and the same faculty can be located in different areas? This leads, of course, to the more general question if it is really possible to find a special place for each faculty. And if one goes a step further, questioning the materialist framework, one might even ask if it is really possible to find a material place for each faculty. Aren't there some faculties, for example the rational ones, that are not bound to a special place in the body or even free from all bodily constraints, as dualists from antiquity up to the present time have argued?

---

11 Perhaps there is even a vertical and a horizontal order among the cognitive faculties, as Jerry Fodor, *The Modularity of Mind: An Essay on Faculty Psychology* (Cambridge, MA: MIT Press, 1983), 21, suggests.
12 Most famously Gall and other phrenologists; see Reflection 3.

The functioning problem. Suppose that we have individuated the basic faculties in a living being and located at least some of them in specific parts of the body. How can we then explain their functioning, that is, the fact that they produce activities? As usual, we can start with a simple explanation and respond that their producing activities is nothing but a causal process. Consequently, an account of their functioning amounts to an account of this process. But what does that mean? Should we take the faculty to be a cause in the most common sense, namely as an efficient cause that brings about an effect? Should we say, for instance, that the visual faculty is a cause that produces seeing as its distinctive effect, whereas the rational faculty is another cause that produces thinking as its effect? If this is the case, then Galen's objection that faculties are simply unknown causes is near at hand, and facing the further objection that faculties are taken to be inner things that act is almost inevitable. A living being will then turn out to be a complex thing with a multitude of inner causes, and it will become questionable why it is really the entire living being that perceives, thinks, and composes music and not a conglomerate of distinct causes. To avoid this consequence one could reject the idea that faculties act like inner causes. But how then do they give rise to activities? Do they work like functionally defined systems, that is, like systems that yield a certain output when they receive a certain input? Do they even work like computational systems, at least as far as the cognitive faculties are concerned? And if so, in what sense can we then say that the entire living being uses these systems?

Needless to say, these questions only hint at some of the key problems that have been eagerly debated both by defenders and critics of faculty theories. The six chapters in this book examine them in detail, thus showing that the concept of faculty is rich and explanatorily fruitful, but also needs to be spelled out within a larger metaphysical and epistemological framework. Moreover, the chapters make clear that there is no such thing as a unified doctrine of faculties. Different historical authors tackled the problems in different ways and reached different conclusions, thereby working out different theories. Even philosophers working in the same tradition, say Arabic and Latin Aristotelians or twentieth-century cognitive scientists, presented rather different solutions to the key problems. Moreover, one should not forget that critics who pretended to reject faculties simply rejected a certain solution to a given problem. This is most evident in early modern philosophers who are well known for their harsh attacks on faculty psychology. A closer look at their texts reveals that they often criticized earlier solutions to the categorization problem, yet without giving up the idea that faculties are indispensable for cognitive processes.[13] In short, they rejected certain theories of faculties, not faculties altogether. It is therefore important not to neglect the negative approaches to faculty theories. They make clear that the controversy over faculties was mostly a debate about different metaphysical or epistemological models for explaining them.

What is so fascinating about doing history of philosophy is that we can discover different ways of conceiving of a certain phenomenon within different theoretical frameworks—there are no unchanging, perennial concepts. This is especially true for the concept of faculty. While acknowledging that we need this concept in order to make sense of the diversity of activities produced by a single living being, philosophers in different periods defined this concept in different ways. When criticizing a given concept they often created a new one; sometimes they also returned to an older one.[14] It would therefore be

---

13 To be sure, some authors (e.g. Spinoza and Hume) subscribed to eliminativism and thought that all cognitive processes could be accounted for without any appeal to faculties. Many others, however, still maintained the realist view that there are faculties and simply tried to replace earlier forms of realism (e.g. scholastic forms that characterized faculties as real qualities) with new forms. See chapter 4, sections 1 and 5.

14 The use of older concepts is especially evident in late twentieth-century authors who were inspired partly by Cartesian rationalism, partly by Gall's faculty psychology. See Jerry Fodor, *The Modularity of Mind: An Essay on Faculty Psychology* (Cambridge, MA: MIT Press, 1983), 14–23, and chapter 6, sections 2 and 3.

misleading to tell a simple developmental or even a teleological story about the history of the concept of faculty. That is why the chapters in this book do not offer a single "grand story." They describe the origin, the development, and the use of different concepts, and they attempt to show how problems related to a given concept were addressed within a specific theoretical context. It goes without saying that they cannot cover all the developments in all philosophical contexts. They focus on some key developments, leaving others aside, thus presenting *a* history of the concept of faculty, not *the* history.

The Reflections inserted between the chapters show that faculties are not only a crucial issue for philosophy, they are also widely discussed in literature (Reflections 1 and 2), science (Reflection 3), and health politics (Reflection 4). Different ways of conceptualizing and evaluating them lead to different roles assigned to them in literature, science, and politics. No doubt, the short Reflections can only hint at some of the roles played by faculties in other disciplines. Metaphorically speaking, these Reflections only open the door to some rooms without exploring all of them with all their corners. But they make clear that there are indeed other rooms and that philosophy is just one room in the large house called "Arts and Sciences"—a room that should be seen as being closely connected with other ones.

Finally, it should be pointed out that this book presents a history of the *concept* of faculty, not of the *terms* that were introduced to express this concept in different languages. That is, it does not compare words like "faculty," "Vermögen," and "faculté" as they were used in different intellectual contexts.[15] Nor does it trace the history of the English word "faculty" back to the Latin word *facultas* and spell out its different meanings. Of course, the development of a concept often goes hand in hand with a terminological development, sometimes even with the creation of technical terms. But a concept is not the same as a

---

15 For a brief overview of the way "faculty" and "Vermögen" were established as technical terms, see Klaus Sachs-Hombach, "Vermögen, Vermögenspsychologie," in *Historisches Wörterbuch der Philosophie*, edited by Klaus Ritter et al., vol. 11, 727–31 (Basel: Schwabe, 2001).

term. It is, loosely speaking, a certain way of thinking about a phenomenon and representing it.[16] It aims at describing and classifying a phenomenon, sometimes even at defining it, and is always part of an explicit or implicit theory that relates many phenomena to each other. Hence it is important to situate a concept within a theory and to look at the way it is connected with other concepts. Only then will it become understandable what it means, how it describes and classifies a given phenomenon. This has an immediate consequence for the study of the history of a concept. One cannot understand the meaning of a concept in a past period unless one looks at the theory of which it was part and in which it played an explanatory role. And one cannot become aware of a conceptual change unless one looks at the transformation of the entire theoretical framework. This is especially true for the history of the concept of faculty. Analyzing this concept in earlier periods amounts to reconstructing comprehensive theories about living beings or human minds, and a conceptual development or change does not become visible unless transformations of the relevant theories are identified. This is the main reason why the chapters in this book focus on the place of the concept of faculty in specific theories, and why they attempt to reconstruct the explanatory role the concept played (or still plays) in them.

3. Acknowledgment

This book would never have seen the light of day without the help and constant support of many people. First of all, I would like to thank all the authors for their enthusiasm and patience during the long process of preparation. In 2012, we all gathered at Humboldt-Universität in Berlin and discussed drafts of the chapters. Many thanks go to the "Exzellenzcluster Topoi" for making this stimulating meeting possible.

---

16 For various models that explain this key function of concepts, see *Concepts: Core Readings*, edited by Eric Margolis and Stephen Laurence (Cambridge, MA: MIT Press, 1999).

Christia Mercer, the series editor, also participated in this event, making detailed comments on the individual chapters and the structure of the entire book. Thanks to her for her valuable advice. Finally, I would like to thank Christina Banditt, Cécile Bonneton, Jennifer Marušić, and Claudio Mazzocchi for their help with the preparation of the manuscript.

CHAPTER ONE

Faculties in Ancient Philosophy

Klaus Corcilius

1. INTRODUCTION

According to the *Oxford Concise English Dictionary* the term "faculty" derives from the Latin expression for "making" or "doing" (*facere*) and primarily means "inherent mental or physical power." According to the *Oxford Dictionary*, to ask what faculties are, then, is to ask what a particular kind of *power* is, namely such a kind of power as inheres in minds or more generally in physical things. This close connection with the concept of power is also characteristic for the history of the conceptualization of faculties in ancient Greek thought. I can think of two main reasons for this. First, there is but one term for powers, dispositions, faculties, and the like in the ancient Greek. This is the term *dunamis,* an abstract verbal noun derived from the verb *dunasthai*, which means "being able to..." or "can"; second, in the formative period of classical Greek philosophy discussions of the concept of power are either prompted by, or overlap to some large extent with, discussions of mental faculties. In what follows I will trace in all brevity a basic outline of the history of the concept of *dunamis* from the time before Plato to later antiquity with a special focus on the discussion of the faculties of the soul. The story that I will tell is that of a linear development. It starts with powers of physical bodies (physiological powers), continues with faculties of the soul that engage in mental episodes (psychological powers), and culminates in the metaphysical application of dispositional analysis (ontological powers). Hippocrates, Plato, and Aristotle will be the landmarks of this story. Toward the end of the chapter I will very briefly touch upon later developments in the conceptualization of mental faculties in Greek philosophy.

2. The Prehistory of the Philosophical Concept of Faculties: Hippocratics

For the ancient Greeks the term *dunamis* and its cognates seem to have been pretty much the same as what "power," "disposition," "ability," or "capacity" and its cognates are for us when we employ these terms in a pretheoretical way: they all somehow denote that some thing or person *can* either do or undergo something. This broad usage is attested throughout the different periods of Greek literature from the beginnings in the Homeric poems down to late antiquity. The concept of power is one of the rather basic notions with which both the ancient Greeks and we ourselves describe and orient ourselves in the world around us. We have a grasp of things not only by attending to what they actually do or undergo at a given moment but also, and even more so, by attending to their *potential* to either do or undergo things in the future, that is, by attributing powers, abilities, and susceptibilities to them. These powers, abilities, and susceptibilities are the dispositional properties of the things around us. They are what these things either can, or are liable to, do or undergo. The importance of powers in this broad sense can hardly be overestimated; it is difficult even to imagine what our world would be like without them. But our familiarity with the use of the concept contrasts unfavorably with our understanding of it. Powers seem so fundamental a part of our cognitive grasp of the world that we hardly ever come to reflect on them.

This is no different in the history of the conceptualization of powers in western philosophy. As far as we can tell, early Greek thinkers did not write about powers *as such*.[1] This should not surprise us, given that the isolation of a concept of such fundamentality requires a rather developed and sufficiently abstract set of conceptual tools, which was not yet available to the early Greek thinkers. The fact that those who first identified powers as an object of systematic study in western philosophy still did not yet clearly distinguish between properties and their bearers may serve to illustrate this point: to the authors of the early Hippocratic writings properties such as hot and cold and so on were not different from the things that paradigmatically possess these properties, like for example fire and ice.[2] So, given that there is no particular thing in the world that would even be a candidate for paradigmatically instantiating the concept of power, one can perhaps see just how much of an intellectual achievement its isolation must have been. Looking at things from that historical perspective makes it a question well worth asking how such an abstract concept *could* be isolated in the first place. How could early thinkers identify the concept of power? The answer presumably has to do with the practical interest of the ancient doctors: the medical profession depends for its existence on systematic and teachable expertise on how to produce health in a patient, and this requires a more or less systematic understanding of the powers both of the human body and

---

[1] See Heinrich von Staden, "*Dynamis:* The Hippocratics and Plato," in *Philosophy and Medicine 29: Studies in Greek Philosophy*, edited by J. Boudouris Konstantinos (Athens: International Association for Greek Philosophy, 1999), 265. Regarding the early history of the concept of power (*dunamis*) I rely on von Staden's impressive study.

[2] See Heinrich von Staden, "*Dynamis:* The Hippocratics and Plato," in *Philosophy and Medicine 29: Studies in Greek Philosophy*, edited by J. Boudouris Konstantinos (Athens: International Association for Greek Philosophy, 1999), 267–68.

of the things that are either conducive, or detrimental, to health. Hence there was a keen interest among ancient doctors in particular powers—which things are powers for what effects[3]—and, from a more methodological perspective, also in powers generally.

The best textual evidence we possess for the explicit methodological interest of ancient doctors in powers can, somewhat anachronistically, be found in *Plato's* writings. In the dialogue *Phaedrus* he credits the historic Hippocrates with the following remarkable statement:

> First, we must consider whether the object regarding which we intend to become experts and capable of transmitting our expertise is simple or complex. Then, if it is simple, we must investigate its power [*dunamis*]: What power it has by nature to be acting upon in relation to what, or what power it has to be acted upon and by what? If, on the one hand, it takes many forms, we must enumerate them all and, as we did in the simple case, investigate how each is naturally able to act upon what and how it by its nature can be acted upon by what. (*Phdr.* 270C/D)[4]

The passage says that all objects of systematic study and instruction are bearers of powers and that their study consists in an investigation of what these powers are. Given our concerns, I shall focus on the following claims:[5]

(1) To have teachable expert-knowledge about something is to know the power (or the powers) that something has.

---

[3] This interest in the character of the particular powers that things possess would by itself not suffice to distinguish the ancient Hippocratic doctors from other early Greek thinkers. Indeed, one of the most important concepts in pre-Socratic philosophy, the concept of nature (*phusis*), is a dispositional concept.

[4] Trans. Nehamas/Woodruff, in Plato, *Complete Works*, ed. J. Cooper (Indianapolis: Hackett, 1997), modified.

[5] The passage has no exact parallel in the Hippocratic corpus such as it has come down to us. On the passage and its relation to the Hippocratic corpus, see Mario Vegetti, "La medicina in Platone: IV. I 'Fedro,'" *Rivista critica di storia della filosofia* 24 (1969): 7–22.

(2) Powers are either for acting on something else or for being acted upon by something else.
(3) To investigate powers is to observe their manifestations, that is, by observing what things actually do or undergo we identify the powers they have.

In what follows I will refer to this set of claims as the *Hippocratic doctrine*.

Plato emphatically endorses the Hippocratic insight. He will make frequent use of, elaborate on, and apply it to new areas of inquiry over and above the powers of physical bodies that the Hippocratics were mainly interested in,[6] most notably to the soul.

3. Plato: Faculties of the Soul

Plato's tripartition of the soul in *Republic* IV is the first explicit systematic philosophical treatment of the faculties of the soul in western philosophy.[7] It is also an extended application of the Hippocratic doctrine, even if Plato does not bother to mention Hippocrates in the relevant passages. But before we can duly appreciate its significance, it is important to point out that Plato adds some refinements to the Hippocratic doctrine. These are, first, that faculties fall under the general concept of relatives and, second, that there is a one to one correspondence between each relative term and its correlate (438a7). Hence, argues Plato,

---

[6] These are physiological powers, although the Hippocratics do not use this designation. Note that in the early medical writings this includes not only the dispositions of basic bodies (elemental bodies) but also their physiological compounds, extending sometimes even to such seemingly psychological faculties as perception, see, e.g., *Hp. Vict.* II 61. But in such cases the focus lies neither on cognition nor on other psychological features, but on the physiology of perception. The powers identified by the ancient doctors are bodily *causal* powers.

[7] Plato does not use the term for power/faculty (*dunamis*) in the relevant passage. But the language he uses makes it very clear that this is what he has in mind. (He uses the *-ikos* ending, which in the Greek indicates a capacity, and he uses the periphrastic construction "that *with which* we do…" (*Resp.* 435b). There is brief mention of some faculties of the soul in terms of their location in different parts of the body in a fragment of the pre-Socratic Pythagorean Philolaos (fragment B 13). But this falls short of an explicit *treatment* of the faculties of the soul.

our faculty of knowledge (*epistêmê*), to name but one example, correlates with just one epistemic object (*mathêma*), which is specific to it and to no other faculty. This one-to-one correspondence recurs on all levels of generality and specificity. Plato illustrates this by using the faculty of thirst (i.e., the desire for drink) as an example. Thirst is correlated to a specific object, drink. Additional qualifications of that object, for example, cold drink, are not correlated to *thirst* as such, but to their own specific, and accordingly qualified, *kinds* of thirst, in this case for "cold drink" and so on (437d8). These refinements, trivial as they might seem, are important. Plato puts, as it were, conceptual order in the realm of powers: he establishes that powers fall under the general concept of relatives[8] and that they have specific and unique objects correlating to them on each level of generality and qualification.

In order to see how the Hippocratic doctrine forms the methodological underpinning of Plato's tripartition of the soul, a few words on Plato's underlying general views about the soul are in order. Plato has what one may call a *psychological* conception of the soul. What I mean by this is that he conceives of the soul as the subject of mental actions and affections. It is the *soul* that experiences episodes of perception, thought, desire, and so on.[9] Similarly, the faculties of the soul are conceived as the powers that the *soul* possesses and that enable the *soul* either to act or to be acted upon in determinate ways. What is important for my concern is that Plato makes the soul distinct from the powers that it has: it is the underlying subject of its faculties and as such it seems significantly different from them. I will say more about this as I

---

[8] Plato does not use a technical terminology that he is invariantly committed to. Many times he avoids abstract designations and uses the concrete terms, as e.g. in our case where instead of "faculty" and "desire" he prefers to speak about more concrete cases, e.g., assent and dissent, appetite, and wish (437b–c). Similarly, in the case of the even more general notion of relatives, he says "the things that are 'such as to be *of* something' (*einai tou*, 438a7–b1); see the discussion of relatives in *Charm*. 167b .

[9] This is, in spite of its Cartesian appearance, *not* quite equivalent to saying that Plato is a "substance dualist": to say that the soul is an autonomous subject of mental episodes does not imply any determinate view about the substantial autonomy of the body (which, it seems, is denied by Plato, who seems to think that the body is dependent on the soul). Plato at no point *argues* for his psychological conception of the soul.

go along. A second important thing to note about the tripartition of the soul in *Republic* IV is that there are strong contextual limitations to it. For one thing, tripartition in the *Republic* regards only desiderative attitudes, that is, the desire for learning, a desire of a more spirited sort, and the desire for the pleasures of food, drink, and sex (437b). But this choice of faculties does not necessarily imply that Plato did not allow in other contexts for other faculties as well, which in fact he did (e.g. cognitive faculties later in the *Republic*). And one could argue that the reason why he does not mention other faculties in *Republic* IV is simply that he is concerned there only with the specific issue whether there are psychic equivalents for the three political classes of the ideal political community previously distinguished in *Republic* II and III.[10] On the other hand, one should also take into consideration the fact that at other places he returns to the tripartite soul in ways that seem to suggest that the tripartition in *Republic* IV was meant to be the canonical way of dividing the human soul.[11] But however that may be, possible contextual limitations should not blind us to the philosophical significance of tripartition: Plato, for the first time, *systematically* discusses the soul as the seat of its multiple faculties. Sometimes he calls them *parts* of the soul (*meros*, 442b, 444b), but we should not be overly impressed by his use of the language of parts in this context. It need not imply more than that he arrived at them by way of some kind of dispositional analysis, which seems confirmed by his use of the much weaker "kinds" to refer to the faculties of the soul.[12] As just mentioned, Plato's starting point for tripartition is the question whether there is a psychic equivalent for

---

10 These are the moneymaking, the auxiliary, and the deliberative classes of the ideal political community. The question is motivated by the analogy between the constitution of the political community and the soul's inner constitution that governs the whole dialogue. On Plato's part, there is no explicit commitment toward an exhaustive treatment of the soul in *Republic* IV. See Christopher Shields, "Plato's Divided Soul," in *Plato's "Republic": A Critical Guide*, edited by Mark L. McPherran (Cambridge: Cambridge University Press, 2010), 147–70. For a philosophical defense of tripartition, in spite of its problems, see Myles Burnyeat, "The Truth of Tripartition," *Proceedings of the Aristotelian Society* 106 (2006): 1–23.

11 *Tim.* 69d–71a.

12 *Eidē*, throughout 434d–441c. A passage in his latest dialogue, *Laws*, which offers "affections" (*pathē*) in addition to "parts" of the soul, seems to show some uneasiness with relation to the use of the language of parts in connection with the faculties of the soul (863b).

each of the three political classes of the ideal city in the individual soul, each of which is responsible for its own specific activity, or not:

> Do we do these things [i.e. learning, being angry, desiring drink, food, and sex] with the same part of ourselves, or do we do them with three different parts? Do we learn with one part, get angry with another, and with some third part desire the pleasures of food, drink, sex, and the others that are closely akin to them? Or, when we set out after something, do we act with the whole of our soul, in each case? (*Rep.* 436a8–b2)[13]

In order to properly address the question and to be able to determine whether it is the soul as a whole that is the subject of each of these actions and affections (learning, being angry, desiring the pleasures of food, drink, and sex) or some part of the soul, Plato introduces a *criterion*. This is the so-called *principle of opposites*:

> It is obvious that the same thing will not be willing to do or undergo opposites in the same part of itself, in relation to the same thing, at the same time. (*Rep.* 436b7–8)[14]

According to the principle of opposites, two faculties cannot be the same if they exhibit opposite manifestations in the same subject at the same time and in the same respect.[15] It is plain that the criterion can work only on the basis of the Hippocratic doctrine, since its applica-

---

13 Trans. Grube/Reeve, in Plato, *Complete Works*, ed. J. Cooper (Indianapolis: Hackett, 1997).
14 Trans. Grube/Reeve, in Plato, *Complete Works*, ed. J. Cooper (Indianapolis: Hackett, 1997).
15 Plato refines the criterion by confronting it with two difficult cases: (1) A person stands still while moving her hands and head; is the person standing still and moving at the same time? Response: the subject of these activities is not the person, but different physical parts of the person, namely her hand and her head. (2) The removal of the second difficulty requires a more subtle distinction: a spinning top revolves around its fixed axis. It hence moves while standing still at the same time. Response: In one *respect*, i.e. in respect of its axis, the spinning top stays put, while it revolves in respect of its circumference. It is noteworthy that Plato removes the second difficulty by invoking the rather subtle distinction between different *geometrical respects* and, perhaps even more subtly, by attaching different motions/states of rest to them.

tion presupposes a previous identification of a plurality of faculties as candidates for being (or not being) the same thing. So the issue the principle of opposites is designed to settle is not how one ought *to individuate* faculties of the soul, but only how to tell apart various faculties that have already been individuated on the basis of the Hippocratic doctrine.[16]

Plato, as just noted, takes it for granted that the soul is the subject of its actions and affections. The *soul* is the "thing" that will not do or undergo contrary manifestations in the same part of itself at the same time and in the same respect. Plato now makes the following observation: appetitive desires occasionally oppose rational desire, for example when we have an appetitive desire for a drink and at the same time also a rational desire that bids us to avoid this very same drink, for example for reasons of health. According to the criterion, then, appetitive desire cannot be the same faculty as rational desire, since both do cooccur at the same time in the same respect and in relation to the same object. It follows that appetitive desire (*epithumia*) is a faculty (or part) of the soul different from the rational faculty. Having established this, Plato applies the same line of thought to a different case, this time illustrated by way of an example. This is the (perhaps historic) story of the Athenian Leontios. Leontios was reported to have scorned himself for having an—apparently voyeuristic—appetitive desire to look at the corpses of executed criminals outside the city walls of Athens. Since he was said to have been angry with himself *because* of his base appetitive desire, his anger is naturally taken as counteracting his own voyeuristic appetite desire. The faculty responsible for anger therefore seems to be a different part of the soul than appetite. But does this make anger a part of the soul on its own? In order to establish that hypothesis it will have to be shown that anger is not the same as rational desire as well. Plato argues as follows. Anger, unlike rational desire, is "hot" and spirited in its nature. Moreover, it occurs in small children who still lack rationality

---

16 Namely in the passage *Resp.* 436a/b quoted above.

and even in animals that lack rationality altogether. From that he concludes that the "angry" part of the soul must be different from both the appetitive and the rational part. Plato thus ends up with a tripartite view of the soul as possessing a rational, an appetitive, and a spirited faculty.

With this dispositional analysis of the soul Plato applies the methodological framework of the Hippocratic doctrine to a nonbodily entity, and he attributes the status of real existence to the results of this procedure. Plato is an outspoken ontological *realist* about powers, believing that powers and faculties, although they are not observable, really exist: "powers," as he says later in the Republic, "are a class of things that *are*."[17]

Plato's tripartition, undeveloped and inchoate as it is, gives rise to some pressing questions:

(1) If the soul possesses a set of diverse faculties each of which is responsible for its own activity, how are we to regard the soul a unitary entity?

(2) If the soul is the possessor of a set of faculties, as the language in the *Republic* strongly suggests,[18] how does the soul relate to its faculties?

(3) How do the faculties of the soul relate to the causal powers of the body?

(4) Does Plato's acceptance of the Hippocratic doctrine not lead to an undesirable multitude of faculties of the soul?

These questions, which are closely related to one another, will be the focus of much what is to follow in ancient and medieval philosophical reflection about the faculties of the soul.

---

17 *Resp.* 477c1–478d5, where he also gives a general characterization of powers, which again very much relies on the Hippocratic doctrine. His argumentation for the existence of ideas in turn relies on the real existence of the cognitive *powers* correlating with them (*Resp.* 477b and following. A passage in the *Sophist* (246e–248a) sketches a somewhat similar argument on the basis of existence claims concerning the soul and the virtues. It additionally—unlike the *Republic*—suggests that the possession of an active or a passive power is necessary and sufficient for existence. But the context suggests that this claim is not Plato's last word on the matter (247e–248a: "For perhaps later to both us and them something else will appear to be the case").

18 *Tria eidē en tēi psukhei*, 440e7, 441a2.

Plato himself addresses only (1) explicitly. From the perspective of a theoretical investigation into the nature of the soul his answer might not seem the most satisfying, but it does seem more satisfying from a practical point of view: it is one of the major contentions of the *Republic* that the unity of the parts of the soul is a *moral task* and achievement of the fully virtuous person. It is very interesting to see that Plato conceives of that unity not in terms of an undifferentiated and static whole, but in dynamic terms as a joint effort in which each of the parts fulfills its own function and thus cooperatively brings about a higher unity: the *Republic* identifies the unity of the soul with the right and harmonious *relation* among the various activities of the different faculties of the soul. This, as will turn out, is a hierarchical relation with the rational faculty at the top and the other faculties at its service.[19]

Since Plato consistently talks as if he thinks that the soul is the underlying subject of its faculties, we can to some extent answer question (2), which concerns how the soul relates to its faculties: Plato did not *identify* the soul with its faculties. This is confirmed by the memorable comparison of the tripartite soul with the sea-god Glaukos in *Republic* X:[20] as the deformations from barnacles, seaweed, and other maritime encrustations he suffers prevent us from properly discerning Glaukos's original and divine form, so the true immortal and rational nature of the soul cannot be properly discerned due to the deformations it suffers from its union with the body. This, to be sure, casts doubt on the claim that it is Plato's last word on the matter that there are three faculties of soul rather than one. But however that may be, for the *embodied* human soul Plato in the immediate sequel of that passage explicitly reaffirms tripartition.[21] What are we to think, then, of the relation of the soul to its faculties? It seems Plato wants to distinguish between an immortal (rational) part of the soul, which is the true nature of the human soul, and the accretions that result from its union with a mortal

---

19 For a clear statement of this, see *Resp.* 588e and following.
20 611b–612a.
21 *Resp.* 612a3–6 explicitly suspends judgment on the question of whether the true nature of the soul is of one or of more forms (*eite polueidēs eite monoeidēs*, 612a4).

body. On that picture the nonrational parts of the soul are strictly speaking not faculties of the true nature of the soul, but additions to its true nature that somehow follow from its embodiment. If this is correct, then there is one sense in which the nonrational parts of the soul are possessed by the rational soul as the true subject of the faculties (namely as the accretions it acquires during embodiment) and another sense in which both the rational and nonrational faculties are possessed by the *embodied soul*.

Regarding (3), the relation of the faculties of the soul to the causal powers of the body, Plato speaks of faculties in terms of powers to either act or to be acted upon. This does of course not have to mean that the causal powers of the soul *exhaust* what the soul can do or undergo, but it does suggest that the question of how the faculties of the soul relate to the causal powers of the body (or body-parts)[22] is not a question about how a supposedly noncausal (mental) entity relates to a physical world that is categorically different from it; rather, for Plato, the faculties of the soul are by their very definition causal agents (they act on, and are acted upon, by things), and this even though they are said not to be materially extended. This points to an important difference between Plato's and the classical modern conception of a physical cause: Plato does not limit physical causality to materially extended bodies. As we know from his discussions in other works, he thinks of the soul as a nonbodily, yet self-moved, entity. He describes it as a self-mover with the causal power to bring about changes in bodies by imparting its own motions to the body to which it is attached.[23] So, even though most of the details of his views about the soul are opaque, what seems clear is that Plato has no deep philosophical problem with the

---

22 Or body-parts: in the *Timaeus*, Plato will attach the three parts of the soul to three different regions of the body. The "lower parts" are the place of the appetitive desire, the breast of spirited desire, and the head of the rational part (69d–71a).

23 *Phdr*. 245c–246a, and to bodies quite generally, see *Tim*. 34 B, 36 E and following; and *Lgg*. X, 895e–896a. On Plato's conception, all physical motion is *causally* accounted for by way of souls as self-moved movers. There is no obvious reason to suppose an inconsistency of Plato's faculty view of the soul with the self-mover view.

relation between the faculties of the soul and their causal basis in the human body, because the psychic and the causal are not yet two separate spheres whose interaction needs to be accounted for by way of bridging principles.[24]

Question (4): If we, with Plato, accept the Hippocratic doctrine as a general principle for the individuation of soul-faculties, the danger of an uncontrolled proliferation of faculties seems to loom. If difference in manifestation and the principle of opposites are the only criteria for generating psychic faculties, as they seem to be for Plato, why should we not, for example, postulate a faculty of memory, another faculty for remembering animals, and another faculty for remembering particular animals like, say, dogs or horses, and yet another faculty for remembering this particular dog Fido, and so on ad infinitum? Surely, this problem seems easy to deal with, but my point here is that Plato's refinements of the Hippocratic doctrine do not provide an explicit principle that helps us to avoid such an uncontrolled proliferation of faculties: the principle of unitary and cospecific correlates excludes only a plurality of correlates for each faculty—it does not exclude an infinite number of faculties. Similarly with the principle of opposition: it excludes opposing manifestations of the same faculty, but that still allows for an infinite number of not directly opposed faculties (e.g., a faculty for perceiving, a faculty for perceiving one object, another faculty for perceiving two objects, etc.). Thus, question (4) seems to formulate a serious worry. And as we will see, Aristotle will raise it against Plato's methodology in partitioning the soul. Plato himself does not address the question.

Apart from the general merit of having conceived of the soul explicitly as the unitary subject of various faculties,[25] two features of Plato's

---

[24] However, the bodily and the psychic seem to be two such separate realms for Plato. Hence, the hard question for him seems to be not so much how the soul interacts with the body, but how it comes to be in a body.

[25] The unity of the *faculties* of the soul with each other, as we have seen, has a somewhat problematical status for Plato, since he conceives of it as an achievement of the fully virtuous person.

account of psychic faculties stand out. This is, first, their *causal* nature: faculties are powers to either act on, or to be acted upon by, something; they are powers to *do* or *undergo* things in the physical world.²⁶ Second, faculties are *possessed* by subjects, and in the case of the faculties of the soul this subject is the *soul*. The soul is the underlying subject of its faculties and as such it is distinct from them.

4. Aristotle: Hylomorphic Faculties of Living Bodies

Aristotle systematizes the work of the Hippocratics and Plato. In his *Metaphysics* he develops an ontology that situates mental faculties within the framework of a general theory of powers. But he also adds new items to Plato's specific ontology of powers. While endorsing Plato's ontological realism about powers, he does not confine dispositional analysis to what Plato and the Hippocratics focused on, namely change. For Aristotle, change requires the interaction of an agent and a patient that are *physically distinct from one other*, so that both active and passive powers for change always relate to something physically distinct from whatever possesses them: powers for change are powers to either act on, or to be acted on by, *something else*.²⁷ Aristotle, to be sure, does accept the existence of powers for change. But he also introduces an additional kind of power different from the kinds of powers known to Plato and the Hippocratics. Indeed, this extension of the use of the machinery of dispositional analysis is one of his most significant metaphysical innovations: Aristotle applies the language of powers and dispositions to the *being* of things. His discussion of *dunamis* (what later came to be called *potentiality*), and *energeia* (*actuality*) in

---

26 Compare *Phdr.* 245c–246a, *Lgg.* X, 895e–896a. Again, this need not mean that this is *all* they can do.

27 It follows that physical self-change or self-affection in a strict sense (as seems to be the case with Plato's self-moving soul) is impossible for Aristotle: for every change there must be some X that actively changes and some Y that is passively changed, and X and Y cannot be the same *strictly speaking* (*Met.* IX, 1046a28–29; *Phys.* VIII 5). Self-change in a looser sense is not a problem for him, though, as long as the agent of the change is a *part* of the body distinct from the affected part (*Met.* IX, 1046a10–11).

the *Metaphysics* prepares us for this new application by claiming that there is a further kind of potentiality besides the more familiar powers for change:

> For "potentiality" and "actuality" extend more widely than those things that are said [to have potentiality or actuality] only according to change. (*Met.* 1046a1–2)

What is this other kind of potentiality? What Aristotle has in mind are potentialities to *be*, rather than *to become or to change into*, an X, such as for example the potentiality to *be* a house, a statue, or a living being. Such potentialities do not directly relate to changes in the course of which what is not yet a house, statue, or living being, comes to be a house, and so on (and hence *ceases* to be what it is now);[28] rather, they relate to the being of the very things whose potentialities they are. The potentiality to be a house relates to the actuality of being a house and similarly in the other cases. That means that potentialities for being do not relate to *something else*: the actual being of the house, for instance, is not a thing different from the thing that is potentially a house. Physically speaking, they are one and the same thing: the house. They are two ontological modes or aspects of the house, the one being its material aspect (say, the bricks, wood, etc. of the house), the other its formal aspect. The latter aspect is also called "essential form," which in the case of the house amounts to something like "certain arrangement of bricks and stone for the sake of shelter and storing goods" in these bricks, wood, and so on. There are, then, two different domains to which Aristotle applies the dispositional framework of the concepts of actuality and potentiality: either to changes, in which case powers for change involve a relation to something physically

---

28 There is the difficult case of powers for *being changes* (action and affection are modes of being as well, *Cat.* 1b27). In such cases, the power would of course relate to a change, but it would do so in a way that is different from the way in which powers for change relate to their correlated changes: X's power to *be* a certain kind of change Y would not constitute a relation to something *else*; rather, X, by becoming Y, would preserve what it already essentially is, namely a change.

distinct from them, or to the being of things, in which case the *relata* are ontological modes or aspects, namely the essential form and the matter of what physically speaking is one and the same thing:[29]

> For some [of the things said to be in actuality are said] as change is said in relation to a power, and the others as essence in relation to some matter. (*Met.* 1048b8–9)[30]

The dispositional analysis of things in terms of their ontological aspects is crucial not only for Aristotle's conception of the soul and its faculties. His entire philosophy of nature and natural things hinges on this metaphysical innovation. His view of the basic structure of natural things is known as hylomorphism (although he does not use this term himself). Hylomorphism is a thesis about natural bodies, which states that every physical object is a form/matter—composite in the sense just explained that every natural body is *ontologically* structured by two correlating aspects: matter (*hylê*), the relatively indeterminate potentiality for its essential form, and form (*morphê*), the determination and actuality of its matter (hence the later coinage "hylomorphism"). According to hylomorphism, the essential form and the matter of natural things are physically *inseparable* from each other, the only way of separating them being precisely hylomorphic analysis.[31] Form and matter are in this sense merely metaphysical aspects, and not physical parts of what physically speaking is but one single thing, the natural body. However, and this is going to be important in what follows, *when considered in isolation from their matter* in hylomorphic analysis, the essen-

---

[29] I am cutting a much longer story short here; readers should consult Stephen Makin, *Aristotle: Metaphysics Theta*, translated with an introduction and commentary (Oxford: Oxford University Press, 2006); Jonathan B. Beere, *Doing and Being: An Interpretation of Aristotle's Metaphysics Theta* (Oxford: Oxford University Press, 2009); Aryeh Kosman, *The Activity of Being: An Essay on Aristotle's Ontology* (Cambridge, MA: Harvard University Press, 2013); and, for the history of the concept of *energeia*, Stephen Menn, "The Origins of Aristotle's Concept of Ἐνέργεια: Ἐνέργεια and Δύναμις," *Ancient Philosophy* 14 (1994): 73–114.

[30] See *Met.* IX, 1046a1–4, and the discussion in 1048a25 and following.

[31] *Phys.* II 2.

tial forms of natural bodies are what these bodies *can do* or *undergo* insofar as they are what they are, that is, they are the *powers* or *faculties* that are possessed by these bodies and that make them the kind of bodies they are.[32] This may seem strange at first sight, since what was supposed to be the *actuality* of the natural body now turns out to be a power and thus something *dispositional*. But Aristotle has a point: the inseparability of form and matter requires that the actuality of the natural body be the *exercise* of its faculties as they occur in the form/matter composite. It thus seems right to conceive of the actuality of the natural body taken by itself and in isolation from the body as certain of the powers the body possesses.

4.1 The Hylomorphic Conception of the Soul

For Aristotle, living things are a specific *class* of natural bodies. That means that hylomorphic analysis applies to them in a straightforward way. The essential form of the living matter is the *faculties* that it possesses and that make it the kind of natural body that it is. The faculties of the soul are thus the essential *functions* of living bodies. The only difference between general hylomorphism and its psychological application is that the essential form of living bodies is called *soul* and the correlated matter is called *living body* (*DA* II 1, 412a11). This last point, that it is not the dead but the *living* body that is the matter of the soul, reminds us of the fact that the form/matter distinction is not a distinction between two physically distinct components of natural bodies, but between two ontological aspects of one and the same natural body. Aristotle goes so far as to deny that the lifeless body, a human corpse for instance, is a human body in the first place. This, even though it might perhaps sound counterintuitive at first blush, makes good sense: a body without soul lacks the essential form that made it the kind of body that it was—it has lost its identity as a human body. Therefore the

---

32 *Dunamis*, see *GC* 322a28–29. See *Met. 1049b8-10*.

corpse is a human body "only in name," similar to the way in which a statue of a human body is a "human body" (*DA* 412b10–24). This *homonymy principle*, as it came to be called later, shows very clearly that psychological hylomorphism conceives of the soul strictly as the essential form of the *living* body; physically soul and body are inseparable from each other.

Aristotle's psychological hylomorphism focuses broadly on all living bodies, and not just on the human soul (as Plato did in his *Republic*). This makes the hylomorphic approach to the soul not psychological but *biological* in scope, as it includes even the life of plants. The hylomorphic soul is the explanatory principle of natural life generally.[33] Apart from its broad focus, the hylomorphic conception of the soul has further significant consequences. This starts with the body/soul relation: since form and matter are ontological aspects of what is physically one and the same thing,[34] the hylomorphic soul is not physically distinct from the living body. The soul is the form in the sense of the essence *of* the living body, and the living body is the matter *of* its form and essence, the soul. Both correlate as physically inseparable aspects of one and the same thing. The question whether souls and the bodies of living things are one, therefore, does not arise:

> We should not ask whether the soul and body are one, any more than whether the wax and the impression (in the wax) are one, or in general whether the matter of each thing and that of which it is the matter are one. For, while unity and being are so spoken of in many ways, that which is most properly so spoken of is the actuality. (*DA* 412b6–9)

Living body and soul are identical in the way in which the essence of some matter is identical with this matter. Consequently, there can be

---

[33] Plato occasionally talks about plants as possessing soul as well (*Tim.* 77b), but that is not characteristic for his psychology.

[34] *Met.* VIII 6, 1045b16–19 makes that point by saying that what is potentially X and what is actually X are "somehow one."

neither a soul in separation from its living body nor a living body in separation from its soul. This has the further interesting consequence that the soul is not a possible subject of mental episodes. As the actuality of the living body the soul does not, and could not, act on the body, since such action, as I have shown, would require an agent physically distinct from the patient. There can be no "psychic process" that is not thereby also a process of the living body. This is the whole point of psychological hylomorphism—that it is in virtue of the soul as the essential form of the living body that the living body *is* such as it is: psychic episodes therefore *are* episodes of the living body.

Regarding the faculties of the soul, psychological hylomorphism makes not the *soul* the bearer and possessor of the faculties, but the *living body*. The hylomorphic soul, as just shown, is not different from its faculties: it is the set of faculties living bodies possess insofar as they are alive, in the same way in which the form of every natural body, when considered in isolation from its matter, is a certain power possessed by that matter. Faculties of the soul are the essential life-functions of living bodies. As such they belong immediately to the body, without any intermediate (psychological) bearer.[35]

There is, however, one exception. This is the part or faculty of the human soul that is responsible for thought. Aristotle says that this part does not have a body specifically attached to it. (Unlike Plato, Aristotle did not think that the brain is a specific organ for thought.) It therefore looks as if the faculty of thought, unlike the other faculties of the soul, *can* be separated from the body after all:

> That, therefore, the soul or certain parts of it, if it is divisible, cannot be separated from the body is quite clear; for in some cases of them the actuality is of the [bodily] parts. But at any rate nothing prevents that some parts [are separable from the body], because they are actualities of no body. (*DA* 413a3–7)

---

35 *DA* 414b19–23.

So there is one part or faculty of the soul that is not hylomorphically related to the living body. But in spite of this exception to the hylomorphic rule, Aristotle's conception of the soul, including its rational faculty, remains different from Plato's in two fundamental ways. For Aristotle, the soul is not distinct from its faculties, and the faculties cannot possibly interact with the body: the hylomorphic ones cannot do so because they are identical with the living body (or some aspect of it), and the rational faculty cannot do so because rationality is completely detached from all physical action and affection.[36]

4.2 The Faculties of the Soul and the Science of Life

Aristotle's definition of the soul in his *On the Soul* (*De Anima*) is not just an exercise in philosophy of mind. It is meant to express the first principle of a comprehensive science of living beings, with a broad focus not only on humans, but on all animals and plants.[37] But to know that the soul is the essential form of the living body is to have only a very general and almost vacuously abstract conception of the soul as a principle of botanical and zoological explanation. What a science of living beings needs is a much more specific account of the soul, one that is capable of *explaining* the phenomena of life, that is, the features that living beings possess insofar as they are alive. To that end, Aristotle, having given his general account of the soul as the essential form of the living body, makes a fresh start:

> We take, then, as our starting point for discussion, that what has soul differs from what has no soul, in that the former displays life. Now this word has more than one sense, and if any one of the following [activities] is present in a thing we say that it lives. (*DA* 413a20–23)

---

36 See e.g. *DA* III 4 and *GA* 736a27–29.
37 *DA* I 1, 402b3 and following.

This is the first step in the study of the soul as the principle of the science of living beings: Aristotle establishes that the soul is coextensive with life. In a second step he identifies four different kinds of life-activities. These are the activities that constitute life in its most basic form. To be able to perform one or a plurality of these activities is sufficient for the ascription of life to a body (*DA* 413a23–25):

(1) thought (*nous*)
(2) perception (*aisthêsis*)
(3) local movement and rest (*kinesis kai stasis kata topon*)
(4) nutritive movement, decay, and growth (*kinesis kata trophên kai phthisis te kai auxêsis*)

It is not immediately clear on the basis of what criteria these four activities (rather than others) are chosen by Aristotle. In any case, in a next step he establishes the soul as the principle in the sense of the *explanatory first cause* of these four life-activities:

> At present we must confine ourselves to saying that the soul is the principle of these [activities] and is divided into these:
> (i′) the nutritive faculty [*to threptikon*];
> (ii′) the perceptual faculty [*to aisthētikon*];
> (iii′) the faculty for thinking [*to dianoētikon*];
> (iv′) local motion [*kinēsis*]. (*DA* 413b11–13)

This list matches more or less exactly the above list of life-activities. The difference is of course the addition of "faculty" in three of the four cases.[38] This is how Aristotle introduces the faculties of the soul in his scientific account of living beings: as explanatory principles for the

---

38 In the Greek this corresponds to the *-ikos* suffix, which linguistically indicates an active power or faculty, ability, or affinity (see note 6 above).

activities of living things. They are the principles whose existence we have to primitively assume if we want to scientifically explain the phenomena of life and living things. At this very early stage of the investigation into the causes of living things, the faculties of the soul are still mere placeholders for the four basic items that are going to play the corresponding explanatory roles in the science of living beings. The search for the definition of each of these principles structures the treatise *On the Soul* (in the same sequence as above in i′–iv′). The resulting definitions, Aristotle says, also provide the most adequate definition of the soul as a whole (*DA* 415a12–13). He can say this because on his conception the soul is identical with the faculties of the living body. Hence, to define the soul's faculties is equivalent to defining the soul as a whole. The discussions in the course of the search for the definitions of the faculties of the soul are difficult, and their interpretation is often controversial. However, Aristotle completes the job and ends up with the following definitions: The nutritive faculty is defined as the capacity for self-preservation, as the capacity, that is, to preserve the body that possesses that faculty (*DA* 416b17–20); the perceptual faculty is defined as the capacity to take on perceptual forms without their matter (424a17–24); and the rational faculty is defined as the capacity to take on intelligible forms (which are essences for Aristotle) (429a10). In the case of the capacity for locomotion things are more complicated. Unlike the other cases *On the Soul* does not offer us a clear definition, but gives a causal account of animal locomotion that involves a plurality of psychic and bodily factors (mainly perception, thought, and desire, *DA* III 9–11) thus showing that there is no specific faculty of the soul exclusively responsible for locomotion.[39] On the whole, the following picture emerges. For Aristotle the soul is the first principle of life. It fulfills that role in virtue of being a set of faculties that are basic for the activities of living things insofar as they are alive. These faculties

---

39 Klaus Corcilius, *Streben und Bewegen: Aristoteles' Theorie der animalischen Ortsbewegung* (Berlin: de Gruyter, 2008), offers a full interpretation of the discussion of animal locomotion in *DA* III 9–11.

serve as partial subprinciples of the different kinds of life-activities, whose definitions do not make reference to each other and together exhaust the being of the soul as a whole. The faculties inhere in living things as the primary cause of their life-activities, and thus also serve as the first principles in the scientific explanation of the phenomena of life. They are: self-preservation (nutrition, growth, sexual reproduction), sense-perception, and thought. Other related activities and phenomena of the biological lives of plants, animals, and humans, such as desire, pleasure and pain, imagination, opinion, and others, should in one way or another be *derived* from these basic subprinciples.[40]

Aristotle's conception of the faculties of the soul raises questions as well. The following seem the most pressing:

(1) Why and on what basis did Aristotle choose precisely these four activities as basic for life?
(2) How do the different faculties of the soul relate to each other?
(3) How do the various faculties of the soul constitute a unitary soul?
(4) How do the faculties of the soul relate to the causal world?
(5) How does Aristotle avoid the uncontrolled proliferation of faculties?

Regarding (1), I think the only way to assess the accuracy of Aristotle's choice is to see how far this set of life-activities gets us in the actual scientific explanation of the phenomena of life. Unfortunately, even though we possess a great deal of his zoological writings, Aristotle's science of life is not much more than a torso. But however that is, we have good interpretive reasons to suppose that Aristotle's criterion in picking out these particular faculties as parts of the soul was their

---

[40] In which way this derivation is supposed to work is a complicated question and a matter of controversy that certainly cannot be pursued here. For a more extensive discussion of Aristotle's faculties of the soul as foundational for Aristotle's science of life, see Thomas K. Johansen, *The Powers of Aristotle's Soul* (Oxford: Oxford University Press, 2012).

explanatory power as first principles in accounting for the phenomena of life.[41] This shows that his conception of the faculties of the soul ranges on a rather high level of abstraction.[42]

(2) How do the different faculties of the soul relate to each other? Aristotle raises this question himself. His answer, though far from satisfactory, is very interesting. It also provides a first approach toward answering the third question regarding the unity of the soul. Aristotle says that the different faculties of the soul, in those animals that possess more than one faculty, are related through *potential inclusion* and *teleological nesting*:[43] the lower faculties are *contained potentially* in the higher faculties in a way similar to that in which simple geometrical figures, for example triangles, are contained in more complex figures like squares. Squares contain triangles, even though they are not visibly—that is, actually—contained in them as discrete parts. This claim of a seamless *fitting* of the lower faculties into the higher ones is meant to account both for the continuity in the ascent to higher soul-faculties in nature generally,[44] and for the integration of even the highest psychic functions in what in each case is essentially one unitary life. This idea of complete functional integration is brought out by the principle of *teleological subordination*, according to which the lower faculties exist "for the sake of" the higher faculties. Thus the basic faculty of nutrition, in living beings that possess more than one faculty, is

---

41 Compare *DA* 402b22–403a2.

42 Their role as first principles of Aristotle's biology makes the faculties of the soul quite different from Fodor's modules of the mind. Jerry Fodor, *The Modularity of Mind: An Essay on Faculty Psychology* (Cambridge, MA: MIT Press, 1983). For Aristotle, the, the faculties of the soul are basic *explanatory* items, in the sense of foundational for a deductively structured science of life. This means that the criterion for their individuation is their explanatory power as first principles in a system of scientific propositions, and not, or not primarily at least, the findings of empirical research. It is true that Aristotle thinks that the parts of the soul are also *causally* basic (*DA* 415b8–28), but that should not be taken as an *empirical* claim about their causal independence on the level of mental episodes. Whether the functions of the faculties of the soul can occur independently from each other in an actual living being (soul/body—composite) is a different (and much more specific) question for Aristotle.

43 What follows is based on the *DA* 414b19–33.

44 This is the basis for Aristotle's idea of a *scala naturae*, i.e. the idea of a more or less continuous ascent from lower to higher species in nature, *PA* 681a12 and following, *HA* 588b4 and following.

not only potentially contained in the perceptual faculty, it also exists for its sake.[45] Similarly with the perceptual faculty: souls of intelligent animals include the perceptual faculty, which is also teleologically subordinated to it.[46] Hence, perception is fully functional in rational animals, but in ways that integrate it in a functional context that is determined by the animal's intellect. This being so the perceptual faculty of rational animals seems different from the perceptual faculty in animals that lack such a superordinate context. Aristotle seems committed to something like the following: when considered in abstraction from superordinate contexts, the faculties of the soul have essences that are *definitionally separable* from each other (their definitional accounts do not make reference to each other), but as integral parts of functionally superordinate contexts they are inseparable parts of a natural whole. This is how the *Nicomachean Ethics* makes this point about the relation between the rational and the nonrational part of the specifically human soul:

> Whether these parts (of the soul) are separate like the parts of the body or anything else that can be physically divided, or whether they are inseparable by nature but two in account, like the convex and concave in a curved figure, does not matter for our present concern. (*NE* 1102a28–32)

The suggestion is that these two parts of the human soul, even though they have other accounts, are *by nature* inseparable from each other in a way similar to the way in which the convex and the concave form part of, for example, a circle or any other curved figure. So it seems that Aristotle thinks that the faculties of the soul are functional aspects

---

45 Conversely, *teleological subordination* of the lower faculties corresponds to the *existential dependency* of the higher on the lower ones: without nutrition there can be no perceivers, and without perception there can be no intellect in nature, etc.
46 Note the structural similarity to Plato's aforementioned ethical conception of the unified soul as the right hierarchical relation among the parts of the soul.

that, although separable *in account*, are somehow integrated in what *by nature* is one indivisible soul.[47]

(3) How do the faculties of the soul constitute a unitary soul? This question has to some extent been answered in (2), even though many critical issues are still unresolved. At any rate, these issues are relevant not only for Aristotle's theory of life. For given that the soul is the essence of the living body, and the faculties of the soul are somehow parts of that essence, the answer to the question of how the faculties of the soul relate to the soul as a whole has a direct bearing on the more general questions of how essences can have parts in the first place, and how these parts relate to the essence as a whole. These are difficult metaphysical puzzles whose significance far exceeds the ambit of philosophical psychology. However, apart from what I just said about teleological nesting and the analogy with the convex and the concave, there is not very much to be found in Aristotle's philosophical psychology.[48] But it is clear that the soul, in spite of being a plurality of faculties, should have unity in some strong sense for Aristotle.[49]

(4) As the foregoing discussion of psychological hylomorphism has shown, the soul is the actuality *of* the living body. This holds also for the faculties of the soul, given that the soul is a set of faculties for Aristotle. Living bodies act and are acted upon in the ways that are characteristic for them *in virtue of the fact that their soul-faculties are parts of their essential form*. Being of such a nature, the faculties of the soul determine the normative equilibrium states of the living body and thus fundamentally explain why living bodies function in the ways they do. Since these are facts that are true *of* the living body—namely that it is a body of such and such a kind tending toward self-preservation

---

47 This view is confirmed by a parallel passage in *EE* 1219b26–36, which adds as a further analogy the inseparability of the straight and the white aspect in (presumably) a straight white object; this brings out more clearly that the essences of the two aspects, when taken by themselves, are separate from each other.

48 There is a significant suggestion in this regard in his *Metaphysics* (*Met.* 1038a9 and following).

49 Aristotle takes the soul not only to be unitary, but even to be the *principle* of the unity of the living body (*DA* 411b9–14).

in this particular way, is capable of perceiving, and so on—and not *things* that have a physical existence besides the body, the question of how the soul interacts with the body does not arise.[50]

However, such a deflationary answer is not available in the case of the thinking part of the soul, since Aristotle, as we have seen, does not believe that there is any specific matter attached to the faculty of thought. Here it should be noted that Aristotle's position is not as naïve as it might seem at first glance. He is acutely aware of the causal, the representational, and the linguistic requirements of human thinking.[51] But he tends to think of them in terms of *necessary conditions* for the exercise of thought and not as *essential* features of the thinking activity. What this means is only that thinking, unlike perception and biological self-preservation, for him is not *essentially* connected to physical bodies. It does not mean that human thinking does not involve causal, representational, and linguistic factors, for he clearly thinks that it involves them. But it does make thought as such a nonphysical entity, and Aristotle insists on thought's nonphysicality in many places. This suggests that he faces a problem that might be not exactly like the Cartesian interaction problem, but somehow analogous to it. This is the question of how it should be possible that a nonphysical entity like the rational faculty, which is neither extended nor moved in a physical sense, connects to the physical world and can—via

---

[50] Sometimes, though, Aristotle talks as if there was interaction between the soul and the body, especially in the places where he talks about the soul as the primary cause of locomotion. However, these passages can, and also *should*, be given interpretations that avoid the attribution of a Cartesian body/soul interaction to Aristotle. His conception of physical causes is much broader than the modern concept of a physical cause (see the discussion in *Phys.* II 7) and he also accepts causes that occur in nature that are entirely unmoved, but still causes of motion in, as he says, a "nonphysical way" (*ou phusikôs*). This explicitly includes the essences of natural things (198a28–29, a35–b4), which in turn suggests that there is no interaction between body and soul for Aristotle, but only an essential relatedness of the body to its soul, by means of which he wants to *explain* why living bodies are moving/behaving in the way they do.

[51] See Jean-Louis Labarrière, *Langage, vie politique et mouvement des animaux: Etudes Aristoteliciennes* (Paris: Vrin, 2004), Philip van der Eijk, "The Matter of Mind: Aristotle on the Biology of Psychic Processes and the Bodily Aspects of Thinking," in *Aristotelische Biologie: Intentionen, Methoden, Ergebnisse,* edited by Wolfgang Kullmann and Sabine Föllinger (Stuttgart: F. Steiner Verlag, 1997), 221–58, and Michael V. Wedin, *Mind and Imagination in Aristotle* (New Haven: Yale University Press, 1988).

human action—even have physical effects in it.⁵² Naturally, there is no room to address such intricate questions here.⁵³ There is, however, another question more pertinent to my present concern with faculties. If rationality is not essentially of a body, why should it be a faculty (power) in the first place? Again, there is no room here to address this question properly, but Aristotle distinguishes a *faculty* of thought, which is passive and capable of receiving every object of thought (by thinking it), from another kind of thought, which is active *by its very essence*.⁵⁴ Whatever it is exactly that Aristotle has in mind here, which is a matter of notorious controversy, the mere fact that he makes that distinction suggests that for Aristotle thought taken by itself may indeed *not* be a faculty. But this, if correct, is a fact about thought taken by itself, to which humans have only very limited access. In *human* souls the thinking part is defined as a faculty. The faculty of human thought, then, seems essential for us humans, but *for thought* as such it seems neither essential to be human nor to be a faculty.

(4) How does Aristotle deal with the proliferation problem? He explicitly raises the problem as a challenge to (presumably) Plato and some of his followers.⁵⁵ This suggests that he must have thought he had somehow come to grips with it. Here is how he poses the challenge:

> The problem at once presents itself, in what sense we should speak of parts of the soul and how many [we should distinguish].

---

52 The nonphysicality of rationality comes out very clearly in *DA* III 4. *Met.* IX 2 and 5 discuss the causal efficacy of human rational powers. See also the general discussion of the relation between the rational faculty of the soul and physical processes in, e.g., *Parts of Animals* 641a32–b10; *GA* 736b28–29; *NE* 1154b21–28.

53 See Klaus Corcilius, *Streben und Bewegen: Aristoteles' Theorie der animalischen Ortsbewegung* (Berlin: de Gruyter, 2008).

54 *DA* III 5, 430a10–25.

55 In the following passage Aristotle distinguishes two groups of thinkers: those who divide the soul into two parts, the rational and the nonrational parts, and those who divide the soul into three parts, the spirited, the appetitive, and the rational. The latter group, as I have shown, is Plato and his followers. The former group is much harder to identify, especially since bipartition into a rational and a nonrational part was accepted not only by Plato and his followers but also by Aristotle himself (in his ethical works); see Paul A. Vander Waerdt, "Aristotle's Criticism of Soul-Division," *American Journal of Philology* 108 (1987): 627–43.

For in a sense there seem to be indefinitely many parts, and not only those that some people mention when they distinguish the rational, the spirited, and the appetitive, or with others the rational and the nonrational; for *if we take the dividing lines by which they separate into these*, we shall find parts standing more apart from one another than these, namely those we have just treated: the nutritive, which belongs to plants as well as to all animals, and the perceptual, which cannot easily be classed as either nonrational or rational; furthermore, the imaginative, which is different in being from all the others, while it presents a great problem with which of those it is the same or not the same—*if one posits separate parts of the soul*; and in addition to these, the desiderative, which would seem to be different both in account and in power from all the others. (*DA* 432a22–b4, emphasis mine)

This is a remarkable criticism, especially since it does not concern the specific psychology of the Platonists (i.e. the parts of the soul they postulated) but the *methodology* they applied in dividing the soul. Aristotle's targets are the criteria they use for partition (1) and the particular *way* in which they conceive of the parts of the soul (2). The argument in (1) has the form of a *reductio*: if we accept the Platonists' criteria of partitioning the soul, we end up with an undesirably high number of parts of the soul. On these criteria, difference in activities, as long as their accounts are different from each other,[56] is sufficient for the postulation of different faculties of the soul. So the Platonists' criteria for partitioning (difference in account) are too loose to prevent an uncontrolled proliferation of parts. And apart from that, Aristotle argues, there are other faculties of the soul that have a greater right to be regarded as parts of the soul; these are the three main parts of the soul he himself distinguished previously in his work *On the Soul*, namely the

---

56 It is possible, and in my opinion also not unlikely, that the criterion "difference in account/being" mentioned above (432b3–4) is also meant to cover Plato's principle of opposites.

nutritive, the perceptual, and the rational parts. Hence, the suggestion goes, the Platonic criteria should be rejected or at least modified. But, then, what are Aristotle's own criteria for partitioning the soul? It seems that Aristotle distinguishes between faculties and parts of the soul. Faculties of the soul are simply all sorts of powers that we can ascribe to the soul in virtue of the fact that we observe different activities in living things. Parts of the soul, by contrast, as I have argued above, are a particular subset of faculties of the soul, namely those that are both explanatorily basic in Aristotle's deductive science of living beings and definitionally separable from one another.[57] On that distinction, Aristotle can easily accept an indeterminate number of faculties of the soul as long as this does not lead to postulating corresponding *parts* of the soul (which should be as few as possible). This means that Aristotle's own criterion for partitioning the soul, while taking on board the Platonic criterion of difference in account, adds an important methodological criterion to it, namely *explanatory primitiveness*. This leads over to (2) the criticism of the Platonic way of conceiving of the parts of the soul: the Platonists precisely did *not* distinguish between parts and capacities of the soul. For them, as the text suggests, every faculty of the soul is thereby also a separate part of the soul. It seems that it is precisely this separation of faculties from one another, including those whose accounts are different but make reference to one another, that produces the *problem* in the uncontrolled proliferation of parts of the soul for Aristotle. Souls can have as many faculties as there are corresponding activities. The problem only arises once these faculties are conceived of as separate parts of the soul, since that makes them explanatorily basic, and to have an uncontrolled proliferation of explanatorily basic parts is a serious problem for any principled account of the soul.

---

[57] This view of his conception of parts of the soul is argued for in K. Corcilius and P. Gregoric, "Separability vs. Difference: Parts and Capacities of the Soul in Aristotle." In *Oxford Studies in Ancient Philosophy* 39 (2010): 81–119.

On the whole, it seems that Aristotle's conception of the faculties of the soul owes much of its systematicity to the fact that it is embedded in his scientific account of living things. Plato's diverse treatments of the soul and its parts, by contrast, lack such scientific background. Consequently, one could argue that Aristotle's methodological arguments, for example against the uncontrolled proliferation of soul-parts, may be slightly off the mark. In terms of the specific ontology of powers, Aristotle's most significant contribution consists in having applied the Hippocratic machinery of dispositional analysis beyond powers for change. This, as we have seen, gives him the means to account for the *essential being* of natural substances in terms of certain of their faculties.

5. Epilogue: Faculties of the Soul in Later Greek Philosophy

In the Hellenistic period much of Aristotle's metaphysical thinking about powers, the soul, and its faculties seems to have been either neglected or forgotten.[58] This tendency has certainly been corroborated by the ontological corporealism that prevailed in the two dominating Hellenistic schools of philosophy. Stoics and Epicureans agreed that the basic entities in the world are bodies, and that incorporeals like powers and faculties are not basic (save the void for the Epicureans, of course). This made them much less favorably disposed toward Plato's and Aristotle's ontological realism with regard to powers. Both schools applied their corporealism to their conceptions of the soul. They applied the following line of reasoning: there is interaction between soul and body; but if there is interaction, the soul *has* to be a body, because only bodies can act on bodies. But from here onward agreement ends. For while the Epicureans held that the soul is an aggregate of discrete

---

58 See Robert Sharples, "The Hellenistic Period: Whatever Happened to Hylomorphism?," in *Ancient Perspectives on Aristotle's "De Anima,"* edited by Gerd Van Riel and Pierre Destrée, 156–66 (Leuven: Leuven University Press, 2009).

and especially fine soul atoms,[59] the Stoics identified the soul with a continuous body, a breath-like and fiery substance they called Pneuma. This difference in the basic structure of the (otherwise not very different) bodies with which they each identified the soul, led, as I will now show, to considerable differences in their conceptions of faculties of the soul.[60]

The Epicureans made the soul an *aggregate* or blend of discrete parts. That might initially suggest a picture on which the different faculties of the soul are simply different material parts of that aggregate. But in this case the physical discreteness of the atomic parts of the soul does not translate into a functional differentiation: the Epicureans *reduced* life and the life-functions to supervening properties of the aggregate of soul-atoms in the body. This reductive move was of utmost importance to them, because it provided physical grounds for their thesis that at the moment of death all our life-functions come to a definitive end.[61] When properly grasped by the Epicurean philosopher, this thesis frees him or her from irrational beliefs such as that being dead is a bad thing and that there is punishment in the afterlife, and so on. It is clear that for this to work properly it is vital that not the atoms themselves—which the Epicureans considered lifeless bodies—but their *aggregates* are the proper bearers of life. And the same applies to the various faculties of the soul. They are physically grounded in the atoms

---

59 They imagined it as being in certain respects like wind and fire, *Ep. Hdt.* 63.
60 See Anthony A. Long, "Stoic Psychology," in *The Cambridge History of Hellenistic Philosophy*, edited by Keimpe Algra, Jonathan Barnes, Jaap Mansfeld, and Malcolm Schofield (Cambridge: Cambridge University Press, 2005), 560–84. What follows is a very brief summary of only the most general features of the Epicurean and Stoic accounts of faculties of the soul. For an account of Epicurean psychology, see Stephen Everson, "Epicurean Psychology," in *The Cambridge History of Hellenistic Philosophy*, edited by Keimpe Algra, Jonathan Barnes, Jaap Mansfeld, and Malcolm Schofield (Cambridge: Cambridge University Press, 2005), 542–59; for (early) Stoic psychology, see Anthony A. Long, "Stoic Psychology," in *The Cambridge History of Hellenistic Philosophy*, edited by Keimpe Algra, Jonathan Barnes, Jaap Mansfeld, and Malcolm Schofield (Cambridge: Cambridge University Press, 2005), 560–84; and for a discussion with a specific focus on the question of the parts of the soul in early Stoicism, see Brad Inwood, "Walking and Talking: Reflections on Divisions of the Soul in Stoicism," in *Partitioning the Soul: Debates from Plato to Leibniz*, edited by Klaus Corcilius and Dominik Perler (Berlin: de Gruyter, 2014), 63–83.
61 *Ep. Hdt.* 65–66.

that underlie them, their functions being properties not of the discrete parts but of the aggregate:

> The primary particles of the elements so interpenetrate one another in their motions that no single element can be separated off nor can its power be active when spatially divided from the rest, but they are, as it were, the many forces of a single body. Just as in the flesh of any living creature there is a scent and a certain heat and flavor, and yet from all these is made one body grown complete. (Lucr. III.262–65)[62]

The different faculties of the soul or life-functions are not different material parts of the soul, but properties that supervene on the particular blend of atoms that constitute the living aggregate.

This differs from Stoic psychology, since their conception of Pneuma is that of a *continuous* material substance that is *itself alive*. Hence, for the Stoics, life is not an emergent or otherwise supervenient property, but a basic constituent of the Kosmos. The Stoics identified Pneuma with the active and divine cosmic principle whose motions pervade the whole Kosmos and thereby impose determinacy, structure, and purposiveness on passive matter, which was the other cosmic principle they believed in. That, however, does not mean that wherever there is Pneuma, there are living creatures, since on their view biological life requires Pneuma of a special kind. The Stoics believed that Pneuma comes in different degrees of tension (*tonos*) and that it endows the bodies it pervades with different qualities accordingly. The result was a Kosmos populated by a hierarchical sequence of increasingly organized and purposeful bodies. This started with lifeless bodies, and continued with plants and animals, with rational animals at the top of the sequence. The Pneuma in lifeless bodies they called their "having" or "holding" (*hexis*), which we may think of in terms of their "material

---

[62] Trans. Rouse/Smith 1924, modified = Anthony A. Long and David N. Sedley, *The Hellenistic Philosophers* (Cambridge: Cambridge University Press, 1987), 14 D 1.

structure," with the caveat that the Stoics conceived of that structure in active terms, that is as the result of the active effort of Pneuma. The Pneuma explaining the cohesion and functioning of plants and other vegetative bodies they called "nature" (*physis*); animal Pneuma, which is responsible for perception and motor impulses, was called "soul" (*psychê*), and the highest degree of pneumatic tension, reached in rational animals, was called *logos* (speech/rationality). Now in *humans* all these different degrees of tension had somehow to come together and form a coherent whole. That was no easy task because, famously, the Stoics were psychological monists, that is, they were firmly committed to the *identity* of all of the parts of the human soul with the rational part. So, for the Stoics there is somehow only one faculty of the human soul, namely the rational faculty, which they called "the commanding faculty" (*hêgemonikon*). That psychological monism committed them to identifying *all* psychic functions with the functions of the commanding part.[63]

As far as the criteria they used in distinguishing the parts and capacities of the human soul are concerned, it is very interesting to see that they relied on both bodily and functional criteria:

> How are the soul's faculties distinguished? Some of them, according to the Stoics, by a difference in the underlying bodies. For they say that a sequence of different breaths extends from the commanding-faculty, some to the eyes, others to the ears, and others to other sense-organs. Other faculties are differentiated by a peculiarity of quality in regard to the same substrate. Just as an apple possesses in the same body sweetness and fragrance, so too the commanding-faculty combines in the same body impression, assent, impulse, reason. (Iamblichus in Stobaeus, *Eclogae Physicae et Ethicae*)[64]

---

[63] This is why they identified emotions and other nonrational attitudes with assent to (bad) judgments.

[64] Ed. C. Wachsmuth (Leipzig: Teubner, 1884), 1. 368, 12–20: Anthony A. Long and David N. Sedley, *The Hellenistic Philosophers* (Cambridge: Cambridge University Press, 1987), 53 K.

With regard to the five senses bodily parts correspond directly with functional parts,[65] but at the same time there is a purely functional division of spatially indistinguishable parts of the commanding part of the soul (they are: impression, assent, impulse, and reason). But the apparent methodological pluralism in the individuation of soul-parts, it seems to me, should not be overstated. Their corporealism and corresponding nominalism with regard to incorporeals give the Stoics strong ontological grounds to prefer a division by bodily parts as more basic.

It is striking that Stoics and Epicureans show considerable agreement in the basic conception of powers that underlies their division of faculties of the soul:[66] faculties of the soul are powers either to act upon something else—namely the body and via the body on the external world—or they are powers to be acted upon by something else. So with regard to their conception of faculties both schools revert back to the original Hippocratic doctrine, and everything indicates that they agreed in rejecting Plato's ontological realism about powers as well.[67] Both put much emphasis on physical bodies as *causally grounding* the faculties of the soul. Their main difference lies in their conceptions of what particular kind of body is supposed to do that grounding: for the Stoics the underlying subject of the soul's faculties is Pneuma, a simultaneously bodily and mental stuff that is primitively alive, whereas for the Epicureans the subject of the soul's functions is the *aggregate* of atoms inside the living body, and whose life-functions are "nothing but" their particular blending.

---

65 Inwood calls this "spatial/functional isomorphism." See Brad Inwood, "Walking and Talking: Reflections on Divisions of the Soul in Stoicism," in *Partitioning the Soul: Debates from Plato to Leibniz*, edited by Klaus Corcilius and Dominik Perler (Berlin: de Gruyter, 2014), 68.

66 For a detailed account of some of the shared holistic (psychophysical) views of Epicureans and Stoics, see Christopher Gill, *Naturalistic Psychology in Galen and Stoicism* (Oxford: Oxford University Press, 2010).

67 The Stoics went so far in their corporealism that they declared even the virtues to *be* bodies. See Anthony A. Long and David N. Sedley, *The Hellenistic Philosophers* (Cambridge: Cambridge University Press, 1987), 29B.

Four hundred years later, and with remarkable methodological self-awareness, the famous doctor-philosopher Galen of Pergamum expresses some very interesting concerns with regard to faculty psychology as a whole. Galen himself advocated a Plato-inspired tripartite psychology, but also shared some of the Hellenistic intuitions about their physical grounding. His doubts, and resulting expressions of agnosticism,[68] are motivated by a basic conceptual fact about faculties. This is the fact that faculties belong to the category of relatives, as was pointed out already by Plato. But relatives, by their very conceptual nature, are unable to provide us with knowledge of how things are *in themselves*. For suppose we know that the soul has a faculty to cause the body to φ. In that case, what we know about the soul is what it is in relation to something else, namely that it is a cause *of* the body's φ-ing. But as long as we do not know *how* the soul does this, then all we know about it is what it is in relation to something else. And this is Galen's point: faculties are mere placeholders for whatever item *actually*, that is, in nonrelative terms, explains what they stand for (ignored by us). This basic fact, he claims, has been neglected by philosophers:

> On this point many philosophers appear to be in some confusion, lacking a clearly articulated notion of faculty. To me they seem to imagine faculties as things which inhabit substances in much the same way as we inhabit our houses, and not to realize that the effective cause of every event is conceived of in relational terms; we may *talk* of this cause as of a particular object, but the faculty arises in relation to the

---

[68] They are described by Pierluigi Donini, "Psychology," in *Cambridge Companion to Galen*, edited by Richard J. Hankinson (Cambridge: Cambridge University Press, 2008), 184–209. For more thorough discussions of Galen's psychology, see Christopher Gill, *Naturalistic Psychology in Galen and Stoicism* (Oxford: Oxford University Press, 2010), and Galen, Claudius of Pergamon, *Psychological Writings*, edited by P. N. Singer and translated with introductions and notes by Vivian Nutton et al. (Cambridge: Cambridge University Press, 2014); for a discussion of his conception of faculties of the soul, see Richard J. Hankinson, "Partitioning the Soul," in Galen on the Anatomy of the Psychic Functions and Mental Illness," in *Partitioning the Soul: Debates from Plato to Leibniz*, edited by Klaus Corcilius and Dominik Perler (Berlin: de Gruyter, 2014), 84–106.

event caused.[69] We therefore attribute as many faculties to a substance as activities. For example, aloe has a cleansing and toning faculty in relation to the throat, and a faculty of binding bleeding wounds, of scarring over grazes, and of drying the moisture in the eyes. But there is no other object apart from the aloe which is performing all these actions. The aloe is what is active, and it is because of its ability to perform these actions that it is said to have these faculties—as many faculties, in fact, as the actions in question. (QAM 769–70)[70]

Galen here argues against what he takes to be a seriously misguided metaphysical conception of faculties: it is a mistake, he says, to treat faculties as if they were intrinsic features of the substances whose faculties they are. But this category mistake is precisely what many philosophers are guilty of in conceiving "of faculties as things which inhabit substances in much the same way as we inhabit our houses." The error is to take faculties, which are relational properties, as parts of the *substance* of the items whose faculties they are. To find out about the true substance of things, he argues, we should attend not to their relative properties, but to the substantial items that *causally ground* them. He illustrates this point with an analysis of the various faculties of aloe and their relation to its substance: aloe cleans and tones the throat, it heals bleeding wounds, it scars over grazes, it dries the eyes. But aloe *is* not that set of relational properties (and hence does not contain them as parts of its substance); aloe is the substance that *causally grounds* these various faculties. Because of this, he claims, faculties are something of an *asylum ignorantiae*:

> But, if the cause is relative to something—for it is the cause of what results from it, and of nothing else—it is obvious that the faculty

---

69 The Greek text seems a bit murky here, but the overall thrust of the argument seems clear enough; see the apparatus in Galen, Claudius of Pergamon, *Scripta Minora*, edited by Iwan von Müller, vol. 2 (Leipzig: Teubner, 1891).
70 Trans. Singer (slightly modified), in Galen, Claudius of Pergamon, *Psychological Writings*, ed. P. N. Singer, trans. Vivian Nutton et al. (Cambridge: Cambridge University Press, 2014).

also falls into the category of the relative; and so long as we are ignorant of the true essence of the cause which is operating, we call it a *faculty*. (Galen, *On the Natural Faculties* I.iv)[71]

Galen's point here is that we individuate things in terms of faculties *only for as long as we are ignorant of the true substance of the cause*. To conceive of things in terms of their faculties, therefore, is tantamount to saying that we do not yet know the true causes. But contrary to what one might expect in view of such remarks, Galen is very much inclined to accept Plato's three-faculty analysis of the soul. But that does not mean that he here follows his master against his own better judgment. On the contrary, Galen believes that—given our ignorance of the true nature of the soul—the only way for us to know anything about it is to analyze it in terms of faculties, and that the Platonic way of individuating the faculties represents the best possible manner of doing this. As we lack knowledge of the true substance of the soul, we simply *have* to proceed on the basis of faculties. But since there are "as many faculties as there are actions," we should subject our dispositional analysis of the soul to strict *methodological criteria*:[72]

So, that part of the soul which we call the rational is desiderative in the broad sense of that term: it desires truth, knowledge, learning, understanding, and recollection—in short *all* the goods. Similarly, the spirited is desiderative of freedom, victory, power, authority, reputation, and honor. The part which Plato calls desiderative *par excellence* has the desire for sexual pleasure, and for the enjoyment of *all* kinds of food and drink. This part

---

71 Trans. Arthur J. Brock (Cambridge, MA: Harvard University Press, 1916).
72 Galen's above point bears an interesting relation to the much later and famous *vis dormitiva* objection against the explanatory vacuousness of powers in Molière's *The imaginary invalid*. (A sleeping pill allegedly induces sleep in virtue of the fact that it has the power to induce sleep, which is basically stating the cause in terms of its effect.)

has no capacity to desire the good, and no more does the rational have the capacity to desire sex, food, or drink, or indeed victory, authority, fame or honor. Nor, of course, can the spirited have the same desires as either the rational or the desiderative. (QAM 772)[73]

Galen defends Platonic tripartition as resulting precisely from such rational methodology. On his interpretation, Plato divided the soul not into an arbitrary collection of faculties, but according to a rational procedure, namely by (1) bundling all of the soul's countless faculties into three basic kinds (this is why Galen says that Plato "speaks broadly" when he uses specific designations for the parts of the soul) that cover *all* sorts of desiderative attitudes, and by (2) accepting only such faculties that exclude one another.[74]

So it is the rational methodology on which it is allegedly based (and the additional fact that he takes it to be empirically confirmed by the findings of anatomy) that make Galen endorse Platonic tripartition. However, when it comes to Plato's additional claims about the immortality of the rational soul—a massive ontological claim about the intrinsic nature of one particular part of the soul and thus clearly without foundation in the foregoing methodology—Galen finds himself unable to follow:

There is, however, a further belief, that of these three "forms" or parts of the overall soul the rational is immortal; and of this Plato seems convinced. For my part, I am unable to make a confident assertion one way or the other. (QAM 772–73)[75]

---

[73] Trans. Singer, in Galen, Claudius of Pergamon, *Psychological Writings*, ed. P. N. Singer, trans. Vivian Nutton et al. (Cambridge: Cambridge University Press, 2014).

[74] Similar to the criteria applied in Aristotle's *De anima*.

[75] Trans. Singer, in Galen, Claudius of Pergamon, *Psychological Writings*, ed. P. N. Singer, trans. Vivian Nutton et al. (Cambridge: Cambridge University Press, 2014).

From the standpoint of Galen's methodology, in this case suspension of judgment makes good sense. As far as his positive views on the faculties of the soul are concerned, his conception of faculties as placeholders for (unknown) true causes seems to entail a rejection of ontological realism about faculties. In this regard he seems to differ from Plato. Galen also had a strong tendency toward identifying the faculties of the soul with the faculties of the body,[76] which makes him seem more of an Aristotelian than a follower of Plato who, as I have shown, makes the *soul* the underlying subject of its faculties. But this is to be taken with the due grain of salt, given that Aristotle shares Plato's realism about faculties.[77]

There is of course much more to be said about the faculties of the soul in ancient philosophy. However, given limited space, I hope this overview conveys a rough idea of the development of the conception and presents the basic options that were on the table—options that were considered and extensively discussed by later Arabic and Latin authors.[78]

---

76 Richard J. Hankinson, "Partitioning the Soul," in "Galen on the Anatomy of the Psychic Functions and Mental Illness," in *Partitioning the Soul: Debates from Plato to Leibniz*, edited by Klaus Corcilius and Dominik Perler (Berlin: de Gruyter, 2014), 96–99, provides a concise discussion of Galen's complex attitude toward physicalism.

77 Galen's criticism against those who conceive of faculties as parts of the substance of things "in much the same way as we inhabit our houses" fits, on the face of it, with the Peripatetic conception of the soul as a set of faculties that constitutes the essence of living things.

78 I am grateful to the editor, Dominik Perler, for his patience and helpful suggestions, to Roland Wittwer for his help, and to Christian Pfeiffer for his critical comments on an early version of this chapter. Thanks go also to Michael Arsenault.

Reflection
FACULTIES AND SELF-DEBATE
Helene P. Foley

Faced with sudden isolation, Homeric heroes and rarely heroines (Penelope in the *Odyssey*) can confront difficult strategic decisions in extended dialogue with a part of themselves. This part or faculty, which can generally be understood as something like "heart" since it seems otherwise to be located in the midriff, is able to listen to reasons for taking action and be susceptible to pleas from emotion, in these contexts generally fear or anger. This faculty can be described with various Greek words: *kardiê, êtor, kêr,* or *phrên*, but above all "*thumos*." Homeric *noos* or mind is a mental faculty that can contrive plans and determine actions as well, but it does not engage in the kind of formulaic internal dialogues I have in mind here. In these particular internal dialogues an I (*egô* or a verb in the first person) invites "heart" (*thumos* is used in the opening address of these typical scenes) to weigh the details of particular difficult circumstances in which the subject finds himself or herself, to consider the views of others as well as prototypical examples of appropriate behavior in comparable circumstances, and then move to action.[1] For example, in *Iliad* 11, 403–10, Odysseus, left alone

---

[1] For background and discussion, see especially Bruno Snell, *The Discovery of the Mind*, translated by Thomas G. Rosenmeyer (New York: Harper, 1960), challenged by Bernard Williams, *Shame and Necessity* (Berkeley: University of California Press, 1993), and Christopher Gill, *Personality in Greek Epic, Tragedy, and Philosophy* (Oxford: Oxford University Press, 1996), with extensive further bibliography.

facing terrible odds in battle, considers in dialogue with his greathearted *thumos* (403) the possibility of withdrawing from the field. Flight due to fear would be bad for himself, he asserts, but to be caught alone after all the rest have scattered would be worse. Then Odysseus asks why his *thumos* is having this dialogue with him (*moi*, 407). For he knows (*oida*, 408, I know) that inferiors flee (as are the other Danaans at this point) and the best in battle stand their ground regardless. This citation of aristocratic heroic behavior apparently constitutes a decision, because Odysseus then stands his ground courageously. What kind of alternative faculty Odysseus's *thumos* represents in this dialogue is not entirely clear. It is a part of the self apparently capable of courage or favoring withdrawal. But unlike *egô* it is not here represented as knowing (at least to the same degree) standards of heroic behavior, nor does it initiate the final move to action.

Two of the other three lengthy internal dialogues on the battlefield introduce to the heroic *thumos* other complex considerations.[2] At *Iliad* 17, 90–103 Menelaus is worried on the one hand about retreating from defending the dead body of Patroclus and Achilles's arms, because his fellow Greeks who have been fighting the Trojans for him might be angry, and concerned on the other hand that many Trojans led by Hector will surround him. Fear and shame beset him on both sides. But, he now adds, no one of the Danaans will be angry at retreat when the enemy is supported by a god. The best of two evils would be to find Ajax and return to battle with him. Menelaus continues to ponder this course in his *phrên* and *thumos* as Hector and the Trojans draw nearer. He then seeks and finds Ajax.

In the most elaborate case (*Iliad* 22, 93–130), Hector, facing Achilles's terrifying approach for their final duel, introduces to his *thumos* the better course of a glorious fight to the death. This course

---

2 The third involves Agenor at *Iliad* 21, 552–70.

seems preferable partly due to his humiliating failure to listen to the strategic advice of Polydamas, which led to the deaths of so many Trojans. Hector next considers attempting to lay down his shield to negotiate with Achilles, but then, like Odysseus and Menelaus, asks why his *thumos* is in dialogue with him. Achilles would very likely not pity him and would slay him unarmed like a woman. He concludes that staying and fighting for glory is better. (The recognition of the better choice typically precipitates the move to action.) Once again, Hector's *thumos* seems to become linked by implication with the rejected (and feminizing—see further below) alternative. In these Iliadic battlefield decisions about whether to stand or retreat, however, both *egô* and *thumos* are jointly faced with weighing the applicability of social considerations familiar from public discussions of the proper behavior of warriors in battle. It seems that *egô*, rather than *thumos*, takes the dominant role in exploring the current context in the dialogue and setting up the final move to action with a second address to *thumos*. Hence, the *thumos* addressed and engaged in this dialogue can be approached or distanced, or susceptible to fear or commitment to others, but does not have a clearly identifiable character of its own.

In *Odyssey* 20, 5–24, however, the parts engaged in a tense strategic internal dialogue develop a richer character. Here the sleepless Odysseus, still disguised as a beggar, has observed the faithless serving maids departing to sleep with the suitors. He ponders in his *thumos* and *phrên*, which have been (typically) stirred by this event, whether he (*egô*) should kill them now or wait until he has taken vengeance on the suitors. His heart (now *kradiê*) growls over their deeds like a bitch who faces someone she does not know over her puppies and therefore wants to fight. Odysseus rebukes his now slightly more gendered or animal-like (due to the simile) "heart," and reminds it that it had faced a worse (literally, more doglike) experience when the Cyclops ate his comrades and it had to endure until *mêtis* (cunning intelligence, presumably another faculty) led it (you) out from the cave when it thought it would die. Odysseus's request to his *kradiê* to endure works. Nevertheless, Odysseus continues to lie tossing and turning over how to take revenge on the suitors like a man roasting a paunch before a fire who is eager to finish the job quickly. Athene then appears, reassures him, and puts him to sleep (*Odyssey* 20, 25–55). Both "heart" and Odysseus in this case share moral outrage at and a desire for revenge against the maids and, in the past, at the Cyclops's behavior, but *egô* seems more strongly linked with invoking the capacity to endure and the *mêtis* that extricated the suffering *thumos* from the Cyclops. (It should be noted though that Odysseus's *thumos* can elsewhere show the same capacity as *mêtis* here. In Odyssey 9, 302 another *thumos* checks his first one as he considers how to defeat the Cyclops and escape.)

In the *Odyssey*, Penelope's *thumos* also continues to wrestle with a difficult decision, but here the formula is adjusted for her different social context. As she describes it to Odysseus still in disguise as a beggar (*Odyssey* 19, 524–34; see also 16, 73–77), her *thumos* is debating, given that her son is now a man and worried about his possessions, whether to stay with Telemachus and keep her house, slaves, and possessions safe and respect her marriage bed and the voice of the people or to remarry the best of the suitors who woo her and offer the most bridal gifts. She does not know whether Odysseus is alive, and her son apparently wants her to make the choice, though he will not force her to do so. The poem represents this female self-debate in the *thumos* in different terms as a choice between responsibilities/commitments to others: to her marriage, son, household, and the voice of the people on the one hand and, implicitly, to preserving Telemachus's future and respecting the urging of her parents to remarry on the other. Both choices are traditional options for women, although normally they would be made for them by men. At *Odyssey* 18, 259–71 Penelope describes the parting instructions of Odysseus, which she has thus far followed. Indeed, Penelope views both her capacity for virtue (*aretê*) and her beauty to have been compromised by the absence of her husband (*Odyssey* 18, 251–53, 19, 124–26). The poem also offers her the *kleos* or fame so desired by the poem's elite in the self-dialogues discussed above, but in this case for remembering her husband and for her trick with the web (*Odyssey* 24, 196–97, 2, 125–26; at 19, 126–28 she claims that her *kleos* would be greater if Odysseus returned). This reported self-debate (which does not immediately lead to action) demonstrates self-restraint since Penelope nowhere expresses her own emotional preferences even though the audience knows that she does not desire to remarry. In this scene among others, the poem assimilates her to a masculine style of self-debate and self-restraint without raising questions about her adherence to a female role.[3]

By and large, tragedy does not describe a struggle between capacities in the self in dialogue in these Homeric terms. The partial exception is the famous monologue in Euripides's *Medea* (1020–80) in which the heroine's *thumos*, which has decided to kill her children in order to take revenge on Jason, is urged to consider the arguments for saving them. The self that wishes to save the children is maternal. Her *egô* is for the first time in the play vulnerable to the compelling presence of the children and to the pleasures that their survival would promise: preparing their marriages, knowing that they would tend her in old age and bury her properly. Medea conceptualizes the self that has determined on vengeance, which requires the killing in order to make Jason suffer, in terms traditional to masculine epic heroes. Like these heroes, Medea has earlier expressed the wish to do good to friends and bad to enemies, retain honor, avoid shame and laughter, and win *kleos* (383, 398, 402–6, 765–67, 797, 807–10; 1049–50). On the one hand

---

[3] See further Helene P. Foley, *Female Acts in Greek Tragedy* (Princeton: Princeton University Press, 2001), 126–43.

Jason has refused to recognize this heroic self in Medea, as well as the power of the marriage oaths they swore by the gods, and has neglected friends/relatives (*philoi*). On the other hand Medea, unlike Penelope, is here unable to enact an authoritative female self since she views it (like most of Attic drama) as weak, cowardly, and inferior (263–67, 1042, 1051–52). Earlier in the play Medea asserts she would rather stand three times in battle than bear a child (250–51). Buying an unknown husband with a dowry and taking a master for her body offers a wife a lottery (232–43). In an unhappy marriage the husband can escape from the house to which a wife is confined (244–47). Women on their own can contrive every evil (407–9) but can only do good for men, as Medea did for Jason (476–87). Dependent on men for happiness and survival, women's vulnerability is extreme. The female chorus, who thinks that women do not normally have the capacity to voice an answer to men (421–30), views its investment in children as consonant with its female identity as married women. Medea does represent her *thumos* as capable of persuasion and pity (1057). Yet by the close of the monologue, Medea's heroic self remains in control in what now seems an overdetermined fashion.

There are many textual difficulties and seeming inconsistencies in this *Medea* passage, and my summary here is open to argument on many counts.[4] This self-debate no doubt deliberately recalls yet in the end differs from and partly reverses those in epic. In each case an internal dialogue with a faculty that is subject to reasoned persuasion and emotion is conditioned by references to specific challenging circumstances and to traditional expectations for social behavior.[5] In the Homeric passages, the male *egô* ultimately initiates

---

[4] See Helene P. Foley, *Female Acts in Greek Tragedy* (Princeton: Princeton University Press, 2001), 243–71, and Christopher Gill, *Personality in Greek Epic, Tragedy, and Philosophy* (Oxford: Oxford University Press, 1996), chs. 2–3, with further extensive bibliography.

[5] Christopher Gill, *Personality in Greek Epic, Tragedy, and Philosophy* (Oxford: Oxford University Press, 1996), makes the argument that Medea, like Homeric heroes such as Achilles, struggles to reconcile a commitment to *philoi* and sense of irreconcilable injury (also from a *philos*) to her heroic sense of identity.

the move to action and thus establishes a kind of authority over his *thumos*, which is nevertheless addressed as and remains a part of himself. Both Odysseus's internal debate with his "heart" in *Odyssey* 20 and Penelope's within hers in *Odyssey* 19 involve a broader set of challenges to cultural identity than those on the battlefield. Odysseus is not only beleaguered, but disguised as an unarmed beggar; his "heart" growls like a bitch with puppies. An equally beleaguered Penelope is being forced to make a decision in the absence of and knowledge about her husband. Medea, by contrast, is directly challenged by a maternal capacity/voice in herself that she has heretofore neglected in favor of a heroic self. Her *thumos*, determined on revenge, retains final authority. All three characters look into the past and future in a far broader sense as they weigh their choices as part of a life pattern. Yet only Medea allows her *thumos* deliberately and with full understanding to initiate her own tragic suffering.

CHAPTER TWO

Faculties in Arabic Philosophy

Taneli Kukkonen

1. INTRODUCTION

In the history of the western doctrine of the faculties, Arabic philosophy figures prominently, for several reasons. First of all, speaking purely in historical terms, the Latin scholastics' reception of Aristotelian philosophical psychology was comprehensively conditioned by the way these materials were portrayed in the works of the Arabic philosophers translated into Latin in the twelfth and thirteenth centuries. Above all, this means that Avicenna and Averroes—but to a lesser extent also thinkers such as al-Kindī, al-Fārābī, and the Arabic medical authors—were instrumental in shaping European thinking on the faculties well into the Renaissance and beyond. Thus to write a history of the faculties without covering the principal Arabic authors would severely distort the picture.[1]

---

[1] This is also increasingly being recognized in the way such histories are written. For recent examples see Dag Nikolaus Hasse, *Avicenna's "De anima" in the Latin West: The Formation of a Peripatetic Philosophy*

Second, and more important, Arabic philosophy, far from being a mere conduit for transmission, contributed decisively to the development of faculty theory. In the first instance, the Muslim Peripatetics executed a thoroughgoing systematization of Peripatetic doctrine, one that consolidated many of the insights of the earlier discussions, for example, folding the Platonic tripartite soul into the overall picture. There were important innovations as well, notably the way Ibn Sīnā, or the Latin Avicenna (980–1037), presented a closely argued theory of the five so-called inner senses, cognitive capacities that bridge the gap between sense-perception and intellection. Avicenna provides a detailed account of how each of the inner senses maps on to the process of abstracting from particulars toward more and more universal content. Beyond this, Avicenna exhibits an avid interest in the way the inner senses, which humans share with higher animals, guide animals in their worldly pursuits, and how the living organism's motive faculties interact with the cognitive ones. Avicenna also thoroughly integrates his theory of the faculties with the available medical and anatomical knowledge of the time.[2]

Avicenna's pivotal role notwithstanding, the following presentation takes a thematic approach to the issues and draws examples from across *falsafa*, that is, classical Arabic philosophy up to around 1200 CE. In presenting the materials in this manner, my intention is not to intimate that all Arabic philosophers would have been saying the exact same thing when it comes to the faculties, since plainly they did not. Nor do

---

of the Soul 1160–1300 (London: Warburg Institute, 2001); and Dag Nikolaus Hasse, "The Soul's Faculties," in *The Cambridge History of Medieval Philosophy*, edited by Robert Pasnau (Cambridge: Cambridge University Press, 2010), 305–19; Simo Knuuttila, "Aristotle's Theory of Perception and Medieval Aristotelianism," in *Theories of Perception in Medieval and Early Modern Philosophy*, edited by Simo Knuuttila and Pekka Kärkkäinen (Dordrecht: Springer, 2008), 1–22; Pekka Kärkkäinen, "Internal Senses," in *Encyclopedia of Medieval Philosophy*, edited by Henrik Lagerlund (Dordrecht: Springer, 2011), 564–67; Luis Xavier López-Farjeat and Jörg Alejandro Tellkamp, eds., *Philosophical Psychology in Arabic Thought and the Latin Aristotelianism of the 13th Century* (Paris: J. Vrin, 2013).

2 A useful overview is provided by Robert E. Hall, "Intellect, Soul and Body in Ibn Sīnā: Systematic Synthesis and Development of the Aristotelian, Neoplatonic and Galenic Theories," in *Interpreting Avicenna: Science and Philosophy in Medieval Islam*, edited by Jon McGinnis and David C. Reisman (Leiden: Brill, 2004), 62–86; see now also Jon McGinnis, *Avicenna* (Oxford: Oxford University Press, 2010), 89–148.

I wish to imply that there would not be considerable intrinsic interest in the individual views of particular Muslim philosophers (patently, there is). My chosen mode of representation, rather, reflects the conviction that there are materials all across classical *falsafa* that should be of interest to nonspecialists in the field, whether contemporary philosophers or historians of philosophy specializing in other periods. This is particularly so when it comes to second-order questions regarding the faculties: what they are, how they are to be identified, and what their relation is to the living being to which they are attributed.[3]

2. Natures and Capacities

The first thing to note about the Arabic discussion is that it builds on the established Aristotelian doctrine of passive and active potentialities. This is squarely in keeping with the precedent set by Aristotle in *De anima*: in fact, the Arabic *quwwa* fully matches the polyvalency of the Greek *dunamis*, which according to context may be translated as "potentiality," "potency," "power," "capacity," or "faculty." In this essay, "faculty" will be the preferred translation in all those psychological contexts in which it can be made to make sense, with "power" and "potentiality" denoting active and passive potentialities more generally in instances where the two can be distinguished (and "capacity" being evoked where they cannot).

The Arabic Peripatetic philosophers were, furthermore, heirs to the post-Andronicus consensus regarding Aristotle's school works, which they viewed as a comprehensive and logically ordered curriculum. Within this overall conception of philosophy, Aristotle's teachings regarding the soul occupy a determinate position as a branch of natural philosophy, that is, the study of those corporeal entities having a principle of motion and rest inherent in them. *De anima* according to this

---

[3] To facilitate further exploration, I focus on primary sources for which an English or other western language translation exists, although I have as a rule made my own translations from the Arabic in order to highlight shared terminology.

understanding provides the general principles for treating all living things, and animals in particular, with some necessary refinements and specifications then occurring across the biological works. Ibn Bājja (d. 1139), the first Andalusian philosopher of note, accordingly begins his exposition *On the Soul* by providing a lengthy exposition of movers and things moved, and of agents of change and things being changed (*Nafs* 19–28). And before launching into a detailed examination of the faculty of nutrition, the first and most elementary of the soul's faculties, he again introduces first the notion of potentiality and actuality in relation to bodies (*Nafs* 43–48). Similarly, Ibn Rushd or the Latin Averroes (1126–1198) in his *Epitome* of Aristotle's *De anima* begins from the most general principles of nature and ends up with the composition of the homoeomerous bodies and of the organs that are the instruments of the living being (*Mukhtaṣar* 3–7). And, when paraphrasing Aristotle's definition of the soul in his so-called middle commentary, Averroes quickly runs through Aristotle's general comments concerning substance and then proceeds from here to natural bodies in order to situate living nature within the broader scheme of things (*Talkhīṣ* 43–44).

The Arabic doctrine of the faculties thus finds its rightful place within the context of Aristotelian natural philosophy; indeed, the doctrine is scarcely comprehensible detached from this wider conceptual framework. The various psychic faculties provide the explanatory mechanism for how living organisms navigate the world around them and the specific activities that characterize their existence (Avicenna, *Nafs* I.1, 4.5–7). In so doing, they merely replicate how things act upon other things and are in turn acted upon (Ar. *fiʿl, infiʿāl*). Above all, this is a matter of interlocking patterns of potentiality and actuality, seeing as how it is in actuality, actualization, and activity that the key to full-fledged being is found.[4] But if natural philosophy is explanatory of how living beings act and are acted upon, so also the psychic faculties

---

4 Averroes, *Commentarium* III, comm. 38, commenting on the Arabic Aristotle's words to the effect that the division of beings according to potentiality and actuality corresponds to the differences in the soul (*De an.* III.8, 431b20–26).

provide a paradigmatic example of how natural capacities are realized and what functions they perform. Avicenna, for instance, begins his exposition of potentiality in the metaphysical part of his encyclopedia *The Healing* by referring to the evidentiality with which acts issue from animals and how they do so by virtue of capacities inherent in them (*Ilāhiyyāt* IV.2.1). This establishes the functioning of living beings as the baseline from which to initiate an examination of how "potentiality, actuality, power, and impotence" work, to cite the title heading of Avicenna's chapter, much in the same manner that Aristotle referred to plants and animals as the evident starting point for discussing what substances might look like.[5]

Herein lies the great attraction of faculty theory for anyone operating within the Peripatetic philosophical paradigm. The explanatory power and scope of the potentiality-actuality dyad is both global on the one hand and remarkably malleable to the needs of the occasion on the other. From the most universal of ontologies to the overall principles of nature and to the realization of a single species's or individual's inherent possibilities, an appeal to potentiality can be made to explain every natural and nonaccidental kind of change. There is similar utility to defining being and form as actuality and activity. From the simplest sublunary existents, which is to say the four elements, all the way to the most complex life forms, and all across the ten categories, being is as being does, according to Aristotle.[6]

Faculty theory provides an excellent example of the conceptual scheme's systemic potential. Rather than mystifying the psychic functions as issuing from some "ghost in the machine" or, conversely, reducing them to mere material aggregates (Ibn Bājja, *Nafs* I, 39.9–41.5, summarizing *De anima* I), Aristotelian psychological theory promises to

---

[5] See Aristotle, *Met.* VIII.1, 1041b5–11, and *De an.* II.1, 412a11–13; a reading of this passage and of the Arabic philosophers' understanding of the concept of potentiality is available in Taneli Kukkonen, "Potentiality in Classical Arabic Thought," in *Handbook of Potentiality*, edited by Kristina Engelhard and Michael Quante (Dordrecht: Springer, forthcoming).

[6] See Aryeh Kosman, *The Activity of Being: An Essay on Aristotle's Ontology* (Cambridge, MA: Harvard University Press, 2013).

present the faculties as capacities whose very exercise constitutes their actualization and thereby contributes directly to their, as well as their subject's, ontological makeup. Coupled with the hylomorphism that characterizes Aristotelianism as a whole, the approach combines an admirable level of vertical integration (each living being's faculties all ultimately serve that organism's life and specific mode of being) with a great deal of modularity when and where the latter is desired (each psychic faculty can be treated in isolation in terms of its form, matter, and moving principles).

Averroes's style of paraphrasing Aristotle is instructive. On Averroes's understanding, in order to gain an understanding of the soul in general, one must first investigate its principal manifestations (the nutritive, the sensory, and the imaginative soul, with disembodied reason failing to make an appearance in this context: see *Commentarium* II, comm. 31). This, again, occurs through identifying the faculties associated with these types of souls: and the faculties in question in turn can only be understood in terms of their proper activities. All of this culminates in the study either of the objects of these acts or else of those things that act on the faculties that are passive in nature: the nature of foods and of sense-objects, and of all those natural kinds with which living beings interact. All this is because, as Aristotle states in the *Physics* (I.1), we must always begin from what is better known to us. In this fashion, psychology becomes firmly embedded in natural philosophy. Organisms as natural entities live out their existence in constant interaction with their immediate environment so that correspondingly, their capacities are defined by what they find themselves relating and responding to (*Talkhīṣ* 54–55). A map of the faculties therefore simultaneously serves as a catalogue of those features of external reality most salient to a given organism's purposes.[7]

---

7 Averroes's treatment noticeably removes the question mark from Aristotle's open-ended pondering at the top of *De anima* as to whether this kind of procedure—backing away from questions concerning the soul's essence all the way to the objects of sensation and thought—might not be appropriate: see *De an.* I.1, 402b6–17.

At the same time, the identification of distinct psychic faculties can act as a bulwark against reductionist physicalism. Based on the old Empedoclean principle of like acting upon like—understood here as the need for any potential to be actualized by something that already actually possesses that quality—Averroes in his *Epitome* of *De anima* (*Mukhtaṣar* 7) argues that a living thing, because it minimally possesses the power of nutrition (see Aristotle, *De anima* II.4, 415a23–26), can only issue from a form-giving faculty (*quwwa muṣawwira*) that itself is of the genus of the nutritive soul. What this amounts to is a kind of argument against the notion of emergent complexity, since it denies the possibility of a complex form and disposition (*shakl wakhulqa*) arising from vital heat alone. A hot thing, when heating something colder than itself, imparts upon the subject a form identical to its own, whereas in the reception of life something similarly living must be assumed to be the active power (*Mukhtaṣar* 23–24). This principle can be taken in many ways—Averroes early on appears to have used it as an argument for a supernatural Giver of Forms, whereas later in life he uses the same principle to argue for the transmission of the substantial form through seed—but what the principle unmistakably does is insist on the irreducible complexity of living substances in accordance with their functions. Based on an Aristotelian argument against Empedocles (*De an.* II.3, 415b28–416a5; see *Commentarium* II, comm. 38), Averroes adduces the example of the plant's ability to grow in several different ways in several different directions at once—both to put down roots and to spread its branches outward and upward—which is something no simple body or elemental mixture could do on account of its mere elemental composition (*Talkhīṣ* 48; *Commentarium* II, comm. 14).

Al-Fārābī (d. 950) in his survey work *The Philosophy of Aristotle* similarly describes a transition from the doctrine of the elements—a topic whose consideration culminates in the fourth book of Aristotle's *Meteorology*, the last book in the established curriculum to treat inanimate principles—to the study of sensible qualities, and does so in terms of the powers inherent in things. The question is whether all such powers can be traced back to the elementary level: al-Fārābī's contention is that they cannot, which means that the study of the more complex causal interactions between things must be relegated to a separate study of living nature (*Philosophy of Aristotle* IX.65–66, and see X.74; see Avicenna, *Nafs* II.2, 62). For psychology and biology to stake their claims as autonomous disciplines, therefore, at least as far as al-Fārābī and Averroes are concerned, they must demonstrate the existence of physical processes sufficiently complex so as not to be reducible to elemental physics. Whether through the primitive motive powers of nutrition and growth, or the passive potentiality of taking on the form without the matter that characterizes Aristotelian perception, the faculties of living beings fulfil this remit.

3. Enumerating the Faculties

All the major Arabic authors adopted Aristotle's basic tripartite division between the vegetative, animal, and rational functions. They furthermore trusted to Aristotle's guidance when it came to distinguishing between the different aspects of plant, animal, and human life. Thus nutrition, growth, and the drive to reproduce characterize all life in general, which makes each of the three into faculties of the vegetative soul; self-motion and sensation are added functions apportioned to animals, meaning that they must issue from faculties of their own; while the faculty of reason itself divides into two, theoretical and practical reason, just as Aristotle had taught. To be alive, meanwhile, is defined as having even one of these functions and activities (see Averroes, *Talkhīṣ* 48; *Commentarium* II, comm. 13).

Presentations of the faculties can certainly be identified that diverge from the standard Aristotelian lines of division, such as the so-called Brethren of Purity's insistence in their *Letters* that God granted plants seven distinct powers, or Abū l-Ḥasan al-ʿĀmirī's (d. 992) curious tripartite division between the rational powers.[8] But even here the same basic principle for faculty differentiation holds that had served philosophers since its original formulation in Plato's *Republic*: psychic forces are distinguished by their different or opposing orientations. The Brethren of Purity, for instance, maintain that each power works differently and bears a different relation to the soul; they also maintain that the human being has innumerable powers known only to God (*Rasāʾil* 2:468). While the conclusion is not particularly helpful, the stated principle is. As Avicenna puts it more clearly and concisely, "each faculty, qua being what it is, is what it is because a principal action proper to it issues from it."[9]

On the basis of the principle that a thing is identified by those actions peculiar to it (thus also, e.g., Miskawayh, *Tahdhīb*, 10–11), the Arabic Peripatetics appear to have been willing to designate the various parts of Plato's tripartite soul as so many "faculties," "powers," or "potentialities" (pl. *quwan*) as well, and to do so with little hesitation. This habit of bringing in the appetitive (desiderative) and spirited (irascible) parts of the soul—sometimes called simply the appetite and the spirit, or the appetitive and spirited "souls" in the plural—under the aegis of the soul's faculties is in evidence in nontechnical works on ethics from writers such as Yaḥyā Ibn ʿAdī (d. 974) and Miskawayh (d. 1030). It may originally have been inspired by, and very likely did find legitimation in, Aristotle's willingness to evoke the notion of contending psychic forces on roughly the Platonic model in the context of

---

8 See *Rasāʾil*, 2:156–57; Susanne Diwald, "Die Seele und ihre geistigen Kräfte: Darstellung und philosophiegeschichtlicher Hintergrund im K. Ikhwān aṣ-Ṣafāʾ," in *Islamic Philosophy and the Classical Tradition*, edited by Samuel M. Stern et al. (Oxford: Cassirer, 1972), 49, counts twenty-five distinct faculties in total in the *Letters* and remarks on their eclectic provenance. Al-ʿĀmirī distinguishes between the capacity to formulate inner thoughts (*iʿrāb ʿan al-ḍamīr*), cogitation (*fikr*), and the capacity for practically oriented opinion (*raʾy*): *Amad* VIII.1–2: on possible analogues and precedents, which do not go very far, see Rowson's comments on the passage in the same work.

9 *Nafs* V.7, 252.16–17. In the *Salvation* (*Najāt* II.6.15, 190.6) the equivalent formulation is put in a more convoluted manner: "the activity of each faculty is proper to the thing that is said to be its faculty."

moral psychology.[10] But it is equally prominent in the *Letters* of the Brethren of Purity (*Rasāʾil* 2:389), a work that is additionally helpful in pointing out a basic harmonization of Platonic, Aristotelian, and Galenic elements that became common in Arabic philosophy. In the *Letters*, as in Galenic thought, appetite is associated with the vegetative soul, spirit with the animal soul, and rationality with the human soul, with the associated Aristotelian faculties explained in terms of their utility in realizing the aims that each kind of soul assumes (*Rasāʾil* 1:313–15; see also Avicenna, *Canon of Medicine* I.6).

The signature doctrine developed by the Muslim philosophers, meanwhile, is that of the so-called internal senses, a set of cognitive capacities designed to account for the higher animal functions without thereby granting animals the power of thought. A first prototype of the doctrine appears in an early ninth-century version and paraphrase of Aristotle's *Parva naturalia* that has its origins in the circle of al-Kindī.[11] In this recension we find reference to the formative faculty (*quwwa muṣawwira*), memory (*ḥafẓ*), and cogitation (*fikr*). The Brethren of Purity also reference the internal senses, although their list looks quite different and may have come about as an amalgam of Platonic and Stoic influences (it includes the capacity for expressing *logoi*).[12] But it was Avicenna's theory, with its lucidly argued set of five internal senses, that proved most influential of all. Avicenna carefully makes the case for each of his five internal capacities (common sense, the formative power, memory, estimation, and imagination), localizing them in the ventricles of the brain in accordance with what medical theory

---

10 See Yaḥyā Ibn ʿAdī, *Tahdhīb*, 15; Miskawayh, *Tahdhīb*, 15–16; Aristotle, *Nicomachean Ethics* I.13 (but see *De an*. III.9, 432a19–b7); for the Aristotelian background and its commensuration with Plato, which is incomplete, see William W. Fortenbaugh, *Aristotle's Practical Side: On His Psychology, Ethics, Politics, and Rhetoric* (Leiden: Brill, 2006).

11 See Rotraud Hansberger, "Kitāb al-Ḥiss wa-l-maḥsūs: Aristotle's *Parva Naturalia* in Arabic Guise," in *Les Parva naturalia d' Aristote: Fortune antique et médiévale*, edited by Christophe Grellard and Pierre-Marie Morel (Paris: Presses Universitaires de Paris-Sorbonne, 2010), 143–62.

12 See Harry Austryn Wolfson, "The Internal Senses in Latin, Arabic, and Hebrew Philosophic Texts," *Harvard Theological Review* 28 (1935): 69–133.

intimated:[13] he elucidates what role each of the five plays in the cognitive processes of humans and higher animals, letting on how relatively sophisticated these nonrational cognitive processes can be and how they allow an animal to be appropriately responsive to its environment (*Nafs* I.5, 43–45; *Najāt* II.6.3, 162–63). Averroes's overtures in the direction of reordering the internal senses and his recommendation to eliminate estimation (*wahm*) from the list as an un-Aristotelian accretion mask a more basic agreement and should not be allowed to detract from the larger picture: Avicenna's overall understanding of the higher animal functions carried the day and formed the starting point to discussions going forward, both in Arabic and in Latin philosophy.[14]

4. FACULTIES AND THEIR OBJECTS

The preceding exposition can make it seem as though the Arabic philosophers took on the Aristotelian faculty scheme without consideration or argument. To an extent this is true: the threefold division between the nutritive, animal, and rational faculties is presented as matter-of-factly in the Muslim philosophers' works as it is in the original Aristotelian treatises. The Platonic parts of the soul, meanwhile (appetite, spirit, reason), appear to have been subsumed under the rubric of faculty theory mainly due to the latter's gravitational pull—this notwithstanding the different argumentative aims of their original presentation. (Plato's parts of the soul are meant to explain psychic conflict, while Aristotle's faculties explain how the diverse life functions of animals work in concert.) This is not to say, however, that the Arabic philosophers would have been lacking in tools for distin-

---

13 See Gotthard Strohmaier, *Von Demokrit bis Dante* (Hildesheim: Georg Olms Verlag, 1996), 330–41.
14 See Helmut Gätje, "Die inneren Sinne bei Averroes." *Zeitschrift der Deutschen Morgenländischen Gesellschaft* 115 (1965): 255–93; Dag Nikolaus Hasse, *Avicenna's "De anima" in the Latin West: The Formation of a Peripatetic Philosophy of the Soul 1160–1300* (London: Warburg Institute, 2001), presents an overview of Avicenna's influence on Latin scholasticism; a similar history has yet to be written for Avicenna and later Islamic philosophical psychology.

guishing between the individual psychic faculties in a more principled manner.

Since Plato and Aristotle, one principal tool for distinguishing between the various faculties was to posit that each existed in order to access a different kind of object. As Aristotle already had indicated, this does not really work for distinguishing appetite and desire from reason, since the upshot then would rather be that there are infinite appetites, one for each and every thing desired (*De an.* III.9, 432a22–b4; Averroes, *Commentarium* III, comm. 41). However, in Arabic Aristotelianism, for each modality of perception at least, a different type of entity ends up being postulated—if not a sensible form as such, then a concomitant "intention" (*ma'nā*), to borrow Avicenna's terminology.[15]

But this appears to result in a radical proliferation of entities or—more properly—kinds of entities. A perspicuously worded illustration comes from al-Ghazālī's (1056–1111) famous autobiography, *The Deliverer from Error*. Building the case for the presence of what al-Ghazālī terms the prophetic faculty in some exceptional individuals, al-Ghazālī contends that

> the human substance in its original, innate nature [jawhar al-insān fī aṣl al-fiṭrati] is created bare and innocent, with no information of God's worlds. These worlds are many and nobody but God can number them, as He says: "No-one knows the Hosts of your Lord but He." (Q. 74:31) A person receives information of the worlds by means of perception: each type of perception is created in order that one of the worlds of existents may be disclosed to him. By worlds I mean the [distinct] genera of existents.
>
> Thus, the first thing to be created in a human being is the sense of touch.... Touch, however, falls decisively short of colours and sounds:

---

15 Intentions encompass things such as enmity: despite these not being primary sensibles, and despite their definition being species-dependent (the perch will perceive the pike as a predator, whereas for the human being it will be prey), intentions really exist in the external world, which results in a rather liberal ontology. See Avicenna, *Nafs* IV.1, 166–67.

these are, as it were, nonexistent as far as the reality of touch [is concerned]. Next, sight is created... then hearing... then taste, and so on until the world of sense-objects is exhausted and discernment [tamyīz] is created, approximately at [the age of] seven years. This is a stage [tawr] whose being is different from the others, since here one perceives things in addition to the sense-objects, things that have no existence in the world of sensation.

From here the human being advances to another stage, as his intellect is created. Thus he perceives the necessary, contingent, and impossible, and things that did not exist in the previous stages.

Beyond the intellect there is yet another stage at which another eye is opened, by which a person sees what is concealed [al-ghayb]. (*Munqidh*, 41)

Excepting the closing reference to prophetic vision, this is all standard Peripatetic fare.[16] But what catches the eye is al-Ghazālī's talk of different worlds, which makes it appear as though the different cognitive faculties address wholly separate ontological domains. Though the figure of speech is plainly metaphorical, it finds support in al-Ghazālī's talk elsewhere of faculties being ignorant of the objects of one another, to the point of sounds being nonexistent as far as sight is concerned, and vice versa.[17] The overall impression thus remains one of each cognitive faculty providing access to a distinct aspect of reality.

Perhaps in the case of the proper sensibles, this proliferation of layers of being does not matter overly much (though it should be said that Aristotle can be read in a less ontologically liberal fashion as well).[18]

---

16 See Taneli Kukkonen, "The Self as Enemy, the Self as Divine: A Crossroads in the Development of Islamic Anthropology," in *Ancient Philosophy of the Self*, edited by Juha Sihvola and Pauliina Remes (Dordrecht: Springer, 2008), 205–24, and "Receptive to Reality: Al-Ghazālī on the Structure of the Soul," *Muslim World* 102 (2012): 541–61; Alexander Treiger, *Inspired Knowledge in Islamic Thought: Al-Ghazālī's Theory of Mystical Cognition and Its Avicennian Foundation* (London: Routledge, 2012).

17 See al-Ghazālī, *Iḥyā'* XXXVI, 2585.5–14; see Ibn Ṭufayl, *Ḥayy*, 7–8; the argument goes back to Miskawayh; see Peter Adamson and Peter Pormann, "More Than Heat and Light: Miskawayh's *Epistle on Soul and Intellect*," *Muslim World* 102 (2012): 521.

18 See Stephen Everson, *Aristotle on Perception* (Oxford: Clarendon, 1997).

For as long as the various sensible qualities can be classed as so many bodily accidents, there is a principled way in which to describe all the cognitive faculties as relating to the same reality and thus of the faculties working in concert to form a comprehensive picture of that self-same reality. But a more radical option presents itself. In the psychological and epistemological works of a line of thinkers stretching from al-Kindī (d. c. 870), the first "philosopher of the Arabs," to at least Miskawayh (d. 1030), who was an older contemporary of Avicenna, we find an argument that has it that because the different cognitive faculties access altogether different objects, sensation cannot in the end contribute much at all—if anything—to the workings of the intellect.[19] Al-ʿĀmirī, for instance, paints the difference in the starkest of terms, indicating that the sensual and rational souls have utterly different objects (*Amad* X.5) and that they are receptacles (*maḥall*) to different sorts of cognition entirely (*Amad* XVII.10).

This forcible separation of perception from intellection has two main consequences. On the ontological plane, the possibility opens up of regarding the sensitive and the rational faculties as belonging to two different orders of reality entirely, with only an accidental connection between them. The Brethren of Purity, for instance, say that by (or through) the soul, which is a mortal and embodied principle, the human being hears, sees, smells, tastes, and touches; it is similarly by the soul that the human being eats, drinks, sleeps, and is delighted or sad. By contrast, it is by the spirit that a person intellects and understands and so on (*Rasāʾil* 1:301). This particular way of distinguishing soul from spirit stands in contrast to the more usual parlance, according to which *rūḥ* was associated with the *pneuma* of Greek medicine and therefore identified with a particular kind of subtle body;[20] it resonates

---

19 Peter Adamson, "The Kindian Tradition: The Structure of Philosophy in Arabic Neoplatonism," in *The Libraries of the Neoplatonists*, edited by Cristina D'Ancona (Leiden: Brill, 2010), 360–64.

20 For materials on the standard spirit-soul distinction see, e.g., Y. Tzvi Langermann, "Abū al-Faraj ibn al-Ṭayyib on Spirit and Soul," *Le Muséon* 122 (2009): 149–58. Besides the works of the philosophers, instrumental to the transmission into Latin was the medically oriented *On the Difference between the Spirit and the Soul,* by Qusṭā Ibn Lūqā (d. 912). For the orphan status of spirit and soul in

more with the Platonizing elevation of the spiritual (*rūḥānī*) in the early phase of Graeco-Arabic translations.[21] But the point, beyond contingent matters of vocabulary, still stands: as soon as one is willing to disassociate altogether the functions of the rational soul from those of the animal soul, then any metaphorical talk of two (or more) souls threatens to turn realist.

On the level of epistemology, this opens up a corresponding explanatory gap that is not easy to bridge. If the sensitive soul merely senses, while the intellectual power takes as its object things beyond the confines of the physical world, then what is the point, for example, of doing natural philosophy? Miskawayh, for one, indicates that what the philosophical training of the soul is meant to do is wean us off sense perception entirely, so that through an engagement with the quadrivium—the mathematical sciences, that is, things separable if not always actually separated from matter—our attentions come to be turned toward the intelligible instead.[22] But this leaves open the question of whether the human being's various faculties operate in concert or in fact are at odds—the lower cognitive faculties pulling in the direction of the sensible world, the intellect being drawn toward the intelligible.

In point of fact, of course, the Platonic tradition had long contended that the rational soul's path to *anamnēsis* does take its start from the sensible universe, notwithstanding the latter's lowly status. In keeping with the general explanatory scheme of ancient Platonism, al-ʿĀmirī contends that sensual impressions contribute something, at least, to the intellectual soul's awakening (*Amad* X.10). Thanks to the fact that the things in the sensual world are likenesses of things in the upper world, we, resembling frogs or fish gazing at the distant stars from the

---

early Islamic theological discussions, and for the interpretation of spirit along corpuscular lines, see now Ayman Shihadeh, "Classical Ashʿarī Anthropology: Body, Life and Spirit," *Muslim World* 102 (2012): 433–77.

21 See Gerhard Endress, "Platonizing Aristotle: The Concept of 'Spiritual' (rūḥānī) as a Keyword of the Neoplatonic Strand in Early Arabic Aristotelianism," *Studia graeco-arabica* 2 (2012): 265–79.

22 See Peter Adamson and Peter Pormann, "More Than Heat and Light: Miskawayh's *Epistle on Soul and Intellect*," *Muslim World* 102 (2012): 521.

banks of a muddy river, may be struck by the dim reflections we see of the glory of transcendent things, so that our own higher part becomes awakened to its own true nature (*Amad* XVI.1–3). This indicates that the two souls have to enjoy a nonaccidental connection, which al-ʿĀmirī indeed confirms in another place (*Amad* XX.6–7). Yet the mechanisms for all this are left quite unclear.

This is why Avicenna's intervention on behalf of a robust abstractive view of the degrees of cognition is so important. Avicenna, too, maintains that the animal faculties assist the rational soul: but this is no idle sentiment; instead, Avicenna has a detailed story in which the internal senses each contribute to the form of a sensible object's being gradually stripped of its matter so that finally, the next closest thing to a pure universal may emerge as the substrate for intellectual cognition. (*Nafs* I.5, 50–51; II.2; II.6.7, 168–71; V.3; *Najāt* II.6.11, 182–83) The literature on Avicenna and abstraction has grown quite extensive in recent years.[23] While many philosophical questions remain as to how the abstractive process gels with the ultimate emanation of the intelligibles (i.e., the true universals needed for scientific knowledge) from the separate agent intellect—a thesis to which Avicenna is otherwise pledged—what suffices for present purposes is that Avicenna can be seen as contributing to what Jon McGinnis has termed a project of naturalized epistemology,[24] a system whereby our cognitive apparatus naturally extracts from our environment not only immediate impressions and particular nuggets of information but also patterns and true

---

23 For a start see Dag Nikolaus Hasse, "Avicenna on Abstraction," in *Aspects of Avicenna*, edited by Robert Wisnovsky (Princeton: Markus Wiener, 2001), 39–72. Jon McGinnis, "Making Abstraction Less Abstract: The Logical, Psychological, and Metaphysical Dimensions of Avicenna's Theory of Abstraction," *Proceedings of the American Catholic Philosophical Association* 80 (2006): 169–83; Cristina D'Ancona, "Degrees of Abstraction in Avicenna: How to Combine Aristotle's *De Anima* and the Enneads," in *Theories of Perception in Medieval and Early Modern Philosophy*, edited by Simo Knuuttila and Pekka Kärkkäinen (Dordrecht: Springer, 2008), 47–71. Al-Fārābī, too, has a fleshed-out theory of abstraction (see Richard C. Taylor, "Abstraction in al-Fārābī," *Proceedings of the American Catholic Philosophical Association* 80 [2006]: 151–68), but his presentation wavers more than Avicenna's, and it does not make equal use of the apparatus of the inner senses.

24 See Jon McGinnis, "Avicenna's Naturalized Epistemology and Scientific Method," in *The Unity of Science in the Arabic Tradition*, edited by Shahid Rahman, Tony Street, and Hassan Tahiri (Dordrecht: Springer, 2008), 129–52.

universals—in Avicenna's words, the essential and accidental properties of things, and the difference between the two. This requires that our various faculties operate in concert in disclosing the inner structure of a single reality. It also entails that sensible reality *has* something approximating an intelligible structure—an important return to Aristotle's intentions, at a crucial juncture in history.[25]

Again, notwithstanding Averroes's impassioned entreaties against Avicenna and other latecomers when it comes to one particular aspect of Aristotelian noetics or another, his overall scheme is more akin to Avicenna's than it is different from it when it comes to these basic points. From the point of view of faculty psychology, what is important to recognize is that the senses, both external and internal, all contribute to the task of concept formation and the eventual grasp of substantial form, and that they do so through a process of consecutive abstraction. The venerable principle of distinguishing faculties by their objects nonetheless retains its relevance. To illustrate, consider Averroes's discussion of whether sensation alone suffices to produce appetite and thus motion. Simple animals such as flies and worms provide the test case: Averroes takes the position that since their movements *ex hypothesi* are a species of animal motion (*ḥaraka ḥayawāniyya*), not merely natural motion, some form of imagination must be thought to intervene between sensation and motion, even if in worms and flies and the like it is of such an indeterminate nature that it is not even distinguished from that which is sensed ("ghayr muḥaṣṣal lā yufāriq l-maḥsūs": *Mukhtaṣar* 97–98).

The argument seems question-begging at first, and the conclusion it reaches flimsy. After all, if the act of sensation is indistinguishable from the imaginative representation, why claim that there *is* a difference? And why should we not admit that sensation produces appetite directly and necessarily, just as the combination of light and warmth directly causes the plant to bend toward the sun? The answer, I believe, lies in Averroes's insistence that all animal action is intentional—that is to

---

25 See Dimitri Gutas, *Avicenna and the Aristotelian Tradition* (Leiden: Brill, 1988).

say, that it is predicated on the intentions (pl. *maʿānī*) that exist in the external world but that differ in significance in accordance with the animal species in question (the pike being predator to the perch, but prey to the fisherman). It is the apprehension of such intentions that drives animal action. Nor are these intentions in Averroes's scheme limited to the estimative objects of Avicenna.[26] Even simple sensations can mean different things to different animals, so that one seeks warmth while another shuns it, and perhaps even the same animal can feel excessively warm one moment while at another, warmth is precisely what it seeks. But if this is true, and it remains true at the same time that the cognitive structure of the sense of touch qua touch is identical to all animals at all times (as the Aristotelian explanatory scheme implies), then a different faculty must be evoked to explain how a single sensation can produce different outcomes.

Averroes goes on to discuss whether appetite builds on imagination the same way that perception builds on the nutritive soul. He concludes that it does not; rather, appetite follows upon any act of perception (*idrāk*), so that the appetite is of the nature of a concomitant accident (*lāḥiqa*: *Mukhtaṣar*, 98–99). The subject of appetite is the vital heat itself that courses through the animal. This psychophysical connection explains a number of phenomena, for instance, why the angry person turns red and the fearful yellow or pale. The fact, meanwhile, that the appetite is consequent upon the whole cognitive apparatus accounts for the fact that one can develop multiple opposing appetites, as when our animal appetites stand opposed to our cogitative appetites (*nuzūʿ fikrī*)—something we all can witness, according to Averroes.

5. The Active Intellect

Another way to distinguish the faculties is by way of their relative activity and passivity. When Avicenna states that the soul is known by

---

26 See Deborah Black, "Intentionality in Medieval Arabic Philosophy," *Quaestio* 10 (2010): 65–81.

the name of *quwwa* due as much to its capacity to undertake actions as to its ability to receive the sensible and intelligible forms (*Nafs* I.1, 6.1–4), for instance, this is said in recognition of the active and passive dimensions of the soul's workings.

The first and simplest way of making this distinction, as per Avicenna's quoted statement, is to separate the soul's motive faculties (which ostensibly are active, given how by definition they initiate motion) from the cognitive ones (which are all receptive and thus appear passive). Avicenna, after all, states categorically that "sense-perception is a kind of being-acted-upon, since it is the reception of the form of the sense-object" (Avicenna, *Nafs* II.2, 66.6), which would make each instance of reception similarly passive. But this is too simple, given that (1) the Arabic philosophers, following Aristotle, wanted to grant our intellectual capacity, notwithstanding its receptivity, some measure of self-directedness,[27] and, conversely, (2) the general consensus among the Arabic philosophers was that the lower-order capacities assigned to animals, whether motive or cognitive, are all essentially passive and reactive—in allowing the animal to find its orientation and place in the world, they guide it in more or less automatic fashion. Al-ʿĀmirī, for instance, categorically states that the sensory soul (*al-nafs al-ḥissiyya*) arrives at its acts of discernment (*tamyīz*) and the consequent actions driven by desire or aggression through being-acted-upon in the perceptual register ("bi-ḥasb al-infiʿāl al-idrākī": *Amad* V.11–12). By contrast, the intellect is active and self-sufficient in its operations (*Amad* V.15–16).[28]

However, the division between passive and active does not map tidily onto the sensation/intellection divide either. For the Brethren of

---

27 Thus, e.g., Averroes (*Commentarium* II.55 and III.3) underlines that intellection is passive *only* inasmuch as it is a mode of reception.

28 Avicenna, for his part, evokes the distinction between rational (*nāṭiqī*) and natural (*ṭabīʿī*) capacities already as a matter of metaphysical principle. The difference, Avicenna says in his *Metaphysics* (*Ilāhiyyāt* IV.2.10–11), is that natural potentialities are moved to actuality immediately and automatically upon the coming together of the appropriate active and passive potentialities, whereas a rational capacity lies within the agent's will to exercise or not.

Purity, for instance, the difference between the sensory perceptual faculties and those they call "spiritual" (*rūḥānī*) lies in the former's association with matter on the one hand (for which see above), while on the other hand what is decisive is the latter powers' ability to manipulate the perceptual imprint (*rasm*) as a whole (*Rasāʾil* 2:414; see also 2:471–72). Thanks to that ability, the imagination counts as a spiritual power as well, notwithstanding the fact that it is plainly tagged to a bodily organ. Avicenna similarly distinguishes the compositive imagination (*al-mutakhayyila*) from the faculty of estimation through the former's being active: so does the famed later theologian-philosopher Fakhr al-Dīn al-Rāzī (d. 1210), who manifestly follows Avicenna on this point and says that the imagination, which in humans is called the cogitative power (*mufakkira*), alone among the internal senses can be called operative (*mutaṣarrif*).[29] But this seems either to make the faculty of imagination into an autarchic, yet randomly operating, image-generator—as in some Aristotelian theories concerning the origin of dreams—or alternatively to turn the imagination into a subsidiary of our rational faculty (an option the animals do not have). Both options are in fact seized by Avicenna and the Avicennan tradition, for whom the compositive imagination works in the former way when associated with animals, in the latter way when guided by human reason.

Motivating the original pronouncements regarding the intellect's self-sufficiency and autarchy was the venerable Platonic notion according to which our reasoning part provides the means of our liberation from servitude to the passions. This was met by the Aristotelian desire to portray intellection as an activity in which we ourselves choose to engage or not to engage (*De an.* II.5, 417a21–b29). However, once the Arabic philosophers had assimilated and digested the full Aristotelian theory of the intellect, with its division between the material/potential and agent intellects, this picture came to be in need of serious revision.

---

29 See Jules Janssens, "Fakhr al-Dīn al-Rāzī on the Soul: A Critical Approach to Ibn Sīnā," *Muslim World* 102 (2012): 568.

After all, on the Aristotelian picture, the active party in intellection is not so much the individual human being as the agent intellect, a separate substance whose role is analogous to that of light in our act of seeing. We may gradually be able to advance to a position whereby our own intellect, which originally is purely material or potential, becomes so acquainted with the intelligibles as to become habitually capable of reconnecting with their transcendent source.[30] But this does little to alter the fundamental implication of the theory, which is that our rational soul is essentially receptive and passive when it comes to our reflection on the higher truths, just as much as our senses are receptive and passive when faced with their proper objects.

Avicenna in his psychology ingeniously turns these liabilities into strengths. The basics of his solution are spelled out in a memorable passage in which Avicenna likens the rational soul to an entity facing two ways at once:

> The human soul, as will become apparent later, is a single substance, relating and drawn to two sides, one of which is below it, the other above; and for each side it has a faculty through which the connection between it and this side is administered. Thus, the practical faculty constitutes the power [the soul] has for the sake of the connection it bears to the side that lies beneath it, i.e., the body and its governance. The theoretical faculty, meanwhile, is what [the soul] has for the sake of its connection to what lies above it—to be acted upon by it, to benefit from it, and to receive [things] from it. There are as it were two faces to the soul: a face that is [turned] toward the body—and this should not receive any influence from any kind [of thing] derived from the nature of the body—and a face turned

---

30 On this signature Arabic theory shared by everyone from al-Kindī to Ibn Rushd and beyond, and on its provenance in Alexander of Aphrodisias, see Marc Geoffroy, "La tradition arabe du *Peri nou* d'Alexandre d'Aphrodise et les origines de la théorie farabienne des quatre degrés de l'intellect," in *Aristotele e Alessandro di Afrodisia nella Tradizione Araba*, edited by Cristina D'Ancona and Giuseppe Serra (Padua: Il Poligrafo, 2002), 191–231.

toward the transcendent principles: and this face should be enduringly receptive to what is there and affected by it. Moral qualities issue from the lower face, while the sciences issue from the higher. (*Nafs* I.5, 47.8–18)

Notice how the workings of the theoretical intellect are described in unabashedly passive terms: the human intellect is supposed to be receptive and to be acted upon, and it is precisely in this that its perfection lies. By contrast, the practical intellect, when it comes under the influence of external factors, betrays its true nature. The practical intellect, precisely because it faces downward, should govern and presume an active position toward everything that lies beneath its station (see *Najāt* II.6.4, 163–65). Thus the entirety of a human being's faculties stand in a nested, hierarchical relation, with the lower always serving the higher (*Najāt* II.6.11, 168). But because the intelligible order is one and the same for all, there is nothing to be *done* with it, so to speak, beyond merely actively contemplating it, whereas the practical realm allows for greater leeway (there being no universal truths when it comes to practical life) and therefore an interpretation of activity that more closely corresponds to our understanding of independent agency.[31]

6. Substrate as Suitedness

Besides different faculties taking different objects (or having different agents activate them) and besides their enjoying differing levels of activity and passivity, Avicenna indicates that the faculties can also be distinguished according to the organs in which they inhere or must

---

31 I have argued in Taneli Kukkonen, "The Self as Enemy, the Self as Divine: A Crossroads in the Development of Islamic Anthropology," in *Ancient Philosophy of the Self*, edited by Juha Sihvola and Pauliina Remes (Dordrecht: Springer, 2008), 205–24, that this is the main lesson al-Ghazālī derives from Avicenna's metaphor of the soul's two faces.

inhere.³² At its most rudimentary, this is shown in the localization of the soul's three parts, with the spirited part having its seat in the heart while appetite issues from the liver (*Nafs* V.7, 252.11–12).

The entailments of this principle are best illustrated through looking at an argument Avicenna gives for distinguishing the common sense (*al-ḥiss al-mushtarak*), which he equates with Aristotelian *phantasia* (see *Najā t* II.6.6, 163.1; *Nafs* I.5, 51.5–6), from the formative faculty (*al-muṣawwira*). Both are to be counted among the internal senses, but one of them can only receive the overall impression of the thing conveyed by the senses, while another must see to its retention. As Jon McGinnis points out, this is based on the general principle that the faculty that receives must differ from the one that retains (*Nafs* I.5, 44.9; *Nafs* IV.1, 165.13; *Najāt* 163.5).³³ This in turn is based on physiology and ultimately elementary physics, given that malleability and permanence are underwritten by different and opposing primary qualities—one by that which is watery, the other by the earthen.

Abstracting from the particulars, one may say that on Avicenna's understanding, material conditions may dictate circumstances under which distinct faculties need to develop, even in situations where naked functionality would privilege a single mode of reception. For example, assume that much of the information we need concerning the natural world all about us can be obtained by monitoring the wavelength frequency spectrum. It would therefore be convenient if we possessed a single mode of perception that was able to scan the entire range of frequencies from the subsonic all the way to the terahertz range. However, no such sense-organ is available to us: instead, we must rely on a complement of organs, and accordingly it makes sense to talk about the accordant embedded capacities as constituting different faculties as well.

---

32 The three criteria are set side by side in *Najāt* II.6.3, 162–63; see Jon McGinnis, *Avicenna* (Oxford: Oxford University Press, 2010), 111–13.

33 See Jon McGinnis, *Avicenna* (Oxford: Oxford University Press, 2010), 112.

In light of this very example, however, this criterion for faculty differentiation seems to pull in a different direction from the one that evokes different objects for different faculties. Consider first that most Aristotelian philosophers interpreted the distinction by different perceptual objects much as al-Ghazālī did: our immediate and primitive recognition of different types of perceptions points to their origin in distinct faculties and therefore to the existence of equally many primitive, irreducible, and nonanalyzable ontological domains as well. Colors therefore come to be regarded as fundamentally different things from sounds, and so on. However, the modern physics of sound and light now tells us that these supposedly discrete phenomena in fact fall along a single spectrum of varying wavelengths, even if the greater part of that scale remains inaccessible to us in terms of direct experience. Does the notion of a psychic faculty then become mere convenient shorthand for denoting our direct epistemic capacities—more properly, for pointing out their limitations—with no ontological import at all? That would seem to follow from Avicenna's pragmatic line of argumentation in the case of the internal senses.

There are in fact hints toward this sort of conception in philosophical texts besides Avicenna's. Ibn Bājja puts it that "the faculties are only defined according to the relation the substrate bears to the disposition, and it is according to this that one faculty is distinguished from another: the sensory faculty, therefore, is the fittedness [that exists] in the sense-organ" (*Nafs* III, 94). Here by fittedness (*istiʿdād*) one means a particular receptivity to some specific types of sensory forms: what Ibn Bājja wants to say is that the eardrum is particularly suited to the reception of sound frequencies, just as the retina is particularly disposed to the reception of colors. Al-Fārābī likewise contends that each faculty "receives [what it does] from the agent according to what lies in its substance and fitness [*iʿtidād*] to receive from it" (*Madīna* XIV.3). The receptivity of each perceptual faculty, then, depends both on its form (what the faculty is set to do) and on its matter (what the material means are by which the task is accomplished): an eardrum, thanks to its physical constitution, is especially suited for the reception of sound, the cornea of visual information, and so on.[34] If sensible forms nonetheless are taken to belong to the same genus, then the different perceptual faculties merely add up to distinct "picking" mechanisms that divide up one and the same reality in a functional and yet ultimately contingent manner.

The notion of the faculties being distinguished according to their bodily organs receives a fresh twist when one takes into account that the intellect, the faculty most proper to human beings, is not supposed to be embodied at all. As a starting point, consider Ibn Ṭufayl's (d. 1185) philosophical novel, *Ḥayy Ibn Yaqẓān*, or *Living, Son of Wakeful*. The larger part of the treatise builds on the microcosm-macrocosm metaphor, mapping out how the psychological faculties are tethered to the structural features of the natural world.[35] Yet all this exposition is in service of a paradox: Ibn Ṭufayl in the end wants to say that the most valuable knowledge we can have of anything, which is an intimate acquaintance with God, is not reached through a standard faculty at all. Seeing as that which is wholly immaterial must also be altogether indivisible, it cannot be reached through the senses, either primarily or incidentally (*Ḥayy* 90.9–92.3; see Avicenna, *Nafs* V.2). And because the intellect's operations do not really befall any part of what Ibn Ṭufayl calls the heart—that is, the hylomorphic soul, the spirit that courses through our veins, or any of the organs with which we sense, perceive,

---

[34] Al-Fārābī furthermore adds that in many cases, what is received is not the external quality, quantity, form, or matter itself, but instead its likeness (*Madīna* XIV.4). This touches on a long-standing debate concerning what exactly is received when the various perceptibles are actualized in the soul qua perceptibles. We need not enter here the question of the coloration of the eye-jelly, which has vexed Aristotelians up until the present day: but it is worth noting, at least, that the question of whether the perceptibles cause corresponding corporeal changes in the recipient or instead only have "spiritual" effects has profound consequences for the way faculty theory is constructed. For Averroes's influential take on the problem see Alfred Ivry, "The Ontological Entailments of Averroes' Understanding of Perception," in *Theories of Perception in Medieval and Early Modern Philosophy*, edited by Simo Knuuttila and Pekka Kärkkäinen (Dordrecht: Springer, 2008), 73–86.

[35] See Taneli Kukkonen, "Body, Spirit, Form, Substance: Ibn Ṭufayl's Psychology," in *In the Age of Averroes: Arabic Philosophy in the Sixth/Twelfth Century*, edited by Peter Adamson (London: Warburg Institute, 2011), 195–214.

and execute our bodily functions (*Ḥayy* 121.3–10)—the intellect can only be called a faculty metaphorically (*ʿalā l-majāz*: *Ḥayy* 5.13).

Essentially the same point is put forward by Ibn Bājja (*Nafs* IX, 130), who claims that if a faculty exists that does not make use of an organ (*āla*), then this is not called soul except equivocally. To call the pure, disembodied intellect a soul is therefore problematic: even when found in conjunction (*ittiṣāl*) with a living body, whether it be that of a celestial sphere or the human as rational animal, the makeup of such a curious compound simply does not follow the standard hylomorphic formula.[36] The oddity of this arrangement could be used to argue for the essential immortality and separability of the human essence after its embodied life, as in Avicenna or al-ʿĀmirī,[37] or it could become a source for a sustained meditation on what it means for the so-called agent intellect to be "form for us" without, however, forcing an interpretation that would make this into a basis for individual immortality. The latter is what we find in Averroes's works.[38] In either case, the fact that the agent intellect in particular is to be regarded as a separate substance seriously compromises any attempt to read Aristotle's notoriously knotty *De anima* III.4–5 as simply another application of standard faculty theory.

If from immateriality, unity necessarily follows, as for instance Ibn Ṭufayl appears to think, then we run into a further difficulty, which is described by al-ʿĀmirī (*Amad* VIII.2) as follows. The essence of the

---

36 See Averroes, *De substantia orbis* II; *Tahāfut* IV, 270–71. Arabic authors inherited from Aristotle and the late antique school tradition the notion that the heavens must be deemed to be alive. But because the heavenly movers (which is to say, the heavenly souls or intelligences) are not embedded in bodies in the standard manner, and because their activities are eternally actual without ever lapsing into a state of potentiality, the "souls of the spheres are not faculties at all," but are only said to possess something akin to them in an accidental way (Ibn Bājja, *Nafs* II, 49). Averroes says that "evidently" the heavenly movers possess understanding and self-motion yet lack nutrition and sensation (*Commentarium* II, comm. 15; see also *Commentarium* III, comms. 60–62).

37 For Avicenna see *Nafs* V.4; *Najāt* II.6.13, 185–89; Thérèse-Anne Druart, "The Human Soul's Individuation and Its Survival after the Body's Death: Avicenna on the Causal Relation between Body and Soul," *Arabic Sciences and Philosophy* 10 (2000): 259–73; for al-ʿĀmirī, *Amad* XIV.4–5.

38 See Richard C. Taylor, "The Agent Intellect as 'Form for Us' and Averroes's Critique of al-Fārābī," *Proceedings of the Society for Medieval Logic and Metaphysics* 5 (2005): 18–32.

rational soul, though it is described in terms of a unity, is multiple from the perspective of its faculties. Thanks to its incorporeal nature, these all nonetheless issue from a single form, which is the intellectual power (*quwwa ʿāqila*: *Amad* VIII.8). The different intellectual faculties are distinguished according to the different acts that the rational soul has to perform (*Amad* VIII.1), yet because each act can be said to contain elements of the others, the level of unity enjoyed by the soul's immaterial substance (*jawhar al-nafs*) is greater than anything found in the perceptual universe (*Amad* XI.1–3).[39] But if inner speech, cogitation, and opinion formation all mutually imply one another, what then is the purpose of their differentiation in the first place?

7. Psychic Unity and Its Limits

Much the same question can be put in any context where the Aristotelian theory of the essential unity of the hylomorphic organism receives emphasis.[40] This is a common thread in Arabic Aristotelianism. The famed translator-scholar Ḥunayn Ibn Isḥāq (d. 873), for instance, already states that the vegetative, sensory, and rational faculties are called so many souls only metaphorically (*Farq* 107). And the Brethren of Purity in their *Letters* underline that the soul is essentially one even if it is called the vegetative soul according to how it is engaged in nutrition and growth, the animal soul according to the way it senses and moves the body, and the rational soul in accordance with cogitation and discernment (*Rasāʾil* 2:386–87).

---

39 By contrast, the sensual soul's or spirit's division into powers happens according to the bodily organs and temperaments that allow for the different modes of embodied perception, evaluation, and action (*Amad* VIII.7). Their connection is sequential rather than synchronous: appetite can only be aroused on the basis of imagining, etc. (*Amad* XI.5).
40 Consider here Averroes, *Commentarium* I, comms. 89–90 (Ibn Rushd, Abū l-Walīd Muḥammad Ibn Aḥmad, *Commentarium magnum in Aristotelis "De anima" libros*, edited by F. Stuart Crawford [Cambridge, MA: Mediaeval Academy of America], 1953, 120–21) against Plato's habit of dividing the faculties according to the body parts: "What then holds the soul together?"

Are the various faculties extensions of the living being in question, then, or are they constitutive of the being itself? Al-Fārābī in the *Philosophy of Aristotle* (X.76) unambiguously states that the psychic faculties are merely instruments (*ālāt*) that allow the animal to navigate its environment successfully and to execute what it must in order to live. For al-Fārābī the principal point is that the living being's various powers and potentialities all come under the aegis of promoting a single form of life: "in animate substances," al-Fārābī contends, "nature is not for its own sake but for the sake of the soul." If a touch of anachronistic terminology is permitted, al-Fārābī in this passage is once again trying to delimit the reach of an emergent materialism. The various natural processes that come together under, say, the bodily functions of the living being, whether those of digestion or of sense-perception, are not self-explanatory but instead originate in the needs of the living being as a whole. This notion of a hylomorphic subordination of a living being's powers and potentialities extends to the internal organization of the faculties as well:

> The ruling faculty of nutrition is like matter for the ruling faculty of sense, whereas the sensory faculty is the form of the faculty of nutrition; the ruling faculty of sense is matter for the imagination, whereas the imagination is the form of the ruling faculty of sense; the imaginative faculty is matter for the rational faculty, whereas the faculty of reason is the form of the imaginative faculty without being matter for another faculty: it is the final form of all the forms that precede it. The appetitive faculty is dependent on the ruling faculty of sense and the imaginative and rational faculties, in the manner that heat exists in fire and is dependent on the substantiality of fire. (*Madīna* X.9)

The picture al-Fārābī evokes is one where each higher-order function actively subsumes the lower ones. This means that the modularity of the faculties only goes so far. In the end, only the highest-order entity exists as a fully fledged, actual, and active being, with the lower-grade functions being at best analytically distinguishable within the overall workings of the organism. This also means that the proper locus of explanation is found at level of the whole animal or plant. A firm line is drawn under all this by Avicenna, who coolly remarks that while loose talk of things such as the animal soul and the human soul is understandable—given that one can identify certain faculties and acts that show a kind of kinship—one can only really address in scientific terms the fully defined species. Such generic notions as "animality" can at best be adduced as genus terms: they may form part of a definition, but never the whole of it (*Nafs* I.5, 40.4–13). This is a rather technical way of saying the same thing al-Fārābī says, namely, that the faculties are primarily a heuristic tool.

An acute reader of Avicenna such as Bahmanyār Ibn al-Marzubān (d. 1066) would notice that on the Avicennan picture, the human soul is not one with its faculties, not even with its intellectual power, since not even the intellectual faculty is contained in the very definition of what it is to be soul: instead, the actuality and active power of the soul is what grounds the faculties. The faculties may be necessary concomitants (pl. *lawāzim*) of the living being to which they belong, but they still come second in the ontological order of things.[41] On the epistemological front, what this means is that the faculties cannot be used to get at a "science of the soul" (Avicenna, *Nafs* I.1, 10.15–18; see Ibn Bājja, *Nafs* I, 39). Instead, what the various cognitive faculties do is reflect the different domains in which the living being must find its footing and with which the organism interacts. This may not appear too alarming from the point of view of the modern reader; indeed, the result may be welcomed by many. The faculties, far from becoming reified, instead point in the direction of the outward reality with which they are associated, whether by way of the ultimate objects that form the subject of

---

41 Meryem Sebti, "The Ontological Link between Body and Soul in Bahmanyār's *Kitāb al-Taḥṣīl*," *Muslim World* 102 (2012): 531–32.

one's attention or the organic materials by which access to external reality is granted (the fitness of the various limbs and organs for their intended purposes—see above).

However, taken to its conclusion, such a conviction may in the end seriously undermine any talk of the faculties. Fakhr al-Dīn al-Rāzī, for instance, in his mature authorship appears at first to take on board the Aristotelian and Avicennian classification, with its distinction between the vegetative, animal, and rational functions and the further division of these powers according to the distinct operations that plants, animals, and humans perform, along with the further differentiation between the appetitive, spirited, and rational forces.[42] However, al-Rāzī then goes on to say that although the philosophers assign each different type of action to a different faculty, all the various kinds of perception and all the different types of action in fact belong to the soul's substance. It is the body parts that are the proper instruments for the execution of the soul's actions: the eye for seeing, the ear for hearing, the tongue for speaking, and so on (*Nafs wa-rūḥ* VII, 67–68). But it is the soul as substance that actively *does* all these things, rather than any ghostly faculties in some ill-defined reified fashion.

Al-Rāzī in his treatise *On the Soul and the Spirit* refers to other works of his for the specifics of how his account differs from that of the philosophers. Jules Janssens has given us one such detailed comparison: striking here is that al-Rāzī refuses to accept the very principle, *ex uno unum*, that had been used to differentiate the faculties according to their operations since the time of Aristotle.[43] After discussing the appetitive, spirited, and rational impulses, which in the literature are often treated as semiindependent forces on the model of Plato's *Republic*, al-Rāzī similarly goes on to say that these are not to be regarded as distinctive things independent in their being (*ashyā' mutabāyina*

---

42 See Jules Janssens, "Fakhr al-Dīn al-Rāzī on the Soul: A Critical Approach to Ibn Sīnā," *Muslim World* 102 (2012): 567–68.
43 See Jules Janssens, "Fakhr al-Dīn al-Rāzī on the Soul: A Critical Approach to Ibn Sīnā," *Muslim World* 102 (2012): 568–70.

mustaqila bi-anfusi-hā). Instead, they are, as it were, three tributaries from a single root, branches from a single tree, streams from a wellspring, and so on: or, as the most telling formulation goes, as a man who performs three actions and is thereby called by three different names: blacksmith, goldsmith, builder (*Nafs wa-rūḥ* IX, 84). This is certainly potentially quite radical. Taken literally, it would signal that the division of the soul's actions into faculties would only be nominal, hinging on the bodily instruments with which the soul works in any given instance. Al-Rāzī's formulation in fact represents a further step in the dualist direction from Avicenna's original psychological scheme, which uneasily straddles the substance and hylomorphic models of the soul. Further investigation into postclassical Arabic philosophy would be needed to establish what becomes of Avicennan faculty psychology after al-Rāzī's appropriation and critique of it; that, however, lies beyond the purview of this chapter.

In sum, I have shown that classical Arabic presentations of the psychic faculties draw liberally on the Aristotelian explanatory scheme of potentiality and actuality. It is in this that their chief strength lies. At the same time, it is evident that what puts pressure on this explanatory scheme is the desire to incorporate the peculiar human capacity for intellection within it. Whether one considers the intellect's proper objects, its active nature, or the question of its substrate, it appears that the boundaries of Aristotelian naturalism are easily breached once one tries to account for how our human power of reasoning is supposed to work. This observation is unsurprising in a sense, given how Aristotle himself had already indicated that the concepts of potentiality and actuality had to be applied in nonstandard ways where our reasoning faculty was concerned.

CHAPTER THREE

Faculties in Medieval Philosophy

Dominik Perler

1. Introduction: Faculty Theories—A Scholastic Failure?

Medieval philosophers in the Aristotelian tradition took it to be evident that all living beings— plants, animals, and human beings—have many faculties that enable them to produce a large variety of activities. Since different types of activities require different types of faculties, a clear order must be established among them. That is why scholastic authors usually distinguished three types of faculties: plants have vegetative faculties only, enabling them to nourish and grow; animals have additional sensory faculties, making it possible for them to perceive and remember things; and human beings have on top of that rational faculties, allowing them to think and will. Each type of living being is to be described with respect to a well-defined set of faculties. Human beings in particular cannot be described and understood as belonging to a distinctive type of living being unless one analyzes what their specific faculties are, how these faculties are used in a given situation, and how they are interrelated. Faculties are their "principles of operation," as a large number of authors, ranging from Thomas Aquinas to Francisco Suárez, unanimously claimed.[1] As soon as we give an account of these principles, we make clear how and why human beings behave the way they do.

Clear and simple as this approach to faculties may seem, it is far from being trivial and uncontroversial. In the seventeenth century, scholastic faculty theories became the target of relentless attack. Many opponents pointed out that appealing to faculties as principles of operation amounted to introducing mysterious entities that act inside a human being. John Locke phrased this widespread critique as follows:

> But the fault has been, that Faculties have been spoken of, and represented, as so many distinct Agents. For it being asked, what it was that digested the Meat in our Stomachs? It was a ready, and very satisfactory Answer, to say: That it was the *digestive Faculty*. What was it that made any thing come out of the Body? The *expulsive Faculty*. What moved? The *Motive Faculty*: And so in the Mind, the *intellectual Faculty*, or the Understanding, understood; and the *elective Faculty*, or the Will, willed or commanded: which is in short to say, That the ability to digest, digested; and the ability to move, moved; and the ability to understand, understood.[2]

Obviously, the problem with scholastic talk about faculties is not that it gives an incomplete or an unsatisfactory explanation of human activities. The real problem is that it provides no explanation at all: it is a

---

[1] Aquinas, *Summa theologiae* (*STh*) I, q. 77, art. 1 [q. = question, art. = article], corp.; Suárez, *De anima*, disp. 3, q. 1 (vol. 2, 56). For an overview, see Joël Biard, "Diversité des fonctions et unité de l'âme dans la psychologie péripatéticienne (XIVe-XVIe siècle)," *Vivarium* 46 (2008): 342–67, and Peter King, "The Inner Cathedral: Mental Architecture in High Scholasticism," *Vivarium* 46 (2008): 253–74.

[2] Locke, *An Essay concerning Human Understanding* II, ch. 21, sec. 20 (ed. Nidditch, 243–44).

pseudoexplanation.³ Instead of analyzing the structure of various acti...
ties, the mechanisms that make them possible, and the specific causes of
each mechanism, it simply posits dubious inner agents. These agents are
then characterized in the very same terms in which the activities in need
of an explanation are described—*explanandum* and *explanans* stand in
a circular relation. Moreover, a number of distinct agents are referred to
as if a human being were a household with many employees, each of
them constantly busy doing his job. What remains unclear, however, is
why there should be many agents, why they should have power to act,
and how they are related to the human being as a whole.

In light of this critique, it is understandable why in the seventeenth
century scholastic faculties were discarded in the junkyard of dysfunctional theoretical objects, along with prime matter, real qualities, and
other obscure things.⁴ But what gave rise to this critique? There seems
to be a simple answer. Scholastic accounts of faculties were closely tied
to a specific metaphysical theory: hylomorphism. Following Aristotle,
most medieval authors took the soul to be the substantial form of the
body and claimed that this form was responsible for a wide range of
faculties understood as "principles of operation." This explanatory
strategy was convincing only as long as the real existence of forms was
accepted. It is therefore hardly surprising that Locke and other opponents of hylomorphism openly attacked faculty theories. In a nutshell,
the rise and fall of faculty theories went hand in hand with the rise and
fall of hylomorphism.

It is tempting to choose this line of argument. However, I think that
we should resist the temptation. The philosophical (as well as historical) story is much more complex. It was not simply the rejection of
hylomorphism that gave rise to serious questions concerning the status

---

3 Note that a similar point was already made by Galen, who took faculties to be nothing but an *asylum ignorantiae*; see chapter 1, section 5.

4 See Steven Nadler, "Doctrines of Explanation: Explanation in Late Scholasticism and in the Me-chanical Philosophy," in *The Cambridge History of Seventeenth Century Philosophy*, edited by Daniel Garber and Michael Ayers (Cambridge: Cambridge University Press, 1998), 513–52, and chapter 4, section 1.

of faculties, their relation to the soul, and their causal power. Scholastic authors who defended hylomorphism were already aware of all of these problems. Some even pointed out the very same difficulties that were later identified by anti-Aristotelians. Faculties caused confusion *inside* the Aristotelian-scholastic tradition, which was anything but homogeneous. Analyzing several key texts in this tradition will help us to see how various authors dealt with faculties, how they attempted to explain their status as "principles of operation," how they worked out answers to some crucial problems, and how their answers gave rise to new problems.

2. William of Auvergne: Faculties as Ways of Acting

Faculties were omnipresent in the early and high Middle Ages when most Christian authors followed Augustine and talked about faculties as essential parts of the soul. But in what sense can the soul be said to have parts? When Aristotle's *De anima* was translated into Latin (first in the twelfth century by James of Venice, then again in the thirteenth century by Michael Scot and William of Moerbeke) and studied in the newly established universities, this problem became one of the most vigorously debated issues of the time.[6] Aristotle had also spoken of faculties as parts of the soul,[7] but at the same time he had stressed that the soul was nothing but the form present in the body as its immaterial principle of life.[8] How can an immaterial principle have parts? Surely, it is

---

5 On the Augustinian tradition, see Pius Künzle, *Das Verhältnis der Seele zu ihren Potenzen: Problemgeschichtliche Untersuchungen von Augustin bis und mit Thomas von Aquin* (Freiburg: Universitätsverlag, 1956), 30–96, and the collection of texts in Alain Boureau, *De vagues individus: La condition humaine dans la pensée scolastique* (Paris: Les Belles Lettres, 2008), 337–47.

6 On the translations of *De anima*, see Robert Pasnau, *The Cambridge History of Medieval Philosophy* (Cambridge: Cambridge University Press, 2010), 794; on their dissemination and reception, see Paul J. J. M. Bakker and Johannes M. M. H. Thijssen, *Mind, Cognition and Representation: The Tradition of Commentaries on Aristotle's De anima* (Aldershot: Ashgate, 2007), and Dominik Perler, ed., *Transformations of the Soul: Aristotelian Psychology 1250–1650*, special issue, *Vivarium* 46(3) (Leiden: Brill, 2008).

7 *De anima* II.2 (413b7–8 and 27–28) and III.4 (429a10); for an analysis see chapter 1, section 3.

8 *De anima* II.1 (412a27–28).

not divisible in the same way as a material thing, say a loaf of bread that can be cut into pieces. William of Auvergne was one of the first scholastic philosophers to point out this difficulty, thus making clear that the problem with faculties was first and foremost a metaphysical one. In his *De anima*, written most probably before 1240, William examined several metaphysical models that explained the status of faculties.[9]

According to a first model, faculties are distinct substances. Each of them is endowed with causal power and therefore capable of producing its own activity. Thus, the sensory faculty is a substance bringing about acts of perceiving and desiring, and the rational faculty is another substance producing acts of thinking and willing. The single substances are united in a human being and together form a complex soul. But this model cannot be accepted, as William is quick to remark, because it presents the soul as a mere heap or "aggregate" of independent entities.[10] How can they form a unity, given that there is nothing that connects them? And how can the body, which consists of many parts, form a coherent, unified thing if there is no principle of unity? If the soul is nothing but a heap of substances, it can hardly be such a principle.

To avoid these difficulties, one may be tempted to choose a second model, claiming that faculties are mere accidents that exist in the soul as their underlying substance. When adopting this model, one can easily contend that there is just one substance, serving as the principle of unity for the body, and nevertheless claim that many faculties exist on this basis. Different types of souls will then have different bundles of accidents, and the human soul, being the most complex, will have the largest number of accidents. However, William also rejects this model.[11] If faculties were nothing but accidents, they could be sepa-

---

9 On the composition and structure of this work, which was probably planned as a part of the *Magisterium divinale*, see Thomas Pitour, *Wilhelm von Auvergnes Psychologie: Von der Rezeption des aristotelischen Hylemorphismus zur Reformulierung der Imago-Dei-Lehre Augustins* (Paderborn: Schöningh, 2011), 28–29.
10 *De anima* 3.2, 88a.
11 *De anima* 3.2, 88a–b.

rated from the soul. It would therefore be possible for the rational faculty to be cut off from a human soul, just as it is possible for a color to be separated from its underlying substance. This contradicts the fundamental thesis that rationality is an essential and hence inseparable feature of all human beings.

How then is a statement like "The human soul is endowed with a rational faculty" or for short "The human soul is rational" to be understood if the predicate term signifies neither a substance, coexisting with other substances, nor an accident that is somehow implanted in a substance? William answers this crucial question by making use of a grammatical distinction between two types of predication.[12] In the case of accidental predication, one does indeed claim that a separable accident exists in a substance. By contrast, in the case of essential predication, one affirms that a substance has an essential, nonseparable feature. William's example for this second kind of predication is "Snow is white." When making this statement, we do not say that snow happens to be white and that it could very well have another color. We rather assume that snow is white in virtue of its essence. The statement is therefore to be understood in the sense of "Snow is essentially white." The predication of a faculty is to be understood in the same way; when we say that the human soul is rational we mean "The human soul is essentially rational." What is predicated in this case is an essential feature, not an accident. That is why no special relation between an accident and a substance (say, an inherence relation) needs to be posited. And should someone be incapable of using the rational faculty because of an illness or a permanent disability, the essential feature would not be lost. Due to accidental circumstances, it would simply be prevented from becoming active. Metaphorically speaking, it would be covered by accidents just as the essential whiteness of snow is eventually covered by accidental pollution.

This distinction between accidental and essential predication leads William to an important conclusion: faculties are not special entities

---

12 *De anima* 3.3, 88b.

that are somehow added to the soul and anchored in it. All of them, even the lower vegetative and sensory ones, belong to the essence of the soul:

> The faculty of a human soul is nothing but the soul itself in them [i.e., in human beings], which operates in virtue of its essence. If one says, for instance, that the human soul can think or know and so on, I claim that the verb "can" does not add anything to its essence, nor does it posit or affirm anything except for its essence, as it is also the case when it is used with respect to the praised creator.[13]

The last clause makes clear that William draws a parallel between God and the human soul. That is, when we talk about God's power, we do not speak about a special accident in God, but about his very essence. Likewise, when we speak about the power, capacity, or faculty of a human soul, we simply talk about its essence. And when we refer to different faculties, we talk about different aspects of this essence. Using modern terminology, one could say that words for various faculties, say "intellect" and "will," do not mark an extensional difference (they refer to one and the same entity) but an intensional one (they refer to it under different aspects). William points this out by calling faculties *officia* and by comparing them to offices a person can take.[14] For instance, someone can act as senator or as consul, depending on the political office to which he has been elected. The expressions "senator" and "consul" are then coextensional (they refer to the same person) and differ intensionally only (they refer to different offices of that person). No matter which and how many offices a person takes, it is always the same person who acts and not a number of mysteriously hidden entities. The same applies to the human soul. No matter which and how many faculties are said to produce activities, it is always the same,

---

13 *De anima* 3.6, 92a. (All translations from Latin are mine.)
14 *De anima* 3.6, 92b.

undivided soul that is being active in virtue of its essence and not a number of hidden entities. It simply takes different roles and hence displays different ways of acting.

In defending this position, William clearly holds an *identity theory*: faculties are metaphysically identical to the soul, a simple thing, and only conceptually distinguished when we refer to different roles it plays. Given this position, William does not fall into the trap Locke later saw in faculty theories when he accused scholastic authors of introducing "many distinct agents" that act inside a human being. In recent discussions, this mistake of turning faculties into hidden homunculi has been called "the homunculus fallacy."[15] But William is not guilty of this fallacy, as he does *not* claim the following: "A human being is thinking, therefore there must be an active entity inside it, namely the intellect, that produces this activity; and this entity must interact with other entities that produce acts of perceiving, desiring, and so on." Faculties are not taken to be agents with their own causal power.

Simple and elegant as this explanatory strategy may seem, two problems remain. First, one may ask how faculties as mere roles or ways of acting can be individuated. How can we enumerate them and establish a list for a certain type of living being, say for human beings? William does not provide a clear criterion. He simply remarks that we should pay attention to "the plurality of the acts and operations that stem from the soul,"[16] thus suggesting that we should look at different types of activities and then assign a corresponding faculty to each of them. It is clear that this will lead to a proliferation of faculties. There will not only be the famous three faculties (vegetative, sensory, and rational) but an infinite number of subfaculties, for instance one for acts of seeing, another for acts of hearing, and yet another for acts of tasting. This

---

15 See Anthony Kenny, "The Homunculus Fallacy," in *The Legacy of Wittgenstein*, 125–36 (Oxford: Blackwell, 1984). M. R. Bennett and P. M. S. Hacker, *Philosophical Foundations of Neuroscience* (Oxford: Blackwell, 2003), 68–85, speak about a "mereological fallacy": the activity of the soul as a whole is mistakenly attributed to a faculty (or even to its bodily implementation) as one of its parts.

16 *De anima* 3.6, 92b–93a.

poses a problem because a general theory explaining basic faculties seems to be elusive.[17] One can only sketch a list of faculties for each and every soul—the more activities we observe, the longer the list will turn out to be.

A second problem concerns the order among faculties. While speaking about roles or "offices," William refers to a specific order among them, which is given by nature. He claims that the rational faculty and its subfaculties are superior to the sensory faculty, which in turn is superior to the vegetative faculty. William even goes so far as to compare the human soul to a monarchy with a hierarchical structure.[18] The will, he says, is like the king, who uses the intellect as his counselor when making decisions and issuing orders. The various sensory faculties are the ministers, who communicate the orders to various parts of the body and receive messages from them. But how can faculties act like political figures if they are nothing but roles played by a single soul? And how can they stand in a hierarchical order if they are all identical with the soul's essence? This essence seems to have a complex internal structure, but William neglects to explain how such complexity is possible.

3. Thomas Aquinas: Faculties as Necessary Qualities

Thomas Aquinas, who knew the identity theory quite well,[19] was fully aware of the complexity problem. He emphasized that one cannot explain the status of hierarchically ordered faculties if one simply identifies them with the essence of the soul. Moreover, he took the identity

---

17 It is therefore not surprising that William never provides a complete list. He eventually mentions fifteen different faculties, but this is not a definitive list, as Thomas Pitour, *Wilhelm von Auvergnes Psychologie: Von der Rezeption des aristotelischen Hylemorphismus zur Reformulierung der Imago-Dei-Lehre Augustins* (Paderborn: Schöningh, 2011), 227, shows.

18 *De anima* 2.15, 85b.

19 It was supported not only by William of Auvergne, but by a number of authors in the mid-thirteenth century. For an overview, see Pius Künzle, *Das Verhältnis der Seele zu ihren Potenzen: Problemgeschichtliche Untersuchungen von Augustin bis und mit Thomas von Aquin* (Freiburg: Universitätsverlag, 1956), 116–70.

claim to be unacceptable for metaphysical reasons. In the first part of his *Summa theologiae*, written between 1266 and 1268, he adduced two arguments against the identity claim.

The first argument relies upon the theory of categories.[20] Faculties (*potentiae*) are nothing but potentialities, that is, potential states that can be actualized and hence become actual. Actual states clearly belong to the category of qualities since they are always particular states of a given substance. Consequently, the corresponding potential states must also belong to the category of qualities, because no categorical change takes place when something potential becomes actual. This can easily be illustrated. Being in flame is an actual state, and being inflammable is the corresponding potential state. Now, being in flame is clearly a quality of a substance, say of a tree. Being inflammable, its corresponding potential state, must therefore also be a quality of that tree. But a quality is not something that is identical to the essence of a substance. It is rather something added to it. So, if a faculty is a potential state and hence a quality, it cannot be identical to the essence of the soul; it must be added to it.

Is this a convincing argument? Someone defending the identity theory could concede that a faculty is indeed a quality and nevertheless point out that it is not simply added to the essence of the soul. There are two types of qualities, this person could say, accidental and essential ones. A color is an accidental quality; perhaps being inflammable is also such a quality (provided a tree could be turned into a fireproof object). By contrast, faculties are basic qualities that are not simply attached to the soul. Instead, they are built into its essence or even constitute it. That is why they are identical to it.

This argumentative move cannot be an acceptable solution for Aquinas. For when he talks about the essence of the soul, he only means that which is expressed in its definition, and all that is expressed is the fact

---

20 *STh* I, q. 77, art. 1, corp.; similarly *Quaestio disputata de anima* (*QDA*), art. 12, corp., and *Quaestio disputata de spiritualibus creaturis* (*QSC*), art. 11, corp.

that the soul is "the first principle of life" for the body.[21] This principle, taken in itself, is only responsible for the fact that a body is alive, but it does not determine *how* it is alive, that is, which activities are carried out and how all of them are made possible by specific faculties. That is why faculties cannot be included in or identical to the mere essence of the soul. Admittedly, they can be based on the essence or perhaps even be caused by it, but these dependency relations are clearly distinct from an identity relation.

This line of argument makes clear that Aquinas defends a *distinction theory*: the essence of the soul and its faculties are distinct entities with their own characteristic features and hence their own definition. Furthermore, the argument shows that Aquinas has a minimal conception of the essence of the soul, in particular of the essence of the human soul. Unlike William of Auvergne, who considered the essence to be a complex thing and compared it to a monarchy with a number of political offices, Aquinas takes it to be a very simple thing. Using another political metaphor, we could say that Aquinas considers the essence of the soul to be nothing more than the first and most fundamental article of the constitution of a state. This article simply affirms that a certain geographical territory is a political state and that all the people living on this territory are inhabitants of that state. How the state is organized, which political offices exist, and which duties are assigned to the individuals holding office need to be regulated in further articles. These articles are based on the first one and do not have any validity without it, but they are not identical to it. Similarly, the faculties regulating the behavior of a human being are based on the soul's essence and do not have any causal power without it, but they are not identical to it.

Aquinas's second argument against the identity claim starts with the same austere conception of the soul's essence. Being the first principle

---

21 *STh* I, q. 75, art. 1, corp., and *Sentencia libri De anima* II.1 (Leonina XLV/1, 68–70). Note that Aquinas explicitly speaks about a definition (and not simply about an account or an explanation) that mentions the "essential principle" of the soul.

of life, it is simply the "immediate principle of operation" that is permanently active.[22] It permanently guarantees that a body remains a living piece of matter, thus acting like a basic power station that keeps the body going. By contrast, the single faculties are not permanently active, as we can tell from our own experience. Sometimes we perceive things around us and sometimes we do not; hence our perceptual faculty is not always active. And sometimes we are engaged in acts of thinking and sometimes we are not; hence our rational faculty is not always active either. But an entity that is not always active cannot be identical to one that is always active, because strict identity requires sameness of all features.

How then are faculties related to the essence of the soul? Aquinas gives a rather obscure answer, affirming that they "flow" from it as their first principle.[23] A number of late medieval and modern commentators took this claim as being tantamount to the thesis that there is a real distinction between them.[24] That is, the essence as the basic power station generates further entities, and the entire soul ends up being a complex thing consisting of many items. These items always coexist, but in principle all of them can be taken apart, for whenever two entities, x and y, are really distinct, x can exist without y and y without x.[25]

Simple as this explanation may appear, it cannot be Aquinas's view, for he makes clear that essence and faculties can never be separated from each other. They form an indissoluble whole. Even when the

---

22 *STh* I, q. 77, art. 1, corp.
23 *STh* I, q. 77, art 6, corp.; art. 7, ad 1; *QSC*, art. 11, ad 12.
24 See John Duns Scotus, *Quaestiones in II Sent.*, dist. 16, q.u. (Opera omnia 13, 24–25); Suárez, *De anima*, disp. 3, q. 1 (vol. 2, 58). In the sixteenth century, the Thomistic position was standardly referred to as the real distinction position; see Dennis Des Chene, *Life's Form: Late Aristotelian Conceptions of the Soul* (Ithaca: Cornell University Press, 2000), 145. A modern Thomist like Pius Künzle, *Das Verhältnis der Seele zu ihren Potenzen: Problemgeschichtliche Untersuchungen von Augustin bis und mit Thomas von Aquin* (Freiburg: Universitätsverlag, 1956), 215–18, still claims that there is a real distinction.
25 This understanding of real distinction became influential after the famous Paris condemnation of 1277, which stated that God, using his sheer will, can separate everything that is really distinct: real distinction entails separability. Stephen Menn, "Suárez, Nominalism, and Modes" in *Hispanic Philosophy in the Age of Discovery*, edited by Kevin White (Washington, DC: Catholic University of America Press, 1997), 228, aptly calls it "the voluntarist axiom."

body decays and the soul, being immaterial and immortal, continues to exist, it is not just the naked essence that keeps existing but the essence together with all the faculties, including the so-called lower ones, such as the digestive and the breathing faculties.[26] Of course, they will then no longer be active. Lacking the appropriate bodily organs, they cannot bring about acts of digesting or breathing. But they will still exist in some kind of hibernation state, and when the body will be restored on resurrection day, they will be reactivated. This is not a mere theological concession to the Christian dogma of bodily resurrection, as one might suspect, but a genuinely philosophical claim that elucidates a crucial point: faculties are not separable entities that can, but need not, be attached to the essence of the soul. They are rather *necessarily* attached to it.

But how is this attachment possible? What kind of entities are faculties if they are forever glued to the essence? As has already become clear, they are qualities, but very special ones, as Aquinas hastens to add, namely *propria*, that is, necessary qualities. Their existence is explained as follows:

> A *proprium* does not belong to the essence of a thing. But it is caused by the essential principles of the species. That is why it stands between the essence and the accident so described. And in this way the faculties of the soul can be said to stand between substance and accident, as it were as natural properties of the soul.[27]

The classic example of a *proprium*, frequently mentioned by medieval authors, is risibility.[28] This quality does not belong to the essence of human beings and is therefore not mentioned in their definition (only

---

26 *STh* I, q. 77, art. 8, corp.; *QDA*, art. 19, corp.
27 *STh* I, q. 77, art. 1, ad 5. See also *QDA*, art. 12, ad 7.
28 This example was introduced by Porphyry in his *Isagoge* 4 (4a) and standardly mentioned in medieval handbooks, for instance in Petrus Hispanus's *Tractatus called afterwards Summule Logicales* II.14 (ed. de Rijk, 22).

"animal" as the proximate genus and "rational" as the specific difference are mentioned), but it immediately follows from the essence. For if a person is rational, she can in virtue of that basic fact understand humorous remarks and laugh at jokes. This example shows that a *proprium* is not an essential quality but nevertheless a necessary one that immediately follows or "flows" from the essence.

But how many necessary qualities are there? And how can we establish a general list? If we start with an observation of single activities, as William of Auvergne suggested, and if we ascribe corresponding faculties to all of them, we will end up with an infinite number of faculties. This is a consequence Aquinas wants to avoid. He is convinced that we can and even ought to present a general list, because it is only such a list that enables us to characterize different types of living beings. Moreover, only a general list allows us to draw a map of all the faculties in a given living being and to point out how they are interrelated. We therefore need to indicate criteria that enable us to individuate various types of faculties.

It might seem quite easy to find appropriate criteria: we simply need to look at the bodily organs in a given living being and assign a faculty to each of them. Since there is a limited number of organs, there will also be a limited number of faculties. However, this cannot be accepted as an adequate answer. One problem is that it does not enable us to individuate rational faculties, which, according to the standard Aristotelian view, are not located in any specific organ. But the main problem is that it gets the priorities wrong. Aquinas unmistakably holds that "faculties do not exist on account of organs, but organs on account of faculties."[29] The faculties have priority because they give a specific function to bodily parts, which, taken in themselves, are nothing but chunks of matter. In fact, organs without faculties are not organs at all. Thus, an eye lacking the faculty of seeing is no longer an eye, just as an axe that loses its woodcutting capacity is no longer an axe. Aquinas

---

29 *STh* I, q. 78, art. 3, corp.

clearly defends the principle of homonymy: an organ lacking a faculty is only homonymously called an organ—it is no longer a real organ.[30]

How then are we supposed to individuate faculties if we cannot simply look at various organs? We should pay attention to the objects that surround us and that trigger various activities. This is most evident in the case of perceptual faculties. Since there are different types of objects in the world that can be perceived as such (colors, sounds, odors, etc.), there need to be corresponding faculties that enable us to perceive these objects. And since there are exactly five types of basic objects, there must be five types of perceptual faculties, neither more nor less.[31] But why should we not further subdivide these objects and assign a special faculty to each of them? Why, for instance, should we not assume that there is a special faculty for seeing red colors, another one for seeing blue colors, and so on? Aquinas takes this subdivision to be superfluous because all colors have the same type of form, namely a visible form, and can therefore be perceived in the same way. The same is true for the four other types of perceptible objects. That is why it is superfluous to posit a faculty for each and every individual object. A special faculty is only required for a certain type of form, not for every individual form.

It is clear that this account gives rise to a number of questions. What does it mean that every type of perceptible object has its own type of form? And why should there be exactly five basic types? To answer these questions, one would have to delve into Aquinas's theory of perception. The details of this theory do not matter here.[32] What is important in this context is the overall explanatory strategy Aquinas

---

30 *STh* I, q. 76, art. 8, corp., and more extensively *Sentencia libri De anima* II.2 (Leonina XLV/1, 74–76). On the origin of this Aristotelian principle, see Christopher Shields, *Order in Multiplicity: Homonymy in the Philosophy of Aristotle* (Oxford: Clarendon Press, 1999).

31 *STh* I, q. 78, art. 3, corp.; on the object criterion in general, see *QDA*, art. 13, corp.

32 For an analysis, see Robert Pasnau, *Thomas Aquinas on Human Nature* (Cambridge: Cambridge University Press, 2002), 180–89, and Dominik Perler, "Perception in Medieval Philosophy," in *The Oxford Handbook of the Philosophy of Perception*, edited by Mohan Matthen (Oxford: Oxford University Press, 2015), 51–65.

pursues. When individuating faculties, he does not use an introspective method, nor does he start with an analysis of bodily organs. Instead, he looks at basic objects in the world, which all have distinctive forms, and then argues that there must be corresponding faculties that allow us to grasp these forms. Or to put it crudely: he first looks *outside*, establishing a list of basic objects, and then turns *inside*, assigning faculties to all of them.

The fact that a large number of faculties can be individuated gives rise to another problem we already saw in William of Auvergne, namely the classification problem. How can we establish an order among faculties? Like William, Aquinas assumes that there is an order, and when describing this order he also occasionally uses metaphorical language. For instance, he affirms that the rational faculty "rules in a political way" over the lower faculties, which "obey" it.[33] How is this language to be understood? Is Aquinas introducing homunculi that stand in a hierarchical order, doing their specific jobs and interacting with each other? It is tempting to understand him in this way since he often speaks about faculties as if they were agents. For instance, when explaining emotions he claims that the lower sensory faculty brings about hate, anger, and many other unpleasant passions and that the rational faculty can "moderate" them, sometimes even overcome them.[34] It therefore seems as if he were talking about more or less independent agents struggling with each other. Yet it would be misleading to interpret him in this way. There is a simple metaphysical reason why faculties cannot be agents: they belong to the category of qualities, and qualities do not act. It is always the substance that acts in virtue of a certain quality. And since the soul is the underlying substance or form for all the qualities, it is always the soul that acts. So, to say that the sensory faculty brings about anger and that the rational faculty attempts to calm it are only loose ways of speaking. To be precise, one should say that the soul

---

33 *STh* I, q. 81, art. 3, ad 2; *STh* I–II, q. 24, art. 1, corp.
34 *STh* I–II, q. 24, art. 2.

brings about anger in virtue of its sensory faculty, and that the very same soul eventually calms the anger in virtue of its rational faculty.[35] This shows that Aquinas and William are not really at odds, despite the fact that Aquinas rejects the identity theory. Faculties are intimate features of the active soul, not separate entities with their own causal power.

Why then does Aquinas frequently use hierarchical language and emphasize that the rational faculty is the highest faculty? Why does he not confine himself to mentioning many faculties, which are all intimate features (or, technically speaking, necessary qualities) of a single soul? His main reason is that a mere list of faculties in virtue of which the soul acts would not explain its functional architecture. That is, it would not explain why and how the soul performs some activities for the sake of other ones. For instance, when describing the activities of a given human being we should not simply establish a list that includes items like digesting food, seeing colors, and thinking about objects. We should also explain how these activities are related to each other and which ones are necessary for other ones. Thus, we should say that acts of digesting provide a solid ground for acts of perceiving, because only properly nourished human beings are able to make full use of their sensory organs and to see things around them. Acts of seeing, in turn, provide a solid and even indispensable ground for acts of thinking, for human beings could never form concepts and never produce judgments if they had no perceptions. Consequently, the faculties that make these activities possible also need to stand in a hierarchical—or even teleological—order. Aquinas does not shy away from using teleological language. He explicitly claims that "faculties of the soul that are prior in the order of perfection and nature serve as the principles for others by

---

35 Since the soul is not an independent entity either, but a constitutive principle of the entire human being, one should say that it is the human being that becomes angry or moderates anger. Aquinas follows Aristotle in claiming that all activities should be attributed to the human being as the real agent; see *Sentencia libri De anima* I.10 (Leonina XLV/1, 51).

being their end and active principle."³⁶ So, despite the fact that the rational faculty is the ultimate one, it is nevertheless first in the teleological order. All other faculties are functionally subordinated to it, that is, they bring about activities for the sake of the optimal working of this faculty.

This crucial point can be illustrated once more with the comparison from political science. If the essence of the soul is like the first and fundamental article of the constitution, which defines the existence of a state, then the rational faculty is the second article to which all subsequent ones are subordinated. This article defines the ultimate goal of the state, say the peaceful cohabitation of all the people living in that state, and the subsequent articles serve to spell out this goal. Whatever political organs and electoral procedures will be described, they all exist for the sake of peaceful cohabitation. That is why a constitution is not just a loose assemblage of articles. It has a teleological structure: the top article sets the goal for all the subsequent ones. Likewise, the top faculty sets the goal for all the lower ones. Of course, this way of establishing a hierarchical order has its price. Aquinas subscribes to a deeply rationalist conception of human beings by subordinating all the sensory activities to rational ones. No matter how sophisticated our perceptions, emotions, and desires may be, no matter how valuable they may be in themselves and how much they may contribute to our well-being, they are (and indeed should be) ultimately subordinated to rational activities.

4. William of Ockham: Faculties and a Plurality of Forms

Aquinas's position was discussed throughout the later Middle Ages and did not remain unchallenged. One of the most outspoken critics was the early fourteenth-century Franciscan William of Ockham. In

---

36 *STh* I, q. 77, art. 7, corp.

his *Commentary on the Sentences*, written between 1317 and 1319, he openly attacked the thesis that faculties are qualities that necessarily go along with the essence of the soul but are nevertheless distinct from it. In fact, he explicitly rejected Aquinas's first argument in favor of this thesis.

As I have shown, Aquinas claimed that faculties are potential states that belong to the same category as the corresponding actual states. Since actual states are qualities added to the essence of the soul, the potential states must also be qualities. Ockham judges this argument to rely on a misunderstanding of the ambiguous term "faculty" (*potentia*).[37] In some cases this term may indeed be used as being equivalent to "potential state." But when it comes to faculties of the soul, "faculty" is equivalent to "potency" or "actual power" and signifies not an actual or potential state, but the cause of such a state. And this cause does not belong to the same category as the state it brings about. This can easily be illustrated with the example I already used to explain Aquinas's position. Being in flame is an actual state, and being flammable is the corresponding potential state. But a faculty is not identical to this potential state. It is rather the power to make something flammable (i.e. to endow it with a potential state) or to put it in flame (i.e. to actualize the potential state). Similarly, a faculty of the soul is not simply a potential state but a power that brings about such a state or that actualizes it. It would be utterly misleading to identify the cause of a state with the state itself.

What then is a faculty, in the sense of an active power or cause? Ockham is not at a loss for an answer. He takes it to be the essence of the soul that brings about a state or even a multitude of states, because the essence is in fact a kind of power station. And all the faculties "are identical with themselves and with the essence of the soul."[38] This means, of course, that it does not make sense to speak about faculties as a

---

37 *Reportatio* II, q. 20 (OTh V, 428). Abbreviations: OTh = Opera theologica, OPh = Opera philosophica.
38 *Reportatio* II, q. 20 (OTh V, 435).

bundle of entities that are somehow glued to the essence of the soul and "flow" from it. Ockham clearly rejects Aquinas's distinction theory and returns to the *identity theory* we already encountered in William of Auvergne: faculties are nothing but roles or ways of acting of the soul. Ockham even uses the same terminology as William, remarking that faculties are "distinct offices" of the soul and claiming that it would be erroneous to take them to be really distinct parts.[39] The soul has no parts. It is a single, undivided cause that brings about many potential states and that actualizes some of them under appropriate circumstances, thus producing actual states.

But why then do we speak about various faculties as if they were causes that act and produce various states or acts? Why, for instance, do we say that the intellect brings about acts of thinking and the will acts of wanting? Ockham makes clear that this is just metaphorical talk. Intellect and will are not inner agents with their own realms of activity. If we examine the terms "intellect" and "will" more closely, we will realize that they are mere connotative terms. That is, they are terms that have a primary and a secondary signification.[40] "Intellect" primarily signifies the essence of the soul and secondarily acts of thinking produced by it; "will" primarily signifies the very same essence and secondarily acts of wanting brought about by it.[41] In spelling out the secondary signification we simply specify the "office" or role the soul plays. We make clear what kind of activity it produces when it acts as intellect or as will. But it is always the same soul that acts.

From a methodological point of view, this appeal to a twofold signification is quite significant. It is an example of Ockham's method of linguistic ascent. Instead of distinguishing different entities or parts of entities he examines the terms we use when talking about the alleged plurality of

---

39 *Reportatio* II, q. 20 (OTh V, 441). Note, however, that Ockham does not refer to William of Auvergne. There are no traces of direct influence.

40 On the definition of connotative terms, see *Summa Logicae* I, 10 (OPh I, 36–37). They form a distinctive class of terms and can never be fully reduced to absolute terms. For an analysis, see Claude Panaccio, *Ockham on Concepts* (Aldershot: Ashgate, 2004), 85–102.

41 *Reportatio* II, q. 20 (OTh V, 438–39); *Ordinatio* I, dist. 1, q. 2 (OTh I, 402).

entities. An analysis of these terms then reveals that it is superfluous to posit many entities and to look for a special relation between them. In particular, it is superfluous to posit faculties that "flow" from the soul and form a network of qualities. Should we do so, we would mistakenly take faculty terms to be absolute terms. According to Ockham, it is in fact a semantic confusion that gives rise to a metaphysical confusion: when we naïvely take mere connotative terms to be absolute terms, we assume that they signify distinct entities. Moreover, we are then tempted to assume that they signify many inner agents that are responsible for different types of activities. As soon as we overcome the confusion we realize that there is no need to make such a strong assumption. We can apply the principle of parsimony and emphasize that there is a single agent, namely the soul, which produces a wide range of activities.[42]

Yet this parsimonious approach to the problem of faculties is not without its difficulties. One could immediately object that it fails to explain why the soul can play different roles or take different "offices." When we simply affirm that there is a single soul that produces acts of thinking as well as acts of wanting, we seem to take this entity to be some kind of black box that mysteriously brings about various activities. The crucial question is, however, why it can bring them about. Aquinas's commitment to faculties as qualities may lead to a proliferation of entities, but it has at least the advantage that it explains why we can assign different roles to the soul. Why, for instance, can it produce acts of thinking? Aquinas's answer is clear: the soul (or more precisely: the human being endowed with a soul) can act in virtue of the intellectual faculty, a necessary quality. A certain type of quality makes a certain type of activity possible. Ockham's rejection of special qualities rules out this kind of explanation. How then can he give an account of the fact that the soul is able to act as intellect or as will?

A solution to this problem can be found in Ockham's description of the cognitive process produced by the soul. As far as the process on a

---

42 In *Reportatio* II, q. 20 (OTh V, 436), Ockham explicitly mentions the principle of parsimony.

rational level is concerned, it consists of acts of thinking and acts of wanting. Through acts of thinking we acquire concepts and combine them in propositions; through acts of wanting we give our assent or dissent to propositions and thereby produce judgments.[43] Now the crucial point is that we first come up with acts of thinking and only then with acts of wanting. There is, as it were, a program built into our rational soul that makes it produce a certain type of activity (forming concepts and propositions), which then gives rise to another type (forming judgments). Ockham explicitly holds that there is a given order:

> The intellect is prior to the will because an act of thinking, which is connotated by "intellect," is prior to an act of wanting, which is connotated by "will," for an act of thinking is the partial efficient cause of an act of wanting, and in the natural course of order it can exist without an act of wanting but not vice versa.[44]

This can easily be illustrated. Suppose, when you look out the window, you see a tree covered with fresh snow. You immediately form the concepts "tree," "white," "snow," and so on, and combine them in a proposition, because your soul immediately acts as intellect when it is triggered by certain sensory inputs. The presence of a proposition then triggers further activity. It makes your soul act as will: it immediately produces an act of affirming, thus giving its assent to the proposition and coming up with the full-fledged judgment "There is a tree covered with fresh snow." The mere act of producing and combining concepts may very well occur without an act of the will, because one can simply entertain a thought without assenting to it. But the act of the will

---

43 In *Quaestiones variae*, q. 5 (OTh VIII, 170), Ockham holds that the will is responsible for the formation of judgments. On the process of concept and proposition formation see Martin Lenz, *Mentale Sätze: Wilhelm von Ockhams Thesen zur Sprachphilosophie des Denkens* (Stuttgart: Steiner, 2003), and Claude Panaccio, *Ockham on Concepts* (Aldershot: Ashgate, 2004).

44 *Reportatio* II, q. 20 (OTh V, 441–42).

cannot occur without an act of the intellect, as one cannot give one's assent unless there is a proposition that serves as the object of the assent.

This example shows that Ockham gives a procedural account of the activities of the soul. It is a single rational soul that performs different roles, but this soul follows a certain procedure when it produces various types of activities. It cannot act as will unless it first acts as intellect, and under normal circumstances it cannot act as intellect unless there are certain sensory inputs.[45] How it acts and how it follows a certain procedure is determined by the simple fact that it is a special cause, namely a cause that is responsible for cognitive processes. In presenting this procedural account, Ockham proves to be a reductionist. Not only is he a *metaphysical* reductionist by reducing faculties to mere roles or ways of acting of the soul but also an *explanatory* reductionist by reducing an account of faculties to a description of the procedural steps of the acting soul.

How far does this reductionism go? The most extreme form would be to reduce all the faculties, including the vegetative and sensory ones, to mere ways of acting of a single soul and to claim that there is an all-embracing procedural account for all the activities produced by this soul. Ockham does not go that far. He rather distinguishes two souls in a human being, namely a rational soul that brings about acts of thinking and willing, and a sensory soul that produces all organic and sensory activities, ranging from breathing and nourishing to perceiving and desiring. Consequently, rational faculties are ways of acting of the rational soul, whereas organic and sensory faculties are ways of acting of the sensory soul, and the two souls are really (and not just conceptually) distinct from each other.

---

45 Of course, there need not be an immediate sensory input for every intellectual activity. There can very well be some acts (e.g. acts of remembering) without an immediate input. While defending an empiricist program, Ockham only wants to claim that there has to be an *ultimate* sensory basis for all acts of the intellect. Marilyn McCord Adams, *William Ockham* (Notre Dame, IN: Notre Dame University Press, 1987), 495, aptly calls this "conceptual empiricism."

Why does Ockham think that it is impossible to reduce all the faculties to ways of acting of a single soul? Why does he take it to be indispensable to posit two souls? He presents a number of arguments.[46] The most concise one runs as follows:

> Numerically one and the same form is not both extended and nonextended, both material and immaterial. But the sensory soul in a human being is extended and material, whereas the intellectual soul is not because it is as a whole in the whole and as a whole in every part.[47]

At first sight, this argument looks quite astonishing. Is a form not the counterpart to matter and therefore something immaterial? How can there be a material form? When calling the sensory soul an extended and material form, Ockham does not want to give up the core thesis of hylomorphism, namely that a natural substantial form is an immaterial principle and hence something that is distinct from matter.[48] He rather wants to point out that there are some forms that always exist in matter, even in particular pieces of matter, and that cannot exist on their own. That is to say, he subscribes to a dependency thesis: some forms always depend on matter and are therefore necessarily located in matter. The sensory soul of a human being is precisely this kind of form. It always exists in a human body and ceases to exist as soon as the body decays. When we are asked where it exercises its activities, we ought to point to a part of the body. Thus, we ought to say that it is in the eyes that it performs acts of seeing, and in the stomach that it performs acts of digesting. This means, of course, that sensory faculties as "offices" of the sensory soul are distributed all over the body: the visual faculty is in the eyes, the digesting faculty in the stomach, and so on. By contrast, the rational soul is not dependent on and active in a bodily organ. Ockham claims

---

46 For a detailed analysis of all the arguments, see Dominik Perler, "Ockham über die Seele und ihre Teile," *Recherches de théologie et philosophie médiévales* 77 (2010): 313–50.
47 *Quodl.* II, q. 10 (OTh IX, 159).
48 See *Summula philosophiae naturalis* I, cap. 15 (OPh VI, 195–96).

that thinking and wanting cannot be assigned to the brain, to the heart, or to any other part of the body. As long as the body exists, the rational soul is present in the entire body, yet without depending on it, and it continues to exist even when the body is destroyed.[49] That is why the rational soul is not only functionally independent (it does not need any particular bodily organ to perform an activity) but also metaphysically independent.

Obviously, Ockham defends a dualist position.[50] Consequently, he gives a dualist account of faculties: some of them are necessarily dependent on and present in the body, whereas others are not. Given this crucial difference, it is impossible to pursue a radical reductionist strategy and to claim that all the faculties are simply ways of acting of a single soul. Instead, some are ways of acting of a matter-dependent soul, while others are ways of acting of a different, matter-independent soul. When describing the first type of faculty, we need to pay attention to the material basis. Thus, when dealing with the visual faculty, we need to look at the eyes, in which it is present, and we even need to take into account the fact that the human visual faculty cannot become active in anything other than a human eye; there is no possibility of multiple realization for a given faculty. By contrast, when describing the second type of faculty, we do not need to look at the material basis, even if it happens to have one. Thus, we do not need to analyze the brain or the heart when we are dealing with the intellectual faculty. It will be enough to see how a given program, be it materially implemented or not, makes the rational soul produce concepts and judgments.

---

49 See *Quodl.* I, q. 12 (OTh IX, 68–71).

50 Strictly speaking, he defends a trialist position because he also posits a "form of corporeality" as a third substantial form that is constitutive of a human being; see *Quodl.* II, q. 11 (OTh IX, 162–64). But with respect to the soul, he clearly separates two realms and thereby paves the way for later forms of dualism that set rational activities apart from all other activities. It may be called "mind-soul dualism" as opposed to traditional "mind-body dualism." On its impact on later medieval debates, see Henrik Lagerlund, "John Buridan and the Problems of Dualism in the Early Fourteenth Century," *Journal of the History of Philosophy* 42 (2004): 369–87, and Robert Pasnau, "The Mind-Soul Problem," in *Mind, Cognition and Representation: The Tradition of Commentaries on Aristotle's "De anima,"* edited by Paul J. J. M. Bakker and Johannes M. M. H. Thijssen (Aldershot: Ashgate, 2007), 3–19.

This separation of two souls and hence of two ways of acting inevitably gives rise to the agency problem. Who is acting, the human being as a whole or two inner homunculi, namely the sensory soul that produces acts of digesting, seeing, imagining, and so on, and the rational soul that brings about acts of thinking and willing? Ockham never gives a detailed answer, but he clearly tries to avoid the homunculus fallacy, for he points out that "there is only one total being for a human being," even though each form or soul has its own existence and its own way of working.[51] There is one total being, because the two forms together create a functional unity, and it is only this unity that is to be called an agent. Ockham even goes so far as to claim that a human being deprived of the rational soul would not be a real living being, neither a rational nor an irrational one. It would be a truncated being, something incomplete that has no functional unity and hence no chance to work properly.[52] Let me explain what he means by drawing a comparison.

When we describe the requirements for a modern democracy, we say that it should have three powers, namely a legislative, an executive, and a judicative one, and we insist on the separation of these powers. Each power has its own task, and each should be independent from the other ones. Nevertheless, the three should cooperate and form a functional unity. Only a democracy in which all three powers are equally respected and in which all three complement each other is a well-functioning democracy. Ideally, each power functions so perfectly that it contributes to the flourishing of the entire state. It is then the entire state that is said to act in virtue of one of its constitutional powers. Thus, we say that it is the state that signs an international treaty in virtue of the decision made by the parliament, its legislative power. Or we say that it is the state that condemns a violation of human rights in virtue of a decision made by its supreme court, the judicative power. Should one of the three powers be

---

[51] See *Quodl.* II, q. 10 (OTh IX, 161).
[52] In *Quodl.* II, q. 10 (OTh IX, 161), he says that it would "not be a complete being that exists per se in a genus."

prevented from acting or even be suppressed, the entire state would become dysfunctional: it would become a truncated democracy.

It is along this line that we can understand Ockham's reference to the human being as "one total being" that has two forms as souls (and an additional form of corporeality). Each form has its own task, and each performs this task with its own power. In a well-functioning human being, each form contributes to the flourishing of the whole. When we describe human actions, we appeal to the entire human being that acts in virtue of his or her forms. Thus, we say that someone perceives and imagines material objects in virtue of her sensory soul, or that she thinks about mathematical objects in virtue of her rational soul. Should one of her forms be prevented from acting or be completely destroyed, the entire human being would be dysfunctional—it would become a truncated human being.

In light of this explanation, it becomes obvious why Ockham does not see any difficulty in combining the following two theses: (1) faculties are roles or "offices" of two really distinct souls, which are responsible for two distinct sets of activities, and (2) all the activities can nevertheless be ascribed to the human being as a whole, which is a functional unity. A combination of these two theses is possible because a human being can always be described on two levels. We can first look at the *subpersonal* level and explain what activities are performed by a soul that plays a certain role as an inner cause. Thus, we can say that acts of seeing are performed by the sensory soul acting in the eyes as visual faculty. But we can also look at the *personal* level and affirm that it is in fact the human being that is seeing, not just the visual faculty, because the sensory soul contributes to the functioning of the entire living being and is well-coordinated with the rational soul. There is no additional entity that somehow holds the two souls together.[53]

---

53 Ockham defends mereological nominalism by claiming that the whole is nothing over and above the sum of all the parts; see *Summula philosophiae naturalis* I, cap. 19 (OPh VI, 208), and a detailed analysis in Richard Cross, "Ockham on Part and Whole," *Vivarium* 37 (1999): 143–67.

5. Francisco Suárez: Faculties as Really Distinct Entities

Ockham's position was eagerly debated in the fourteenth century and gave rise to a nominalist movement that shaped philosophical debates up to the early modern period.[54] The Jesuit Francisco Suárez closely studied and assessed it in his *De anima* and in the *Disputationes metaphysicae*, both written in the last three decades of the sixteenth century.[55] While conceding the originality and boldness of this position, he clearly disagreed with it and repeatedly criticized two of its fundamental claims.

First, Suárez stresses that there is just one soul in a human being or in any other living being.[56] He accuses Ockham of mistakenly taking the material presence of some activities to be a sign of a material soul. This becomes evident in Suárez's assessment of the argument Ockham presented to prove the real distinction between sensory and rational soul. As I have shown, he affirmed that the first soul, which is always present in and dependent on matter, cannot be identical with the second one, which is completely matter-independent. Suárez judges this argument to be invalid because it does not take into account the fact that it is only a given set of activities of the soul, not the soul itself, that depends on matter. Digesting, growing, seeing, and so on are in fact all activities that can only take place in a bodily organ. But this merely shows that the soul is not a self-sufficient entity that can exercise all of its functions in splendid isolation. For some of them it needs bodily organs. Should it be present in a mutilated body, it could no longer

---

54 On the controversy over Ockham's position, see Sander W. de Boer, *The Science of the Soul: The Commentary Tradition on Aristotle's De anima, c. 1260–c. 1360* (Leuven: Leuven University Press, 2013); on the nominalist movement, see William J. Courtenay, *Ockham and Ockhamism: Studies in the Dissemination and Impact of His Thought* (Leiden: Brill, 2008).

55 Suárez wrote *De anima*, a set of questions dealing with Aristotle's *De anima*, while he was teaching at the University of Segovia (between 1571 and 1574). He started revising this text later in his life but never finished it. It was published after his death in 1621; see the editor's introduction in *De anima* (vol. 1, xxxvii–lxxiv).

56 *De anima*, disp. 2, q. 5 (vol. 1, 322–30).

produce all of its activities, and should it be completely detached from the body (as it will in fact be after death), it could no longer bring about any sensory activity whatsoever. But the soul itself would not be mutilated or even destroyed.[57] It would simply be prevented from becoming fully active. Ockham and his followers overlook this crucial point and therefore erroneously believe that there is a matter-dependent sensory soul that comes and goes with the existence of a body. As soon as we correct this mistake, we can very well subscribe to the thesis that there is a single, immaterial soul that has several functional levels or "grades" (*gradus*), as Suárez claims.[58] Some of the grades require a body, while others do not. Moreover, one can also understand why there is a hierarchical order among various activities. Why, for instance, do acts of seeing presuppose acts of digesting? This is because acts of seeing can only occur in well-functioning eyes, which require a well-nourished body, which in turn requires acts of digesting. The full functioning of the soul on a superior level always presupposes its functioning on a subordinated one. But in the end, it is one and the same soul that acts on various levels.

Suárez also rejects Ockham's second major claim, namely that faculties of the soul are really identical to its essence and nothing but "offices" or roles it plays. This cannot be true, Suárez holds, because "otherwise this proposition would be true: 'The intellect loves, the will thinks.'"[59] This is not simply an objection concerning the use of language, but a criticism intended to point out an explanatory gap in Ockham's account. Ockham clearly held that there is just one rational soul that plays different roles. In characterizing these roles, he contended that the soul acts as intellect when it produces acts of thinking, and as will when it brings about acts of wanting, including emotional acts. But why is each of the roles responsible for a distinctive type of activity? As long as one

---

57 On the status of the separated soul, see *De anima*, disp. 14, q. 1 (vol. 3, 444–52).
58 *De anima*, disp. 2, q. 5 (vol. 1, 328–30).
59 *De anima*, disp. 3, q. 1 (vol. 2, 62).

does not give an account of these roles, one might as well exchange them and say that the intellect is responsible for acts of wanting and the will for acts of thinking. "Intellect" and "will" seem to be empty labels. There is nothing—no real feature—that distinguishes them, because in reality there is just a simple, rational soul.

At first sight, this objection seems to overlook a crucial point. Ockham clearly gave an account of the two roles by referring to a certain procedure: the rational soul first acts as intellect, producing concepts and propositions, and then as will, giving its assent or dissent to the propositions. It follows some kind of inner program when it brings about activities. Yet this reply gives rise to a follow-up question. What accounts for the fact that the soul first acts as intellect and then as will? As long as one does not provide an explanation one simply takes the soul to be some kind of complex machine with a fanciful inner program. That is why Suárez insists that it is important to indicate a *fundamentum in re* for the different roles it plays, and he does not hesitate to point out the appropriate foundation: it is a special quality that it is responsible for each role. Just as Aquinas, he holds that the soul has many necessary qualities, each of them making a special role or way of acting possible. Faculties are precisely these qualities, and the soul as a whole is an "aggregation" consisting of all of them and a substantial form.[60] To be sure, this does not mean that the soul is a loose bundle of entities. The single substantial form is the indispensable basis for all the faculties. That is why the soul as a whole is a well-defined unity.[61] Nor does the aggregation thesis mean that the faculties are randomly put together. Suárez assumes that there is a hierarchical order among them, and like Aquinas he characterizes this order in teleological terms: the vegetative faculty produces activities for the sake of the sensory faculty, which in

---

60 *De anima*, disp. 3, q. 3 (vol. 2, 132).
61 On the details of this claim, see Marleen Rozemond, "Unity in the Multiplicity of Suárez's Soul," in *The Philosophy of Francisco Suárez*, edited by Benjamin Hill and Henrik Lagerlund (Oxford: Oxford University Press, 2012), 154–72.

turn becomes active for the sake of the rational faculty.[62] It is only a well-coordinated functioning of the faculties at all three levels that makes the full range of human activity possible.

But how exactly are the hierarchically ordered faculties related to the basic substantial form? Using Aquinas's language, Suárez claims that they "flow" from it, but he goes beyond the traditional position by claiming that they are really distinct from it. In fact, he claims that they are things (*res*) that are really distinct both from each other and from the underlying substantial form.[63] Why does he defend this strong claim? His main reason is that they produce distinctive activities, that is, activities that have their special defining character (*ratio*). If the caused activities are distinctive, the causing faculties must be distinctive as well. Suárez illustrates this with the already mentioned example of thinking and loving. These are distinctive activities that need appropriate causes. Should each of the two causes not have a special defining character, it could not bring about a special activity. The decisive point is, of course, that Suárez assumes that there being a special defining character implies that there is a really distinct thing. Or for short: what has its own *ratio* must be a distinct *res*. This has far-reaching consequences, as Suárez's own example shows. Not only is the rational faculty really distinct from the sensory and the vegetative ones but also this faculty is in turn to be subdivided into two really distinct things, namely intellect and will. On each of the traditional three levels of faculties there are many *res*, and all of them are really distinct from the underlying substantial form.

This thesis makes clear that Suárez's reference to the soul as an "aggregation" is to be taken seriously. The soul as a whole is not just a functional aggregate, but a metaphysical one. It consists of many really (and not just conceptually) distinct entities and is therefore a complex thing. To use a political comparison again, we could say that it is like a federal

---

62 *De anima*, disp. 2, q. 5 (vol. 1, 330).
63 *De anima*, disp. 3, q. 1 (vol. 2, 62–64).

state with many provinces. All the provinces are really distinct from one another and have their own governments. They form a unity, because they are all subordinated to the same federal government. But despite this dependency, they all have autonomy and regulate their own administrations, their own schools, and so on. If we want to understand how the state as a whole works, we need to look at the activities of all the provinces and their interactions.

Obviously, this model goes beyond Aquinas's distinction theory, even if Suárez pretends to simply follow his famous predecessor. When claiming that faculties "flow" from the essence of the soul, Aquinas only meant to say that they are entailed by it. He did not appeal to a relation between really distinct entities that are put together in some kind of federal union. Nor did he intend to introduce a relation of efficient causation, for what is entailed by the essence of the soul is not literally produced by it. It is exactly at this point that Suárez radicalizes Aquinas's model:

> For the soul has an efficiency with respect to its faculties, which are really distinct from it. Therefore, as the soul is God's effect, so a faculty is a new effect brought about by the soul itself. But where there is a new effect, there is a new action.[64]

The soul literally produces its faculties, which in turn produce many activities. Thus, the soul literally produces intellect and will as two distinct *res*. Only when these entities are in place can there be activities of thinking and willing. And once these entities are produced, they need to be kept in existence, just as all the creatures need to be maintained in existence by God's constant activity. That is why there is not only an initial production of faculties but also constant causal activity.

But what is the character of this activity: is it necessary or contingent? Could faculties *not* be caused? Could there be a "naked soul"

---

64 *De anima*, disp. 3, q. 3 (vol. 2, 124–25).

without any faculties? Suárez concedes that under normal circumstances it is always the whole package—the substantial form together with all the faculties—that comes into existence. But the existence of a "naked soul" is not ruled out, as Suárez points out by presenting the following scenario:

> Imagine that God creates the substance of the soul and prevents faculties from emanating from it.... In that case the substance of the soul would remain without intellect and will. But if God removed the impediment at a later moment and left to the soul its nature, the faculties would certainly flow from it as in the first moment of generation or creation.[65]

Scenarios invoking divine intervention always refer to situations that are logically and not physically possible. Suárez's hypothetical case should therefore not be understood as a warning that a malicious God could intervene at any moment and prevent us from having intellect, will, or other faculties. All he wants to point out is that it is logically possible to have a "naked soul" without faculties. This possibility is of crucial importance because it shows that faculties are not necessarily attached to the underlying substantial form. Or, to be precise, there is no absolute necessity, but only hypothetical necessity. That is, *if* God does not intervene in nature, *then* the faculties are necessarily produced by the substantial form. But in principle, the natural production could be stopped. Of course, even if it were stopped the substantial form would not be absolutely naked, because it would still preserve its power to produce faculties. But this power would be in vain. Metaphorically speaking, the form would be pumping and pushing, but nothing would come out of it.

This shows again that Suárez has a model of efficient causation in mind—a model that conceives of faculties as distinct things that may

---

65 *De anima*, disp. 3, q. 3 (vol. 2, 130).

or may not come into existence. And it is clear that the form of the soul could exist without faculties. Could it also be the other way around? Could faculties exist without an underlying form? If there is a real distinction between them, as Suárez explicitly holds, this seems to be inevitable. For according to the standard scholastic view, if two entities, x and y, are really distinct, x can exist without y and y without x; there is mutual independency.[66] But Suárez does not hold this view.[67] He affirms that the faculties can never exist without a substantial form as their foundation, not even when God intervenes. The reason for that is an "essential dependency," as he puts it, which is a one-way dependency: the substantial form can exist without the faculties, but not vice versa.[68] Quite obviously, Suárez weakens the criteria for real distinction. What is important for him is not mutual independency, but distinct production. Faculties are really distinct from the underlying form because they come into existence by a "new action." Once they are in existence they remain distinct, even if they stay in existence only together with the underlying form that brought them about. To return to the comparison with the federal union, one could say that the provinces cannot exist without the federal government. Nevertheless, they are really distinct units. Once they have been created by an initial act of federal legislation, they remain distinct, each preserving its own realm of power.

---

66 Ockham is an unequivocal proponent of this view; see *Ordinatio* I, q. 4 and 6 (OTh II, 115 and 193). On the demarcation of real distinction from other types of distinction, see Marilyn McCord Adams, *William Ockham* (Notre Dame, IN: Notre Dame University Press, 1987), 16–29.

67 In *Disputationes Metaphysicae* (*DM*) 7.2 (Opera 25, 261b) he holds that separability is a sign of real distinction. But two items can be really distinct even if they never appear to be separable. Suárez mentions matter and form as two components that never appear to be separable; see *DM* 7.1 (Opera 25, 250b).

68 *De anima*, dist. 3, q. 1 (vol. 2, 78–80), and a detailed analysis in Christopher Shields, "Virtual Presence: Psychic Mereology in Francisco Suárez," in *Partitioning the Soul: Debates from Plato to Leibniz*, edited by Klaus Corcilius and Dominik Perler (Berlin: de Gruyter, 2014), 199–218. Note, however, that the earlier Vivès edition reports a slightly different opinion. There Suárez does not completely rule out a separation of the faculties. He states that it is at least supernaturally possible and draws a parallel to the Eucharist, where qualities can be supernaturally separated from the underlying substance. See *De anima* II.2 (Opera 3, 576a).

What then is the agent at stake: the underlying substantial form, the faculty, or both? The fact that Suárez takes a faculty to be a real thing with its own *ratio* suggests that he takes the faculty to be the agent. In the case of nonliving things, he does indeed make this strong claim.[69] Thus, he says that the heating power in water is a thing that is really distinct from the water itself and that this power does the heating, not the water. Should the heating power be separated from the water (say, through divine intervention), it could still be active and heat. In the case of living beings, however, Suárez is more cautious. He stresses that no faculty can act by itself. The underlying form also needs to be active. He presents three arguments for this claim, all based on experience.[70]

The first argument refers to the phenomenon of attention.[71] We all experience that we do not perceive objects in front of us unless we pay attention to them. Thus, when we are presented with a visual object, it does not suffice to have a visual faculty in order to see it. We also need to focus on that object. But the attention is not an activity of the visual faculty, for this faculty simply assimilates an object; it is a passive faculty. What is required is an active principle that directs the attention to the visual object, which then enables the visual faculty to assimilate it. This principle is precisely the substantial form of the soul. Hence, both this underlying form and the visual faculty are required for an act of seeing.

The second argument, which also concerns attention, refers to the dominant use of a faculty. It can happen that a faculty completely dominates over another one so that the suppressed one becomes inactive. For instance, a person who stares at something does not hear

---

69 *DM* 18.5.4 (Opera 25, 629a).
70 *DM* 18.5.2–3 (Opera 25, 628a–b).
71 This argument, which was meant to reject an account of perception as a merely passive phenomenon, was already presented by a number of Aristotelians before Suárez, among them Fracastero and Zabarella; see Cees Leijenhorst, "Attention Please! Theories of Selective Attention in Late Aristotelian and Early Modern Philosophy," in *Mind, Cognition and Representation: The Tradition of Commentaries on Aristotle's De anima*, edited by Paul J. J. M. Bakker and Johannes M. M. H. Thijssen (Aldershot: Ashgate, 2007), 205–30.

someone speaking to him; the intensive use of the visual faculty suppresses the use of the auditory one. If faculties were independent agents, this could not happen, because each of them would become active as soon as it was presented with the appropriate object. So the simple fact that a faculty can, as it were, be switched off shows that the underlying substantial form must be involved. It must create an order among the faculties, thus making it possible that there will be a dominating and a dominated faculty at any given moment. Consequently, this ordering principle must be active as well as the faculties.

Finally, the third argument points out that there is a dependency among the faculties. When the intellect conceives of something as a good and desirable object, the will is immediately triggered to want it; wanting clearly depends on thinking. And when the visual or auditory faculty perceives an object, the intellect is immediately triggered to think about it; thinking depends on perceiving. So there must be a coordinating principle for all the faculties—a principle that fixes the dependency relations and that makes it possible to have a well-ordered series of activities. Since each faculty is only capable of producing its own type of activity, the coordination task cannot be ascribed to a specific faculty. It is the substantial form that, as it were, stands in the background, holding all the faculties together and establishing an order among them. If it were not constantly active, no well-ordered activities would be possible.

This last argument obviously appeals to the functional architecture of the soul, which was traditionally emphasized by all Aristotelians. But Suárez, unlike his predecessors, does not simply hold that there are different levels and dependency relations built into the soul. He emphasizes that these relations need to be constantly monitored. That is why there needs to be an ordering principle that is constantly active. To return to the political analogy with a federal union, one could say that it is not enough to have a constitution that defines the range of power of all the provinces and that prescribes how they should interact. A federal government is also required to make sure that all the laws concerning the interaction are observed. For instance, it needs to ensure that school diplomas issued in one province will be accepted in another, so that students can move from one province to another. Only federal supervision guarantees a well-functioning interaction between the provinces. Similarly, only a constant regulating activity of the substantial form guarantees a harmonious interaction between all the faculties.[72]

In speaking about two types of agents—the substantial form and the faculties—Suárez avoids the extreme position according to which faculties are completely independent agents. Nevertheless, he turns them into agents, as his final remark shows: "For the form is like a universal principle, whereas the necessary faculty is like a particular one—a more fitting and appropriate one."[73] This statement has a striking similarity with the *concursus* theory that Suárez presents when explaining the relation between God as first cause and natural things as secondary causes.[74] Neither God alone nor natural causes alone bring about effects in the world. The two rather need to "concur," that is, to collaborate. God needs to be active as the universal principle that constantly produces all natural things and determines the dependency relations among them. Natural things also need to be active, namely as secondary causes that produce the effects they are designed to produce according to the natural order. Likewise, there are two types of causes in a soul, and only when they "concur" can there be successful activity. The substantial form as the universal principle of activity needs to produce the

---

72 Note, however, that there is no interaction in the sense of an efficient causal relation between the faculties. They do not act upon each other but are only coordinated in such a way that the activity in one faculty gives rise to an activity in another one. The only entity that literally acts upon the faculties is the underlying form. Suárez speaks about a "sympathia potentiarum," contrasting it with strict causation. See *DM* XXII.2, n. 29 (Opera 25, 818) and an analysis in Josef Ludwig, *Das akausale Zusammenwirken (sympathia) der Seelenvermögen in der Erkenntnislehre des Suarez* (Munich: Karl Ludwig Verlag, 1929).

73 *DM* 18.5.3 (Opera 25, 629a).

74 *DM* 22.2.15–46 (Opera 25, 814a–822b); for an analysis, see Dominik Perler and Ulrich Rudolph, *Occasionalismus: Theorien der Kausalität im arabisch–islamischen und im europäischen Denken* (Göttingen: Vandenhoeck und Ruprecht, 2000), 208–11.

faculties and establish an order among them. The faculties then need to produce the activities they are meant to produce according to the natural order. Neither the substantial form alone nor the faculties alone can bring about the full range of human activity, but if they work together they will successfully produce all that is distinctive of a human being.

6. Conclusion

With Suárez we have finally reached an author who claims that faculties are inner principles of activity. Is he therefore the ideal target of the theory ridiculed by Locke—a theory that introduces a swarm of little agents running around like servants in a fancy household? It would be unfair to ascribe to him this position, for he never speaks about autonomous agents that act inside a human being. As I pointed out in the previous section, he claims that the substantial form needs to be involved in every activity. He even affirms that the substantial form is "the source of all actions and natural motions" in a living being.[75] Since this form is always combined with matter (at least before death), it is always the form-matter compound, that is, the entire human being, that acts. Hence Suárez does not hesitate to claim, just like Aquinas and many other scholastic authors, that it is the human being that digests, sees, thinks, and so on, not an inner agent. However, the entire human being cannot act unless there are distinctive inner principles that enable it to bring about various activities. That is why one ought to say that the human being digests in virtue of the digesting faculty, or that it sees in virtue of the visual faculty. Technically speaking, the human being is the *principium quod* of an action, while the faculty at stake is the *principium quo*. To give a full account of a given activity, one needs to refer to both principles.

---

75 *DM* 15.1.7 (Opera 25, 499b).

What about Locke's second complaint, namely that scholastic authors give a pseudoexplanation of activities when they appeal to faculties? Is Suárez not guilty of this mistake when he claims that a human being sees in virtue of the visual faculty? Quite obviously, he describes the faculty in the very same terms in which he describes the activity and therefore seems to give no real explanation at all. But here, again, it would be inadequate to ridicule him, because he spells out the faculty by referring to a causal process. Why, for instance, should we refer to a visual faculty in a human being? We should do so, Suárez would say, because there are visible objects that cannot trigger acts of seeing unless there is a power that brings these acts about: each activity needs an appropriate source. The fact that we describe this source in the same terms as the activity does not make the explanation vacuous. It merely shows that a certain type of effect requires a corresponding type of cause. Suárez would even say that we neglect to give a real explanation of our activities if we do not appeal to faculties. For in that case we simply assume that there are many activities that miraculously arise in us, but we fail to spell out the specific principles or powers making these activities possible. Only when we draw a map of inner principles, which are all anchored in a substantial form, do we not simply affirm *that* activities arise, but also give a reason *why* they arise. This reason has a solid foundation because we refer to things (*res*) that are really present and constantly active within us. The map of inner principles is therefore a map of indispensable causes—the better we draw this map, the better we can tell a causal story about all our activities.

These replies on behalf of Suárez make clear that he has strong metaphysical reasons for positing faculties as "principles of operation." However, the replies also show that his reasons are deeply rooted in an ambitious metaphysical program. It makes sense to speak about faculties as inner causes only as long as one assumes that the soul is a complex causal network and that it includes a cause for each type of activity. Moreover, one needs to assume that all the causes are present in and dependent on a substantial form, which is the ultimate cause and functions like a power station that constantly provides the intermediary causes with all the energy they need in order to produce activities. As soon as one questions these assumptions, the entire explanation collapses. It is therefore not surprising that Locke and many other anti-Aristotelians saw no explanatory force in late scholastic theories of faculties. Since they rejected substantial forms as well as the manifold causal relations between inner principles and activities, they took it to be pointless to draw a map of faculties. The decisive point is, however, that drawing this map is not necessarily a vain enterprise. It only starts to look dubious when its metaphysical foundation is no longer accepted.

But why was the metaphysical foundation eventually rejected in the seventeenth century? Why did hylomorphism, which had provided a robust explanatory framework for many centuries, lose its attractiveness? There is no doubt that a number of external factors (e.g. the rise of corpuscularian theories in physics and the use of physiological methods in investigations of the soul) played an important role. But there were also *internal* factors that led to a transformation and eventually to a dissolution of scholastic theories of faculties.[76] Two factors in particular proved to be of crucial importance.

The first factor was a thoroughgoing change in the conception of the soul as substantial form. Up to the thirteenth century, many Aristotelians took this form to be an abstract principle responsible for the actual existence, as well as the unity and the functioning, of the body. When talking about faculties, they intended to speak about manifestations of this principle, and they all agreed that there is just one principle in a human being or in any other living body. That is why they never attempted to subdivide the soul or to individuate special soul-parts that are located in parts of the body—an abstract principle is not a concrete thing that comes in different portions. It may play different

---

76 While focusing on a different set of problems, Helen Hattab, *Descartes on Forms and Mechanism* (Cambridge: Cambridge University Press, 2009), points out that internal factors were as important as external ones for the revision and eventual abandonment of hylomorphism as the common explanatory framework.

roles, as William of Auvergne pointed out, but roles are not countable parts or portions of the soul. Thomas Aquinas still subscribed to this view by calling the soul the first principle of actuality. But he drew a distinction between the "naked principle" and the faculties as mere qualities. Even if these qualities necessarily go along with the principle, he claimed, they are distinct from its essence. This led to a first internal division and consequently to a first separation of two metaphysically (and not just conceptually) distinct areas of the soul: the "core soul" is not identical to the "peripheral soul" that consists of faculties. Ockham rejected this distinction but introduced another one, namely between two substantial forms that have their own domain of activity. In emphasizing that these forms are causes that literally produce activities, each having its own domain, he turned the abstract principle into a pair of concrete agents. Of course, he took it for granted that the two forms are components of the entire living being and that they complement each other. But he assumed that there is an interaction between two distinct agents. Suárez returned to the thesis that there is just one substantial form in a living being, but he widened the gap between this form and the faculties by claiming that faculties are really distinct things. In further claiming that the form literally produces these things, which in turn produce activities, he presented the soul as a network of agents. In a nutshell, he turned the abstract principle into a complex aggregation of concrete things. It is therefore not surprising that Suárez carefully analyzed how every single thing acts and how it is connected with the rest of the aggregation. In doing so he reified faculties and deeply transformed the original conception of form.[77]

This reification went hand in hand with another crucial change. Traditionally, Aristotelians referred to the soul as cause, and in doing so they had the model of formal causation in mind: the soul is what makes a living being the very thing it is and that guarantees its identity

---

[77] This tendency to conceive of forms in terms of concrete things can also be found in other parts of metaphysics and natural philosophy, as Robert Pasnau, "Form, Substance, and Mechanism," *Philosophical Review* 113 (2004): 31–88, convincingly shows.

over time. In modern terminology, one could say that they appealed to the soul as the inner structure of a living being, and causal explanations were meant to be structural explanations. Thus, when claiming that the soul actualizes the body and endows it with a number of faculties, they simply wanted to say that it is responsible for the fact that a natural body has a number of structural features. And when explaining different sets of faculties in different types of living beings, they wanted to point out structural differences among natural bodies. But they did not have efficient causation in mind, that is, they did not intend to say that the soul is like a craftsman sitting inside a human being, creating various agents and coordinating them. That is why Aquinas emphasized that the relation of "flowing" between the soul and its faculties is not to be understood in terms of efficient causation. The soul is not literally a source that produces faculties, which flow from it like streams of water. This changed with Suárez, who claimed that the soul is in fact an efficient cause: the faculties flow from it because of an action it performs. Efficient causation is the basic form of causation for Suárez, for causation is nothing but a form of "pouring" (*influxus*); whatever causes something else "pours" concrete existence and power into it.[78] This means, of course, that the soul is not simply an inner structure but an agent that "pours" existence into faculties, thereby producing them as distinct things. Since these things are principles of operation, they are in turn able to produce activities. It was only this transition from formal to efficient causation that enabled Suárez to conceive of the entire soul as a network of acting things.

It is tempting to describe this change as a process of deterioration: the original conception of substantial form and formal causation was gradually abandoned and replaced with a dubious conception of inner efficient agency. But such a pessimistic assessment would only capture part of the story. It would miss the crucial point that it was precisely

---

78 See *DM* 12.2.4 (Opera omnia 25, 384). For a detailed discussion of the priority of efficient causation, see Stephan Schmid, "Suárez on Efficient Causality," in *Suárezian Causes*, edited by Jakob Leth Fink (Leiden: Brill, forthcoming).

the emphasis on efficient causation and on a network of well-defined inner causes that stimulated early modern thinkers to overturn the traditional model and to work out an explanation of faculties that dispensed with the notion of form and appealed to inner processes that could be spelled out in purely mechanical terms. Moreover, it gave rise to models that attempted to explain the relationship between faculties in terms of the interdependency of various inner mechanisms. As with many theoretical shifts, the transformation of an old conceptual framework gave rise to a new one and thereby proved to be innovative.[79]

---

79 I am grateful to Jennifer Marušić, Stephen Menn, and Stephan Schmid for detailed comments on earlier drafts of this chapter. Sections 4 and 5 are based on material discussed in Dominik Perler, "What Are Faculties of the Soul? Descartes and His Scholastic Background," *Proceedings of the British Academy* 189 (2013): 9–38.

Reflection
FACULTIES AND IMAGINATION
Verena Olejniczak Lobsien

Early modern men and women viewed imagination with suspicion. They did not doubt that there was such a faculty, sometimes also referred to as phantasy or, more disparagingly, as fancy, and that human beings were equipped with it. It was indispensable for poets, painters, priests, and prophets. In many respects, however, imagination appeared as a double-edged if not downright dangerous gift. Thus, it was considered responsible not only for bad dreams, hallucinations, and misplaced love but also for religious heresies. It led people into error and laid them open to abuses of belief. It tempted them toward magic or varieties of hermeticism from astrology through cabbala to witchcraft, or caused them to believe themselves victims of the dark arts. Madness was laid at its door. Combined with humoral dysbalance, it could turn a person into a visionary or a suicide. Unrestrained by reason, it would facilitate possession by spirits as well as their casting out by self-professed exorcists. Given its fallibility and vulnerability to various kinds of evil, a general distrust of the faculty seemed advisable.

Still, some products of the imagination were held in high esteem. Poetry flourished under the Tudor monarchs as never before. Sir Philip Sidney emphatically defended its right to "feign" truth; Edmund Spenser asserted its power to "fashion a gentleman or noble person in vertuous and gentle discipline."[1] Shakespeare's theatre provided living proof of the Elizabethans' favorite commonplace that turned all the world into a stage. These literary achievements did not necessarily rely on image-making in the narrow sense. After all, artists making use of the symbolizing power of words could feel justified by traditional rhetorics and its analysis of those functions of speech—metaphorical in the widest sense—capable of conjuring up and placing before the mind's eye things past, absent, even unheard-of, new, perhaps impossible, lacking definable shape.

Nonetheless, common opinion in the Renaissance was and remained, on the whole, wary of imagination and its creations. Warnings usually called to mind the need for a hierarchical order among the faculties, with reason superior, controlling and regulating the others. To mention only a few of the relevant philosophical, medical, and theological voices: Unlike the elder Giovanni Pico della Mirandola in his oration *De hominis dignitate* (1486), who tended to rate enthusiasm, poetic fury, and divine inspiration highly as preparing the way toward transcendence, his nephew Gianfrancesco Pico remained resolutely critical of phantasy. Interested in moral evils and their remedies, in his *De imaginatione* (1501) he warns of its dangers to those who give credence to conceits rising while the body is asleep and reason dormant. In his influential treatise *The Passions of the Minde* (1601), the Jesuit Thomas Wright approached the faculty with the ethos of a medical man, trying to mediate between Aristotelian and Stoical positions. He regards imagination as a catalyst prone to strengthen the affects to which it becomes attached. As it "colors" the whole world according to the quality of the respective passion, it will also intensify its destructive sides. Imagination, for Wright, works like a distorting optics or tinted spectacles affecting the temper as well as the worldview of those who give rein to it instead of harnessing its

---

1 "A Letter of the Authors," in Spenser, *The Faerie Queene: Book Two*, 225.

impulses to reason.² Loss of rational guidance, but above all a lack of spiritual control is also at the heart of Samuel Harsnett's account of the diseased, melancholy imagination at work in the cases of witchcraft and exorcism he exposes in his *Declaration of Egregious Popish Impostures* (1603).

All these authors speak the language of contemporary humoral pathology and psychophysiology. In their inclination to regard the imagination as portal for all kinds of diseases, however, they fall behind positions formulated by authors rooted in the Neoplatonic tradition such as Marsilio Ficino. Harsnett's spiteful allusion to "Porphyrius, Proclus, Iamblicus, and Trismegistus, the old Platonicall sect that conversed familiarly and kept company with devils,"³ indicates his conscious opposition to this school of thought. Still, not everybody shared his disgust, and Neoplatonisms persisted in early modern culture, in the occult regions of hermeticism as well as in the brighter world of courtly civilization with its strong undercurrents of Petrarchism. Here, Ficino's conviction, as articulated in his *Theologia Platonica* (1482),⁴ that phantasy, due to the human soul's overall relatedness to the divine and immaterial, could at least serve as a mediating agent between the lower, bodily realms of sensation and the higher, rational and spiritual faculties of the mind, seemed to be borne out.⁵ In this view, phantasy, interacting with imagination,⁶ is capable of discerning and judging the "intentions" of the images conveyed to it while at least adumbrating (*auguratur*) "the substance of beauty, goodness, and

---

2 See Thomas Wright, *The Passions of the Minde* (1601), reprint (Hildesheim: Olms, 1973), 82–83; 91–92; 94–96.

3 In Frank W. Brownlow, *Shakespeare, Harsnett, and the Devils of Denham* (Newark: University of Delaware Press, 1993), 239.

4 Marsilio Ficino, *Platonic Theology*, vols. 2 and 4; references give book (roman numerals), chapter, paragraph.

5 Marsilio Ficino, *Platonic Theology*, XIII, 4, 4.

6 Marsilio Ficino, *Platonic Theology*, XIII, 2, 18, and 25, as *congregatrix* of the senses identified with the *sensus communis*.

friendship" in which they share.⁷ Yet even Ficino would grant that, though phantasy might aid the mind on its ascent and, through bordering on understanding, communicate with the *mens*, it was beset by numerous dangers and pitfalls and could seriously hinder the soul in its attempt to soar beyond matter.

Poets, too, were aware of the official reservations against imagination, but they went further. Their works certainly reflected its ambivalences. Thus, Shakespeare's *Macbeth* appears to affirm the dangers of witchcraft. Macbeth believes the weird sisters because they speak to his secret wishes and phantastic ambitions. To turn such prophecies into actuality, realizing what seems possible, if beset with moral and mental terrors, obviously equals a traffic with demons. An air of spiritual evil personified also surrounds the more openly didactic presentations of the faculty in the work of Spenser. In book 2 of *The Faerie Queene* the protagonists Guyon and Arthur travel through the Castle of Alma and visit the chambers of the mind in this extended conceit of the human body ruled and inhabited by the soul. Here, the faculties of the sensitive soul reside, quite literally, in individual cells, resembling Carthusian monks. Of the interior senses, reason remains curiously nondescript and anonymous, while memory (ancient "Eumnestes") receives extra attention, with his youthful helpmate Anamnestes fetching, carrying, stowing away, and guarding remembrances of things past. But it is imagination, "Phantastes," that is described with curious detail and the fascination of an allegory not wholly translatable.⁸ The knights find his chamber crowded with the inmate's "idle fantasies...: / Infernall Hags, *Centaurs,* feendes, *Hippodames,* / Apes, Lions, Aegles, Owles, fooles, lovers, children, Dames";⁹ and buzzing with the noise made by his "Devices, dreames, opinions

---

7 Marsilio Ficino, *Platonic Theology*, VIII, 1, 2–3.
8 Spenser, *The Faerie Queene*, II.ix.50–52.
9 Spenser, *The Faerie Queene*, II.ix.50.7–9.

unsound, / Shewes, visions, sooth-sayes, and prophesies; / And all that fainéd is, as leasings, tales, and lies":[10]

> Emongst them all sate he, which wonned there,
> That hight *Phantastes* by his nature trew,
> A man of yeares yet fresh, as mote appere,
> Of swarth complexion, and of crabbed hew,
> That him full of melancholy did shew;
> Bent hollow beetle browes, sharpe staring eyes,
> That mad or foolish seemd: one by his vew
> Mote deeme him borne with ill disposed skyes,
> When oblique *Saturne* sate in the house of agonyes.[11]

Phantastes's chamber appears full of confusion, ornamented with the garishly colored figments of his own exorbitant production. His visitors must feel as if they have stepped into a TV shop with everything turned on full blast. If the room is overcharged with visual and aural stimuli, its inhabitant seems hardly less irritating. Placed in a disturbing context that seems to be an extension of his own being, he is a sedentary figure resembling a caricature of melancholy, complete with *facies nigra* and downcast gaze. Spenser's Phantastes is a child of Saturn, pathologically fixated on the images that torment him. He is a maniac, flooded by visions beyond his control and devoid of divinity. There may be Ficinian echoes in this "blackened" personification of creative fancy, but they negate precisely what Ficino sought to analyze as a precondition of higher insight.[12] Here, imagination does not promise spiritual excellence. The allegorical figure bears no traces of the saturnine temper's potential to transcend its earthbound gravity. But then, this is

---

10 Spenser, *The Faerie Queene*, II.ix.51.7–9.
11 Spenser, *The Faerie Queene*, II.ix.52.
12 See Marsilio Ficino, *Platonic Theology*, XIII, 2, 2 and 33 (in vol. 4, 120–25, 162–65).

merely what the contemporary topical repertoire would have suggested. It is also a way of distancing conventional ways of thinking about a faculty that could indeed appear irritating and incalculable, even uncanny. In the canto's narrative, however, this visit in Phantastes's chamber remains virtually without function. The artistic and visionary potential of imagination is fully realized in a different manner, as a correlative of the overall structure of *The Faerie Queene*, brought to apocalyptic fruition in the *Mutabilitie Cantos*. But this is done, as it were, behind Phantastes's back. As a whole, the dynamics of Spenser's epic, together with explicit authorial reflections on the power of allegorical poetry to lead its readers toward divine truth, bear out what Phantastes denies.

Shakespeare, too, seems to build up a stereotyped version of imagination in order to render it invalid. Like Spenser's, his own poetic performance works against the cliché. Thus, in *A Midsummer Night's Dream* the famous topical account of the faculty's achievements and idiocies is given to Theseus, a character not conspicuous for his imaginative sensitivity:

> The lunatic, the lover, and the poet
> Are of imagination all compact.
> One sees more devils than vast hell can hold:
> That is the madman. The lover, all as frantic,
> Sees Helen's beauty in a brow of Egypt.
> The poet's eye, in a fine frenzy rolling,
> Doth glance from heaven to earth, from earth to heaven,
> And as imagination bodies forth
> The forms of things unknown, the poet's pen
> Turns them to shapes, and gives to airy nothing
> A local habitation and a name. (5.1.7–17)[13]

---

[13] Shakespeare, *A Midsummer Night's Dream*. All quotations refer to the Holland edition by (act.scene.line).

Yet, in his patronizing summary of the delusions caused by the "tricks" of "strong imagination" (5.1.18), Theseus suggests more than he says. Indeed, his speech, like the comments of the more insightful Hippolyta, shows that imagination cannot be reduced to "fancy's images" (5.1.25) and that it may indeed "apprehend / More than cool reason ever comprehends" (6). Imagination's scope is recognized to be wider than reason's. It seems capable of sensing, perceiving, perhaps grasping more than the latter, even though the truth of what it "sees" may appear doubtful. It is precisely the orthodox suspicions leveled against imagination that are both evoked and called into question. Hippolyta's diagnosis of the "transfiguration" the lovers' minds have undergone during the night discerns in it "something of great constancy; / But howsoever, strange and admirable" (5.1.26–27).

This is in turn true of the play itself, due to the ways in which it blends states of mind and levels of being. The lovers' experience of losing and regaining each other is referred to as a mere dream. On stage, however, we are shown that it is really a consequence of Titania's and Oberon's marital discord and their magical interference in the humans' lives. The action certainly demonstrates that love is a kind of imaginative madness, a passion that tends to elude rational control, capable of turning a bush into a bear (see 5.1.22) as well as an ass into a royal paramour. But it also shows, in the artisans' performance, how lack of imagination may render a serious dramatic effort—to boot, an attempt at presenting a play that shows the terrible results of an error of the imagination—inane and ridiculous. Lastly, imagination proves indispensable to the happy ending precisely as an effect of magic demanding to be taken at face value. Demetrius's eyes remain enchanted, while the spell is lifted from the others. Imaginative deceit thus appears essential for theatrical success, for reconciliation among supernatural forces, for desire to persist and for love to issue in nuptial procreativity. To this end, an element of unavailability and unaccommodated difference has to remain; in Hippolyta's words: a trace of the wonderful that causes admiration. It is what the lovers seem to sense when, in their waking speeches, they stress the dubiousness of their restored realities: "Methinks I see these things with parted eye, / When everything seems double," Hermia observes (4.1.188–89), while Helena feels that she has found Demetrius by chance, "like a jewel, / Mine own and not mine own" (190–91). There may be the joy of recognition here, but there is also the fundamental insecurity of a double vision caused by imaginative excess, a way of looking at "these things" that retains a sense of them being less and more than they seem. The experience transcends the comprehension of "cool reason" as well as humans' sensual capacities (see 4.1.208–11).

That phantasy, motivated by desire but perhaps inspired by something it has no name for, may be capable of moving beyond the limits of the sensitive soul also seems to be a guiding idea in *Twelfth Night*. While the festive exuberance its title alludes to is at first hardly prominent, the comedy comes into its own in the final scene, where we witness not only a pairing off of lovers but also a blissful reunion of twins separated by shipwreck. Ultimate happiness is again brought about by imagination, furthermore based on a complicated web of dissimulation. The effect of a re-cognition engineered by imagination, literally a "seeing otherwise,"[14] which enables us to conceive of one thing in terms of another, now appears as a consequence of disguise and courtly rhetoric. Viola, the heroine, played by a boy actor, finds herself in a situation where she not only passes as a boy but is sent by the man she loves to woo another woman in his name, charged with the task of representing and impersonating him to her rival. This multilayered gendering has been much discussed, although critics usually overlook that it, too, is wholly based on imagination. It is

---

14 For a more comprehensive version of this argument see Verena Olejniczak Lobsien and Eckhard Lobsien, *Die unsichtbare Imagination: Literarisches Denken im 16. Jahrhundert* (Munich: Fink, 2003).

motivated by Viola imagining herself in an intimate relationship with Orsino, driven by Orsino's imagination of his mouthpiece "Cesario" addressing Olivia in his stead, and it goes wrong due to the sheer perfection of Viola's role-playing, which causes Olivia to fall in love with the messenger's performance, sensing the imaginative and poetical strength Viola cannot dissemble as successfully as her gender, and attributing it to the virility of the all-too-persuasive youth. As perfect courtier and allegory of herself, Viola fails spectacularly, because her imaginative virtuosity is given one twist too many. Olivia's admiring "You might do much" (1.5.266) aptly sums up the cause of Viola's highest hopes and deepest despair. By driving the achievements of imagination to extremes and by complicating the confusions it causes, Shakespeare's play draws our attention to the profound ambivalence of the faculty, its verbal foundation, and its astonishing potential.

That this potential may ultimately tend toward good becomes obvious only fairly late. Viola is threatened by a duel because she is taken for her twin, but the fearful prospect implicit in the challenge uttered by her brother's friend Antonio (who also has fallen prey to his own impassioned imagination, having devoted himself "to his [i.e. Antonio's] image" and its promise of "venerable worth"; 3.4.353, 354) only causes her to anticipate the happy reunion with her brother: "Prove true, imagination, O prove true, / That I, dear brother, be now ta'en for you!" (3.4.366–67). That Sebastian is still alive and may easily substitute for his sister in Olivia's affections while the twins are reunited and Viola finally assumes her rightful place in Orsino's heart—that imagination will, indeed, "prove true"—is shown in the finale. Now, all participants are undeceived, or rather: productively alienated as, once again, they have to learn new ways of seeing otherwise. All have to imagine their lovers different from what they thought they were, with strange suggestions of omnipresence (see 5.1.221–22) and Neoplatonic overtones in the reunion of the twins, whose identities appear fused into an androgynous whole; to Orsino's amazement—"One face, one voice, one habit, and two persons / A natural perspective, that is and is not" (5.1.209–10). As the sense of relatedness deepens to mutual participation in view of the twin's extreme similarity— "An apple cleft in two is not more twin / Than these two creatures" (5.1.217–18)—imagined oneness begins to bridge the former differences.

Some potential for gender trouble remains. Orsino finds it difficult to perceive Viola as his future wife before he has actually seen the former Cesario in his/her "woman's weeds" (5.1.267). Only then will she/he truly become "Orsino's mistress, and his fancy's queen" (5.1.378). But this is, of course, imagination's truth. The precariousness of the faculty is once again called to mind together with the extent to which it is tied to the soul's signifying powers. In Shakespeare, the human capacity for creating and perceiving symbols, verbal, gestural, or sartorial, in "form and suit" (5.1.229), will shape a world in which, in Feste's words, "Nothing that is so, is so" (4.1.8). But it is imagination that will put it in motion and turn the fool's negative into a unified poetic whole and "something of great constancy."

Renaissance poetry strongly valorizes a faculty underrated in nonliterary discourse. It demonstrates the dynamic of phantasy as capable of reaching beyond a mere representation of things absent and toward new and often amazing unities. It shows how symbolic imagination enables double insight and creates a conceptual surplus in a more than sensual seeing-otherwise. In authors like Shakespeare, it makes a difference.

CHAPTER FOUR

Faculties in Early Modern Philosophy

Stephan Schmid

1. Introduction: New Science, New Metaphysics— and Old Psychology?

Many early modern philosophers became increasingly impatient with, if not overtly hostile to, Aristotelianism, which determined the curricula of Europe's universities until the end of the seventeenth century. Instead, they were attracted by the prospect of mechanist physics as it was pursued by such outstanding figures as Galileo Galilei and Johannes Kepler. Their achievements in applying mathematical methods to natural phenomena promised a form of firm and certain knowledge, giving reason to hope that the endless and allegedly futile disputes among scholastic philosophers could finally be brought to an end.[1] In line with this "new science," many early modern philosophers

---

[1] In his *Discourse on Method*, part VI, Descartes defends his philosophy as providing knowledge that is useful for life. See *Oeuvres de Descartes*, edited by Charles Adam and Paul Tannery (Paris: Vrin,

preferred a broadly mechanist worldview to Aristotelian hylomorphism.

One might expect that this metaphysical transition from a hylomorphist to a mechanist worldview would immediately challenge the positing of faculties, which figured prominently in Aristotelian physiology and psychology, as chapters 1, 2, and 3 have shown. Given that faculties were widely conceived as "parts" of the soul, which was in turn described as the form of a living being, one might suspect that faculties could play no role whatsoever after the hylomorphist framework was rejected. There simply seems to be no ontological room for such kinds of things.

Surprisingly, though, faculties kept on playing a significant role in both early modern psychology and logic. Early modern authors were often fairly traditional in assuming that rational beings have a mind by which they are endowed with various faculties, like the will, and intellect or understanding, as well as various additional faculties that include perceptual faculties, like the outer and sometimes even inner senses such as imagination and memory. Moreover, early modern philosophers even strengthened the role of faculties by assigning them a significant role in their logics by which they tried to replace the Aristotelian syllogistics that they considered to be sterile and futile. As they argued, Aristotelian syllogistic served at best to reorganize knowledge one already had but was utterly inappropriate for gaining new knowledge and certainty.[2] In their view, logic should explore and provide means of acquiring new knowledge. In accordance with this, they thought that a satisfying logic should give a precise description of our cognitive faculties and their various interrelations so as to learn which cognitive processes were reliable, and led to clear and

---

1983–1991), vol. 6, 61 (= AT VI, 61); English translation in *Philosophical Writings of Descartes*, edited and translated by John Cottingham, Robert Stoothoff, Dugald Murdoch (vols. 1–2), and Anthony Kenny (vol. 3) (Cambridge: Cambridge University Press, 1984–91), vol. 1, 142–43 (= CSM I, 142–43).

[2] Descartes famously accused Aristotelian philosophy of being futile and sterile; see his *Letter to Picot* (AT IXB, 17–18; CSM I, 188). This line of critique was often repeated by his successors; prominently by Locke, *An Essay concerning Human Understanding* IV, ch. 17, secs. 4–8 (ed. Nidditch, 670–81).

distinct ideas, and also which were misleading and thus to be handled with some care.³

We thus face an apparent paradox: while authors of the early modern period quite unanimously rejected Aristotelian hylomorphism, in which the scholastic discourse about the soul and its faculties was rooted, they did not shy away from employing traditional talk of faculties. Instead, they kept on explaining cognitive processes in terms of faculties and even relied on faculties in devising their anti-Aristotelian logics. This cries for a resolution. What allowed these authors to do so? Or how did they conceive of faculties so that they fit in their non-Aristotelian metaphysical frameworks? It is this puzzle I want to address here by taking a closer look at four different early modern conceptions of faculties as they are suggested by René Descartes, Nicolas Malebranche, Baruch de Spinoza, and David Hume.

The main focus of this chapter is the metaphysical question of what these authors have taken faculties to be. Yet, since the heroes of this chapter were not always as explicit about their metaphysical views on faculties as one might wish, one will often have to derive their underlying metaphysical assumptions from the way they talk about faculties and employ them in their theories of mind. In order to do so systematically, it will be helpful to take heed of some important distinctions between different levels of analysis and to have a rough overview of the different metaphysical positions that one could adopt regarding faculties.

The first distinction is between semantic and metaphysical questions. Semantic questions concern broadly linguistic items and ask for their meaning. Metaphysical questions are about all kinds of things and concern their nature and (way of) existence. For our task of

---

3 For a comprehensive investigation of Descartes's "faculty logic" see Stephen Gaukroger, *Cartesian Logic: An Essay on Descartes' Conception of Inference* (Oxford: Clarendon Press, 2002). For its reception by other Cartesians, Locke, and the universities of the Dutch Republic in the late seventeenth century see Paul Schuurman, *Ideas, Mental Faculties, and Method: The Logic of Ideas of Descartes and Locke and Its Reception in the Dutch Republic, 1630–1750* (Leiden: Brill, 2004).

reconstructing the metaphysical conception of faculties in Descartes, Malebranche, Spinoza, and Hume, the semantic question is in an important sense prior to the metaphysical one since we can only settle the metaphysical question as to what an author takes faculties to be once we are clear about the semantic question of what he takes the talk of faculties to mean.

Regarding the semantic question one can adopt at least three possible positions. Consider a simple sentence in faculty-language, such as "Lisa has an intellect." Perhaps the most natural way to understand this sentence is to conceive of it as saying that Lisa has a particular faculty or ability called "intellect." Call this the *realist stance* concerning the language of faculties. Instead, one can also take a merely *nominalist stance* toward this sentence, holding that "Lisa has an intellect" only means that Lisa has intellectual thoughts. On this stance, when employing faculty-language one does not attribute a special sort of entity to a person as one does according to the realist stance. Rather, one simply classifies her mental episodes.[4] Despite their differences, exponents of these two stances agree that sentences about faculties are genuinely true or false. This distinguishes them from defenders of an *expressivist stance*, who think that the sentence does not make a true or false statement about Lisa, but rather licenses a range of inferences one might draw about her or as expressing certain expectations about her (such as that she will answer you if you ask her something).[5]

Now, although semantic in nature, these different stances toward faculty-language are not metaphysically neutral. This brings me to the second class of important distinctions, which are associated with the metaphysical question about faculties. According to the nominalist and expressivist stances, there are no such things as faculties. Thus, by

---

4 My labeling of these stances follows the distinction between nominalism and realism in the debate about universals where realists take our predicates (or at least some of them) to refer to entities sui generis, while nominalists take them as mere means to classify particulars.

5 A famous proponent of an expressivist understanding of our talk of dispositions in general and mental capacities in particular is Gilbert Ryle, *The Concept of Mind* (Chicago: University of Chicago Press, 1949), ch. 5.

adopting either of these stances one is committed to *metaphysical eliminativism* concerning faculties. Conversely, sticking to a realist stance toward faculty-language commits one to *metaphysical realism* about faculties, according to which there are such things like faculties. There are, though, multiple ways to spell out this commitment. On the one hand one can just accept that faculties are irreducible primitive entities and thus endorse a *substantial realist position*, which again can be defended in two versions. On one version—call it the *actualist version*—the irreducible powers are always exercised and thus permanently manifested. In contrast to this, one might also accept a *potentialist version* of substantial realism concerning faculties, which allows for unactualized primitive powers. It is primarily this potentialist version of substantial realism that early modern authors ascribed to Aristotelians and to which they were resolutely hostile. This, however, did not deter many of them from abiding by a substantial realist conception of faculties—although they were anxious to make clear that they did not accept the "bare faculties" of their scholastic predecessors.[6] The other possibility for defending metaphysical realism about faculties is to hold a *reductionist position* about them by arguing that they are derivative entities, which can be reduced to other things. Reductionists typically argue that cognitive capacities are ultimately identical with or realized by purely categorical properties or states of thinking subjects.

So much for the various systematic options for spelling out the metaphysics underlying our talk of faculties. It will be the task of this chapter to figure out which of them were adopted by Descartes, Spinoza, Malebranche and Hume. Needless to say, this selective focus on these four authors omits many other important figures of the early modern history of faculties such as John Locke, Gottfried Wilhelm Leibniz, Thomas Reid, Christian Wolff, and numerous other less known authors of this period, who composed countless guidebooks and treatises on

---

6 A representative rejection of "bare faculties" can be found in Leibniz's "Preface to his *New Essays*," in *Philosophical Essays*, edited and translated by Roger Ariew and Daniel Garber (Indianapolis: Hackett, 1989), 294.

the art of thinking.[7] As I will show, however, the respective theories of Descartes, Malebranche, Spinoza, and Hume span a large part of the space of systematic options of theories of faculties, which all early modern authors had to rely on when they joined their scholastic predecessors in describing or explaining mental operations in terms of faculties.

2. RENÉ DESCARTES: MECHANISMS AND COGNITIVE FACULTIES

The impact of Descartes's radical break with Aristotelian hylomorphism and his strong orientation toward the mechanist physics of his time is comparable to almost no other philosophical innovation. In contrast to his Aristotelian predecessors Descartes defended a corpuscularian worldview according to which natural things are made out of infinitely divisible and movable corpuscles that obey purely mechanist laws of motion. This had the advantage of making the natural world exhaustively describable in terms of a mechanist physics that did without any reference to the sorts of causal powers or irreducible dispositional properties invoked by Aristotelians. Instead of explaining natural changes by appealing to manifesting powers, Descartes was eager to stress that such processes can be accounted for simply in terms of the categorical geometrical properties of corpuscles and the universal laws of motion.[8] Descartes's confidence in mechanist explanations was indeed so high that he boldly held that *all* natural or material phenomena were to be explained on mechanist grounds—even the behavior of nonrational animals, which he conceived as complex machines or automata, whose

---

7 The most popular and influential of them might have been the *Logic or Art of Thinking* by Antoine and Pierre Nicole. But until far into the eighteenth century, authors like Anthony Shaftesbury, Francis Hutcheson, and Richard Price were used to developing their moralist views by analyzing our faculties and the origin of our moral emotions. See the collection of Lewis A. Selby-Bigge, *British Moralists: Selections from Writers Principally of the Eighteenth Century*, 2 vols. (Oxford: Clarendon Press, 1897).

8 See for instance his rejection of Aristotelian qualities in *Le Monde* (AT VI, 25); English in *The World and Other Writings*, edited and translated by Stephen Gaukroger (Cambridge: Cambridge University Press, 2004), 18–19.

behavior could be explained according to the same mechanistic principles as every other bundle of corpuscles.⁹ Accordingly, Descartes thought the Aristotelian assumption of an animal soul with vegetative and sensitive faculties to be as unnecessary and futile for explaining a creature's behavior as the assumption of other dispositional qualities. An animal is but a machine, whose metabolism, growth, and various reactions to the environment result from complex mechanist processes of pressure and collision of its corpuscularian parts. Descartes even claimed that the physiological processes in humans were solely due to mechanistic principles. On his diagnosis, Aristotelians only postulated a soul with lower faculties to explain the physiological processes of living beings because they had too little knowledge of anatomy and were thus ignorant of their real causes.¹⁰

Though hostile to the Aristotelian assumption of a vegetative and sensitive soul, Descartes never questioned the existence of a soul altogether. He took it to be an incontestable fact, something we are assured of by our constant self-consciousness, that we have a soul by virtue of which we are able to perform various cognitive operations. In his *Meditations*, Descartes even argued that we not only *have* a soul—as Aristotelians would put it—but that there is a sense in which we *are* a soul or a thinking thing (*res cogitans*), that is, a "thing that doubts, understands, affirms, denies, is willing, is unwilling, and also imagines and has sensory perceptions."¹¹

Instead of completely rejecting the Aristotelian assumption of a soul, Descartes transformed its conception. On the one hand he relegated the lower faculties of the soul, which were supposed to account for a living being's vital operations, to the physiological structure of the body, and retained only its highest part, the rational soul or mind.¹²

---

9 See his *Letter to Reneri for Pollot, April or May 1638* (AT II, 39–41; CSMK III, 99–100).
10 See his *Description of the Human Body* (AT XI, 224); English in *The World and Other Writings*, edited and translated by Stephen Gaukroger (Cambridge: Cambridge University Press, 2004), 170.
11 *Second Meditation* (AT VII, 28; CSM II, 19).
12 So Descartes conceived of the soul no longer as a principle of life in general but simply as principle of thinking in particular. See his *Fifth Replies* (AT VII, 356; CSM II, 246, and AT VII, 352; CSM II, 243).

On the other hand he freed the rational soul or mind from the hylomorphist framework and conceived of it as a complete and independent substance, and not as an ontologically dubious self-subsisting form (*forma per se subsistens*), as scholastic philosophers following Thomas Aquinas tended to do.

Descartes's relegation of the lower faculties to the body and realm of physics surely helped him to avoid certain problems about the unity of the soul that scholastic authors were struggling with, as became plain in chapter 3. At the same time, however, he was faced with a new problem of unity: how can we account for the unity of a human being, which is not only an immaterial soul but also made out of flesh and bone? How can an aggregate of two distinct substances, mind and body, compose a unified human being?[13] What is more, this problem seems to have a bearing on Descartes's theory of faculties, which, apart from the intellect and will, also includes the traditional bodily faculties of imagination, memory, and perception, which all require a body in order to be exercised.[14] But how can an immaterial soul have bodily faculties such as imagination or perception if it is really distinct from the body where the sensory organs reside? Descartes wrote:

> I find in myself faculties for certain special modes of thinking, namely imagination and sensory perception. Now I can clearly and distinctly

---

13 The unity of mind and body in Descartes is widely discussed in the secondary literature. For an entry point into the debate see Marleen Rozemond, "Descartes, Mind-Body Union, and Holenmerism," *Philosophical Topics* 31 (2003): 343–67.

14 For imagination and perception, see his *Sixth Meditation* (AT VII, 78; CSM II, 54). For the body-dependence of memory, see his *Regulae*, rule 12 (AT X, 414; CSM I 41–2) and *Fifth Reply* (AT VII, 356–7; CSM II, 247). Note, however, that apart from this body-dependent memory, there are some texts in which Descartes accepts a form of body-independent intellectual memory, which the soul can still exercise after it is separated from its body; see his *Letter to Mesland*, August 6, 1640 (AT IV, 143; CSMK, 151) and his *Letter to Huygens*, October 13, 1642 (AT III, 580). Unfortunately, apart from saying that intellectual memory depends "on some other traces which remain in the mind itself" (in his *Letter to Mesland*, 2 May 1644, AT IV, 114; CSMK, 233), Descartes does not further explain how there can be traces in the immaterial mind, nor does he make clear to what extent bodily and immaterial memory belong to the same genus; see also Véronique Fóti, "Descartes' Intellectual and Corporeal Memories," in *Descartes' Natural Philosophy*, edited by Stephen Gaukroger, John Schuster, and John Sutton (London: Routledge, 2000), 591–603.

understand myself as a whole without these faculties; but I cannot, conversely, understand these faculties without me, that is, without an intellectual substance to inhere in. This is because there is an intellectual act included in their essential definition. (*Sixth Meditation*, CSM II, 54; AT VII, 78)

According to Descartes then, even the traditionally bodily faculties like imagination or perception belong to the immaterial soul because they give rise to cognitive or intellectual acts or certain modes of thinking, which can by definition only be performed by a thinking thing. Hence, imagination, perception, and memory are (at least in part) clearly mental faculties, which pertain to the soul, and not to the body. They are body-dependent faculties inasmuch as they can only be exercised if the soul is conjoined to a body, but this does not make them bodily or nonmental faculties.

Now, this conception of traditionally bodily faculties like imagination and perception as body-dependent faculties provides Descartes with the means to reconcile his dualism with our self-understanding as unified human beings. The fact that I have body-dependent faculties explains why "I am not merely present in my body as a sailor is present in a ship, but that I am very closely joined and, as it were, intermingled with it, so that I and the body form a unit."[15] True, this is no substantial unity constituted by formal causation, in the way scholastic philosophers tended to conceive of the human being.[16] But the human being is a functional unity, constituted by efficient causation: I am "intermingled" with my body to the extent that my body and my mind stand in constant causal interactions such that I can change some of my body's parts at will and the motions of my body can act immediately on my mind to produce sensory ideas with distinctive qualitative content.[17]

---

15 *Sixth Meditation* (AT VII, 81; CSM II, 56).
16 See Suárez, *Disputationes Metaphysicae*, 15.6.2 (Opera 25, 518b).
17 For more on Descartes's account of the mind-body union see Marleen Rozemond, *Descartes's Dualism* (Cambridge, MA: Harvard University Press, 1998), ch. 6.

Descartes's account of the mind-body union in terms of operationally dependent faculties is revealing for his conception of faculties in general. Although Descartes accepted a variety of different cognitive faculties like the intellect, will, power of judgment, imagination, memory, and perception, he took them all to be ultimately reducible to different operations of the intellect and will. This is perhaps most obvious with regard to the faculty of judgment, which Descartes explicitly analyzes in his *Fourth Meditation* as resulting from the interplay of the intellect, whose function it is "to enable me to perceive the ideas which are subjects for possible judgments,"[18] and of the will, or my "ability to do or not to do something (that is, to affirm or deny, to pursue or avoid)."[19] By means of these two faculties we can assent to the ideas perceived by the intellect and thus form a judgment or withhold from doing so.

The case is less clear with regard to the body-dependent faculties of imagination and perception, since Descartes explained to Gassendi "that the powers of understanding and imagining do not differ merely in degree but are two quite different kinds of mental operation."[20] Yet, this should not mislead us. In the foregoing quoted passage of the *Sixth Meditation*, Descartes tells us that purely mental faculties, such as the will and intellect, belong to me essentially, whereas body-dependent faculties belong to me only accidentally. ("I can clearly and distinctly understand myself as a whole without these faculties.") This makes sense on Descartes's view, since I can only have perceptions and imaginations on the (contingent) condition that I am united to a body. Consequently, the categorical difference between body-dependent and purely mental faculties, which Descartes explained to Gassendi, does not indicate a deep metaphysical distinction between two kinds of faculties. The operations of body-dependent faculties are nothing but the operations of our will and intellect, which happen to be conjoined

---

18 *Fourth Meditation* (AT VII, 56; CSM II, 39).
19 *Fourth Meditation* (AT VII, 56; CSM II, 39).
20 *Fifth Replies* (AT VII, 385; CSM II 264).

with a (suitable) body. Unlike acts of pure understanding, in which the mind "in some way turns towards itself and inspects one of the ideas which are within it," in imagination and perception the mind "turns towards the body and looks at something in the body which conforms to an idea understood by the mind or perceived by the senses."[21]

This finally allows us to characterize Descartes's metaphysics of faculties. As I have pointed out, such a characterization depends on how we understand Descartes's talk of faculties. Descartes takes pains to stress that, unlike some of his scholastic predecessors, he does not hold a reified conception of faculties, according to which they are distinct parts of the soul, which operate as discrete agents: "the faculties of willing, of understanding, of sensory perception and so on... cannot be termed parts of the mind, since it is one and the same mind that wills, and understands and has sensory perceptions."[22] Accordingly, assigning certain faculties to the soul is not to be understood as referring to autonomous parts of the soul. It is rather to say that the one soul can engage in various kinds of activities. Yet, this is not to deny that the soul has various capacities. As Descartes saw it, all our cognitive capacities or faculties are to be reduced to the will and intellect, and even these two faculties are not completely independent from another: Since we can only will something, if we perceive an idea of the thing we want, the will cannot operate without the intellect. And conversely Descartes held that we can only entertain affirmative thoughts, beliefs, or judgments, if we willingly assent to the idea perceived by the intellect. For Descartes then all our cognitive faculties ultimately rely on the two interrelated faculties of intellect and will, so that there is no question that there are two basic faculties that rational beings are endowed with. The only question is: What exactly are these basic faculties? They are not distinct things or *res*, but they are not fictive things either. They are

---

21 *Sixth Meditation* (AT VII, 73; CSM II, 51). See also *Treatise on Man* (AT X, 176; CSM I, 106), *Dioptrics* (AT VI, 140; CSM I, 172), *Second Replies* (AT VII, 160–1; CSM II, 113), and *Conversation with Burman* (AT V, 162, CSMK III, 344–45).
22 *Sixth Meditation* (AT VII, 86; CSM II, 59).

irreducible mental capacities with which God has endowed us:[23] properties, in virtue of which we can perceive ideas and reach decisions, which result in beliefs or bodily movements.[24]

In my taxonomy of positions about the metaphysics of faculties, Descartes is a *substantial realist* concerning cognitive faculties. In the face of his hostility to the Aristotelian assumption of powers, this might strike one as incoherent. Does Descartes's endorsement of primitive cognitive faculties not amount to taking refuge in the very same kind of uninformative explainers that he accused his Aristotelian predecessors of? There are two possible answers that I can think of on Descartes's behalf.

First it is to be noted that Descartes only accepted primitive faculties or abilities in the mind, and it was only with regard to the natural world that he explicitly rejected the Aristotelian assumption of irreducible dispositional properties. Moreover, Descartes is parsimonious in only accepting two irreducible mental capacities, intellect and will, to which all our cognitive faculties reduce: the intellect as the capacity to perceive ideas and the will as the capacity to assent, dissent, or withdraw from any ideas proposed by the intellect, and to form other propositional attitudes toward our ideas (like acts of doubt and acts of wishing),[25] as well as to spontaneously cause bodily movements.[26] Furthermore, these basic faculties are no pure potentialities that enjoy a penumbral way of existence before their manifestation. They are rather permanently actualized. After all, Descartes holds that the mind is an essentially and thus permanently thinking

---

23 See his *Fourth Meditation* (AT VII, 58; CSM II, 40, and AT VII, 60; CSM II, 42).

24 The talk about properties is important here: Unlike many scholastics Descartes does not conceive of these properties as distinct *res*. Consequently, his assumption of two basic faculties does not threaten the unity of the soul any more than the unity of a rabbit is threatened by the fact that it has various properties like a certain age, mass, and color.

25 See his *Fourth Meditation* (AT VII, 57; CSM II, 40) and his *Principles of Philosophy* I.32 (AT VIII-A, 17; CSM I, 204).

26 See his *Passions of the Soul* I.18 (AT XI, 343; CSM I, 335) and I.41–44 (AT XI, 359–62; CSM I, 343–45).

thing.²⁷ Accordingly, our intellect is constantly performing acts of thinking. The same holds for the will, since for Descartes we even manifest our will by withholding a decision.²⁸ As a result of this, Descartes can be seen to hold an *actualist version* of substantial realism concerning cognitive faculties, which does not rely on unactualized powers or bare potentialities whose ontological status has struck many as dubious. Instead, cognitive faculties, for Descartes, are more like constantly manifest forces.

This comes along with an epistemic advantage, which is the second remark I can make in support of Descartes's theory: being permanently actualized, Descartes's cognitive faculties are no epistemologically suspicious "occult qualities." Since our faculties belong to our nature and are constantly manifested, we have direct knowledge of them. Simply in being always conscious of our acts of thinking, we are automatically aware of our faculties such that "when we concentrate on employing one of our faculties, then immediately, if the faculty in question resides in our mind, we become actually aware of it."²⁹

On account of this, Descartes's substantial realist conception of cognitive faculties as permanently active forces seems to have a range of advantages compared with the theories of his scholastic predecessors. However, Descartes's account does not come for free. It is only to be had at the price of a dualist treatment of dispositional properties. While the faculties relegated to bodies are given a reductive account in terms of geometrical properties and laws of motion,³⁰ cognitive faculties rely

---

27 See his *Letter to Hyperaspistes*, August 1641 (AT III, 423; CSMK 189), his *Letter to Gibieuf*, January 19, 1642 (AT III, 478; CSMK III, 203), and his *Letter for [Arnauld]*, June 4, 1648 (AT V, 193; CSMK III, 355).

28 See his *Letter to Mesland*, February 9, 1645 (AT III, 173; CSMK III, 245).

29 *Fourth Reply*, AT VII, 246–47; CSM II, 172. Note, however, that Descartes is not always clear about the fact that our faculties are evident to us. In other passages Descartes considers the possibility that our ideas of extra mental objects might be caused by a faculty unknown to us (see AT VII, 39; CSM II, 27, and AT VII, 77; CSM II, 54). And even according to the quoted passage, our knowledge of our faculties depends on their actualizations. As we will see, Malebranche exploited this claim to argue for the opacity of our nature.

30 It is an issue of debate whether Descartes denied powers to bodies or whether he took moving bodies to be endowed with their own force. For a defense of the former—reductionist—view, see

on irreducible powers of intellect and will. This renders Descartes's picture of the world much less uniform than the one proposed by Aristotelians. We are not only asked to accept two kinds of fundamentally different substances like material bodies and immaterial souls, but we also have to admit that these things possess two fundamentally different kinds of properties. While bodies have only geometrical categorical properties, minds have irreducible faculties or powers in virtue of which they can engage in various cognitive operations. As I will show in the sections that follow, not all early modern philosophers were happy with such a dualist account of dispositional properties in the material and mental realm.

3. Baruch de Spinoza: Faculties as Powerful Ideas

Spinoza's philosophy is often presented as the monist answer to Descartes's dualism. While Descartes took there to be two types of substances, namely minds and bodies, Spinoza famously held that there is only one single substance, namely God or nature. According to this view, singular things, like tables, rocks, horses, and humans, or even minds and thoughts, are but modifications or modes of the single all-embracing substance, conceived under one of God's infinitely many attributes (of which we human beings know only two: thought and extension). On account of this, mental and material things are not really distinct from one another, but rather "one and the same thing, but expressed in two ways";[31] and so the mind and its corresponding body, too, are but the same mode conceived under different attributes.

Daniel Garber, *Descartes' Metaphysical Physics* (Chicago: University of Chicago Press, 1992), 297; for the latter—realist—view, Tad Schmaltz, *Descartes on Causation* (Oxford: Oxford University Press, 2008), 116.

31 Spinoza's *Ethics*, scholium to proposition 7, part 2, henceforth: E2p7s. Further abbreviations I use to refer to Spinoza's *Ethics* are: "a" for axiom; "app" for appendix; "c" for corollary; "d" for definition, if it occurs after the first number, else for demonstration; "le" for lemma; and "pref" for preface. Passages from the *Ethics* and from Spinoza's other works are quoted from *The Collected Works of Spinoza*, edited and translated by Edward Curley (Princeton: Princeton University Press, 1994), to which indicated page numbers refer.

It is observed far less frequently that apart from his commitment to ontological monism, according to which there is only one substance, Spinoza also held an explanatory monism, according to which everything has to be explained in basically the same way. This is particularly manifest in his plea for naturalism:

> Nature is always the same, and its virtue and power of acting are everywhere one and the same, i.e., the laws and rules of nature according to which all things happen, and change from one form to another, are always and everywhere the same. So the way of understanding the nature of anything, of whatever kind, must also be the same, viz. through the universal laws and rules of nature. (E3pref, 492)

This explanatory monism, according to which everything must be understood through the same universal laws, was explicitly directed against Descartes, who taught that the human being—in virtue of its mind and freedom—was not fully subject to the laws of nature, and who argued in the same vein that the dispositions of our bodies were to be analyzed fundamentally differently from those of our minds. According to Spinoza's explanatory monism such a dualist treatment of phenomena is precluded since it seems to introduce illegitimate bifurcations of reality, whose parts are to be explained in radically different ways.[32] Spinoza's aspiration to provide unified explanations is also evident in his treatment of faculties—in particular with respect to their classification and their metaphysics.

In spelling out his theory of mind, Spinoza addressed all four major faculties discussed by Descartes, that is, memory, imagination, will, and intellect. Unlike Descartes, however, he suggested reorganizing their scope of operations significantly. This is most evident with regard to

---

32 Michael Della Rocca, *Spinoza* (London: Routledge, 2008), 5–8, has pointed out that Spinoza's naturalism and his explanatory monism with its drive for unification can be seen as a consequence of his acceptance of the principle of sufficient reason, which rules out there being any brute or unexplainable explanatory differences.

the will and intellect, which Spinoza claimed "are one and the same" (E2p49c). But Spinoza also conceived of the scope of imagination in a surprisingly broad way and argued that it encompassed both sense perception (E2p40s2) and memory (E2p18s). Since we will encounter a similarly broad conception of imagination in Hume, I will omit a detailed discussion of Spinoza's conception of imagination and rather focus on his treatment of the will and the intellect.[33] In doing so, we will not miss a crucial element of Spinoza's theory of faculties, since, in line with his explanatory monism, Spinoza accounts for all faculties in ultimately the same way.

What brought Spinoza to hold that "the will and the intellect are one and the same" (E2p49c)?[34] It is not difficult to guess that Spinoza's startling doctrine of the unity of will and intellect was primarily directed against Descartes's theory of judgment. As seen above, Descartes takes judgments to be a joint product of the intellect and will, where the intellect is responsible for providing their cognitive content and the will for their assertoric force. According to Descartes, simply perceiving an idea by the intellect is not sufficient for believing that the object or state of affairs represented by the idea exists or obtains. Instead, a belief requires an additional act of assent (or dissent) performed by the will. The view that our capacity to believe something depends on the will is called *doxastic voluntarism*—and it has been widely criticized on the ground of the simple experiential fact that it is not up to our will to believe something or not. Spinoza joins this line of criticism and argues (in E2p49) that an idea necessarily involves an affirmation so that we cannot entertain an idea without affirming it (to at least some degree). Interestingly, Spinoza tried to corroborate this view with

---

33 Illuminating accounts of Spinoza's conception of imagination can be found in Peter Weigel, "Memory and the Unity of the Imagination in Spinoza's *Ethics*," *International Philosophical Quarterly* 49 (2009), 229–46, and Martin Hagemeier, *Zur Vorstellungskraft in der Philosophie Spinozas* (Würzburg: Königshausen und Neumann, 2012).

34 For a detailed account see Stephan Schmid, "Spinoza on the Unity of Will and Intellect," in *Partitioning the Soul: Debates from Plato to Leibniz*, edited by Dominik Perler and Klaus Corcilius (Berlin: de Gruyter, 2014), 245–70.

the help of the example of a fictional idea, which at first sight rather seems to be problematic for his view, for we often have fictional ideas without affirming them. Or so at least it seems. Spinoza wrote:

> I deny that a man affirms nothing insofar as he perceives. For what is perceiving a winged horse other than affirming wings of the horse? For if the Mind perceived nothing else except the winged horse, it would regard it as present to itself, and would not have any cause of doubting its existence, or any faculty of dissenting, unless either the imagination of the winged horse were joined to an idea which excluded the existence of the same horse, or the Mind perceived that its idea of the winged horse was inadequate. And then either it will necessarily deny the horse's existence, or it will necessarily doubt it. (E2p49s, 489)

If we had no other ideas that we affirmed and that ruled out the existence of winged horses (like the idea that nobody has ever observed such a horse) our idea of a winged horse would indeed bring us to believe that such creatures existed. This is in fact exactly what we observe in children, who often take their acts of imagination at face value. This defense of the view that even fictional ideas involve some affirmation is telling in at least two respects.

First, it makes plain that contrary to Descartes, who held an atomist theory of belief, according to which we can (at least in principle) assent to ideas quite independently from our other beliefs, Spinoza endorsed a holistic theory of belief, according to which the degree of our affirmation of ideas is essentially determined by our other ideas. So, having strong beliefs about the genre of fictional movies prevents me from taking my sensory ideas produced by watching such movies at face value. Now, this does not mean that such sensory ideas have no effect at all. They can cause a variety of emotional responses and even bring us to look at things in new perspectives. However, as long as they are held in check by stronger ideas incompatible with them, I will neither take them at face value nor will they determine my serious considerations about the world.

This brings me to the second noteworthy point entailed by Spinoza's account of belief. Spinoza holds that believing something or affirming an idea is a matter of degree. That is, although ideas essentially involve some degree of affirmation, they do not necessarily all involve the same degree of affirmation. Instead, their degree of affirmation varies in such a way that some ideas will influence our cognitive economy more forcefully than others.[35] What is important here is that the degree of affirmation an idea involves is sensitive to its degree of adequacy. According to Spinoza an idea's adequacy depends on its relatedness with its reasons or causes (see E2p28d), such that by having an adequate idea of X, we fully understand X. As a result of this, adequate ideas enjoy a greater degree of credibility than inadequate ideas, which are, to a certain extent, inexplicable to us. Consequently, adequate ideas will, *ceteris paribus*,[36] involve a higher degree of affirmation than inadequate ideas.

These might well be good or even convincing reasons against Descartes's voluntarist theory of belief. But they do not yet refute the distinction between will and intellect altogether. Why should we not only deny that an independent faculty of will plays a role for explaining the assertoric force of beliefs or judgments but also—and more controversially—reject the assumption of a separate will in general?

The answer to this question is both simple and highly complex at the same time. It is simple to the extent that it is easily given: the assumption of a faculty of will apart from the intellect is untenable

---

35 This answers an important objection of Diane Steinberg, "Belief, Affirmation, and the Doctrine of Conatus in Spinoza," *Southern Journal of Philosophy* 43 (2005): 149, and "Knowledge in Spinoza's *Ethics*," in *The Cambridge Companion to Spinoza's Ethics*, edited by Olli Koistinen (Cambridge: Cambridge University Press, 2009), 162, who argues that Spinoza cannot coherently explain our denial of fictional ideas in terms of our affirmation of ideas that are incompatible with the fictional ideas since incompatibility is a symmetrical relation such that there is no reason why our fictional idea should be excluded by our other beliefs and not vice versa. Spinoza's view about degrees of affirmation provides such a reason: some ideas involve a greater degree of affirmation than others.

36 More explicitly the *ceteris paribus* clause is: unless the adequate idea is not overtrumped by a more powerful affect arising from a more powerful object than we are (of which there are infinitely many according to E4a1).

according to Spinoza because (1) it relies on a metaphysically distorted view of ourselves and our place in nature so that (2) the phenomena, which the assumption of an independent will is usually supposed to explain, are better explained differently. It is highly complex to the extent that spelling out this answer leads us straight into the thicket of Spinoza's philosophy and hence by far exceeds the scope of this section. Let me therefore elaborate only very briefly on the two components of the simple answer.

The assumption of a faculty of will—at least in the way it was traditionally adopted—is flawed in Spinoza's view since it is at odds with his doctrine of necessitarianism according to which everything happens by metaphysical necessity (E1p29).[37] On account of this, every mode of thinking, including "each volition[,] can neither exist nor be determined to produce an effect unless it is determined by another cause, and this cause again by another, and so on, to infinity" (E1p32d). Nonetheless, most people ignore this fact because "all men are born ignorant of the causes of the things." As a result they "think themselves free, because they are conscious of their volitions and their appetite, and do not think, even in their dreams, of the causes by which they are disposed to wanting and willing, because they are ignorant" of them (E1app, 440). Lacking an explanation of their volitions (of whose true causes they are ignorant), they finally come to postulate a (free) will. The assumption of a will therefore is, on Spinoza's diagnosis, but an illusion arising from our ignorance of the real causes of our volitions and appetites and should consequently be rejected.

This leads us to the second part of Spinoza's rejection of a distinct faculty of will. As we can glean from his diagnosis of our illusion of free will, Spinoza does not deny that we have volitions or appetites. He only thinks us to be ignorant of their real causes. Spinoza's famous *conatus*

---

37 For an account of Spinoza's necessitarianism see Don Garrett, "Spinoza's Necessitarianism," in *God and Nature: Spinoza's Metaphysics*, edited by Yirmiyahu Yovel (Leiden: Brill, 1991), 191–218, and Dominik Perler, "The Problem of Necessitarianism (1p28–36)," in *Spinoza's Ethics: A Collective Commentary*, edited by Michael Hampe (Leiden: Brill, 2011), 57–79.

doctrine offers what he takes to account for our volitions and appetites: "Each thing, as far as it can by its own power, strives to persevere in its being" (E3p6). This *conatus* "is nothing but the actual essence of the thing" (E3p7). As he explained:

> When this striving [*conatus*] is related only to the Mind, it is called Will; but when it is related to the Mind and Body together, it is called Appetite. This Appetite, therefore, is nothing but the very essence of man, from whose nature there necessarily follow those things that promote his preservation. And so man is determined to those things. (E3p9s)

Our volitions then, are nothing but expressions of our essential striving for self-preservation. They are not additional acts produced by a distinct faculty of will. This is quite a radical suggestion. Let me illustrate it with an example. Suppose Bob wants to build a house while Sue dreams of a house. Philosophers usually say that the mental states of Bob and Sue are, despite their similar contents, different in kind: Bob's wanting to build a house is a conative act, while Sue's dreaming of a house is a cognitive act. Yet this is not how Spinoza would see it. On his account, Bob and Sue entertain the same imaginative idea of a house. That Bob's idea qualifies as a volition and not, like Sue's idea, as a daydream, is simply due to the fact that these ideas are differently related to their respective strivings for self-preservation and their other beliefs. Being a volition, Bob's idea of a house is a direct expression of his *conatus*. Perhaps Bob has seen that he has saved enough money for a house, and thinks that having his own house would be good for his self-preservation, and accordingly forms the idea of a house that he strives to realize. Not so with Sue. Her idea of a house does not involve a tendency on her part to realize it. Instead, her dreaming about a house might help her to relax. And unlike Bob's idea, Sue's idea of a house is not caused by a belief that owning a house conduces to self-preservation, but by her need for relaxation.

Although inevitably sketchy, this brief outline of Spinoza's theory of the will and intellect provides enough material to finally address the metaphysical question as to what faculties are for Spinoza. Unlike all other authors discussed here, Spinoza is surprisingly explicit with regard to this question. His answer is rather deflationary:

> Faculties are either complete fictions or nothing but metaphysical beings, or universals, which we are used to forming from particulars. So intellect and will are to this or that idea, or to this or that volition as "stone-ness" is to this or that stone, or man to Peter and Paul. (E2p48s)

Faculties for Spinoza are not entities over and above particular ideas—just as universal predicates designate nothing, according to Spinoza's staunch nominalism, over and above their particular instances. In the same way that we should not take our talk of stone-ness or humanity to refer to general entities, which explain why particular stones and humans are as they are, but simply as a handy means to refer to all stones and humans at once, we should not conceive of talk about faculties as referring to genuine entities, which somehow cause our ideas. Instead, talk of faculties is but a way of classifying different ideas with respect to their content and the functional role they play in our cognitive economy. For Spinoza ideas of imagination are inadequate ideas, which represent external objects through bodily affections (see E2p17s), while ideas of the intellect are adequate ideas, which represent either common features or essences of things in a completely intelligible way.[38] And ideas qualify as volitions to the extent that they dispose us (by mediation of

---

[38] The fact that the intellect for Spinoza is simply constituted by adequate ideas explains why he strictly distinguished between the mind (which he describes as a bundle of ideas that contains adequate and inadequate ideas; see E2p15 and 3p3d) and the intellect (which is only constituted by adequate ideas). For more on this see Yitzhak Melamed, "Spinoza's Metaphysics of Thought: Parallelisms and the Multifaceted Structure of Ideas," *Philosophy and Phenomenological Research* 200 (2011): 25–26. Note, however, that when Spinoza claims that "the will and the intellect are one and the same" he takes the term "intellect" also to comprise inadequate ideas.

our background beliefs and our striving for self-preservation) to realize their objects.

Returning to the taxonomy introduced in the first section, Spinoza clearly took a nominalist stance on our talk about faculties and thus held an *eliminativist conception* of faculties. Faculties, for Spinoza, are not genuine entities that play a causal role in the production of ideas. Talking about them is but a way of referring to different classes of ideas, which are gradually distinct from each other through their degree of adequacy and their contribution to our behavior. This should not be too surprising. For unlike Descartes, who conceived of the mind as a distinct substance, Spinoza held a bundle theory of the mind, according to which our mind is but a bundle "composed of a great many ideas" (E2p15). Such a bundle theory of mind already rules out a realist conception of faculties for the simple ontological reason that a bundle is not the right sort of thing to figure as a performer of mental acts nor, consequently, as a bearer of a faculty, that is, a property in virtue of which it can perform these acts. Being only bundles of ideas, Spinoza's minds do not *have* or *bear* ideas in the way in which a substance distinct from these ideas could have or bear ideas. Consequently, they cannot literally *perform* mental acts or *form* ideas. They can at best contain ideas, which by themselves entail subsequent ideas and influence others.

This sheds light on yet another feature of Spinoza's theory of mind: the Spinozist mind is no inert collection of ideas but rather a highly dynamic bundle, constituted by ideas that are all endowed with a certain degree of power by means of which they produce further ideas and eventually determine our actions or prevent other ideas from taking their effect. Thus, despite his defense of an eliminativist conception of faculties, Spinoza does not do so because he rejects dispositional properties or powers in general. Quite to the contrary, ideas, which constitute the mind and its subclasses of ideas that our faculty language refers to, are themselves powerful entities that strive to bring about their entailments and to destroy ideas they are incompatible with.[39] They do so in virtue of being particular things, which—according to the *conatus* doctrine—strive, as much as they can, to persevere in their being.[40]

This brings us back to Spinoza's explanatory monism. By interpreting our talk about faculties as a way of classifying intrinsically powerful ideas, Spinoza radically departed from Descartes's hybrid strategy of accounting for bodily capacities in a different way from mental capacities. Given Spinoza's parallelism between mind and body (see E2p7), the dynamic structure of the mind is mirrored by the structure of its corresponding bodies, such that "the Mind's striving, or power of thinking, is equal to and at one in nature with the Body's striving or power of acting" (3p28d). Given Spinoza's general *conatus* doctrine, bodies or singular modes of extension display their own striving for self-preservation, too,[41] such that the operations and capacities of material things arise from its powerful constituents in the very same way that our cognitive activities result from our intrinsically powerful ideas. Modes of extension are endowed with an intrinsic power that allows them to act on other bodies and to resist adverse affections.[42]

Spinoza's eliminativist conception of faculties relies on the assumption of intrinsically powerful modes of thought and corresponding modes of extension, which give rise to different processes that we are used to describing as growth, digestion, imagination, memory,

---

39 This provides a deeper explanation for the fact that ideas essentially involve an affirmation: an idea's affirmation of its object is but an expression of its individual *conatus*. For an elaboration of this point, see Diane Steinberg, "Belief, Affirmation, and the Doctrine of Conatus in Spinoza," *Southern Journal of Philosophy* 43 (2005): 154.

40 Spinoza's theory of powerful ideas relies on his general metaphysics of power that is nicely worked out by Valtteri Viljanen, *Spinoza's Geometry of Power* (Cambridge: Cambridge University Press, 2011).

41 Thus, Spinoza rejects Descartes's view that extended things are entirely passive. He does so explicitly in his *Letter 81 to Tschirnhaus*.

42 In his brief digression about physics in the second part of his *Ethics*, Spinoza points out that bodies are characterized by a distinct share of motion and rest (E2le1), which one might understand as basic shares of force of motion and force of resistance (for a defense see Don Garrett, "Spinoza's Theory of Metaphysical Individuation," in *Individuation and Identity in Early Modern Philosophy: Descartes to Kant*, edited by Kenneth Barber and Jorge Gracia (Albany: State University of New York Press, 1994), 73–101, and for an elaboration Valtteri Viljanen, "Field Metaphysic, Power, and Individuation in Spinoza," *Canadian Journal of Philosophy* 37 (2007): 407–11).

hallucinations, understanding, or wanting. This is surely an impressively unified picture of nature. But does it not fall prey to Descartes's objection that nonmental powers are ultimately obscure and unintelligible entities? Spinoza would deny this: the intrinsic powers of individuals are neither primitive nor unintelligible. Rather, the power of a particular thing or mode is due to the fact that this thing or mode is (according to E1p25c) by itself but a determinate expression of God's power—and "God's power is his essence itself" (E1p34). And being a substance, God can be conceived through himself (see E1d3), and is thus the intelligible thing par excellence.

While this response rebuts Descartes's objection, it also makes clear that the power of finite things for Spinoza is actually God's power. It is no surprise, therefore, that Leibniz assimilated occasionalism, which holds that that the only thing that can figure as a real cause and bring about effects is God, to Spinozism.[43] Yet, since it was not Spinoza who made history as *the* occasionalist of the early modern period but Nicolas Malebranche, it will be worthwhile to see how Malebranche conceived of cognitive faculties.

4. Nicolas Malebranche: Thinking in a Supernaturalist World

Unlike Spinoza, Malebranche deeply admired Descartes and followed him in many respects.[44] With regard to metaphysics, Malebranche was an ardent adherent of Cartesian dualism of body and mind and even defended the *bête-machine* thesis.[45] Similarly, he agreed with Descartes's

---

43 See for instance his *On Nature Itself*, sec. 15, 165.
44 Malebranche explicitly expressed his indebtedness to Descartes (as well as to Augustine) in *Letter to Arnauld* (in his *Oeuvres complètes*, edited by André Robinet [Paris: Vrin, 1958–84], vol. 8, 998, henceforth cited as OCM).
45 See *Recherche de la vérité* (henceforth *RV*) II.1.5.1 (OCM I, 215; English translation in *The Search after Truth*, translated by Thomas M. Lennon and Paul J. Olscamp [Cambridge: Cambridge University Press, 1997], 101, henceforth cited as LO) for Malebranche's defense of dualism, and *RV* VI.2.7 (OCM II, 389–99; LO 492–7) for his defense of the *bête-machine* thesis.

mechanist conception of the physical world and maintained that the nature of bodies consisted in extension alone such that they only have categorical properties, which were to be completely described in geometrical terms.[46] Malebranche also took sides with Descartes in methodological issues and held that science ought to aim at firm and unmistakable knowledge; we ought to pursue Descartes's method of doubt and accept only what we cannot rationally doubt.[47] In line with this, the examination of our cognitive faculties played a crucial role in his philosophy. This is particularly evident in his monumental *Recherche de la vérité* (1674–75), where Malebranche developed his philosophy by examining the mind.

At the same time Malebranche was an ardent defender of Augustine, who endorsed a view called "supernaturalism." According to supernaturalism there is no principled distinction between natural and divine phenomena like miracles. Instead, a supernaturalist holds that what we ordinarily take to be natural (like the physical processes of our world, say) is in the end as unintelligible without reference to the operations of a transcendent God as are supernatural phenomena.[48] Following Augustine, Malebranche applied supernaturalism to all sorts of issues, from questions of moral values to the phenomena of causation and cognition.[49] Regarding causation, Malebranche argued for "occasionalism," according to which finite things cannot be real causes that bring about their effects on their own, but are merely "occasional causes" that serve as occasions for God to perform the causal work we usually attribute to finite things. Regarding cognition, he argued that our knowledge consists in a form of divine illumination. In his view, we can only

---

46 Malebranche gives an outline of his mechanist physics in *RV* VI.2.4 (OCM II, 321–45; LO 453–66).

47 For further discussion of Malebranche's Cartesianism, see Andrew Pyle, *Malebranche* (London: Routledge, 2003), 5–7.

48 Augustine states his supernaturalism succinctly in his *City of God* XXII, chs. 4–5.

49 Malebranche even advertised his philosophy as demonstrating "that we must think only of God alone, see God in all things, fear and love God in all things." (*RV* VI.2.3, OCM II, 320; LO 452.)

have genuine knowledge about things because God lets us participate in his ideas of these things and thus illuminates us.

Given Malebranche's indebtedness to Descartes, it is hardly surprising that he largely adopted his classification of faculties and their interrelations. Like Descartes, he took the mind to be endowed with two fundamental faculties—that is, will and intellect—which, in conjunction with the mind's union with the body, jointly give rise to all other faculties such as judgment, perception, imagination, and memory.[50] However, on account of his supernaturalist theories of causation and cognition, Malebranche could not equally adopt Descartes's metaphysics of faculties. Since God is the only source of causation and cognition, our mental activity must be due to God, too, and this rules out treating our will and intellect as primitive powers as Descartes suggested. What then are faculties in a supernaturalist world as Malebranche envisaged it?

In accordance with occasionalism, Malebranche is careful to stress that faculties are, like all other natural causes, "not real causes, but causes that might be called *occasional*."[51] By this he avoids the hybrid account suggested by Descartes according to which powers of material things (like an animal's power to digest) are to be reduced to purely categorical geometrical properties, while cognitive faculties are just accepted as irreducible powers. In Malebranche's view, such a resort to primitive cognitive faculties is no less objectionable than the Aristotelian resort to primitive vital powers, which Cartesians so decisively dismissed:

> They criticize those who say that fire burns by its *nature* or that it changes certain bodies into glass by a natural *faculty*, and yet some of them do not hesitate to say that the human mind produces in itself the ideas of all things by its *nature*, because it has the *faculty* of thinking. But, with all due respect, these terms are no more meaningful in

---

50 For this at least Malebranche argued in turn in *RV* I.2.1 (OCM I, 49–50; LO 7–8, on judgment); III.1 (OCM I, 379–89; LO 197–202, on perception and imagination); and II.1.5 (OCM I, 212–29; LO 101–9, on memory).
51 *RV* I.4 (OCM I, 65; LO 18).

their mouth than in the mouth of the Peripatetics. (*Elucidation* X, OCM III, 144; LO 622)

If it is illegitimate to accept primitive powers of material things, it is no less illegitimate to allow for irreducible powers of immaterial things. In accordance with this our notion of cognitive faculties is to be given an equally reductive analysis as the powers of material things:

> But just as it is false that matter, although capable of figure and motion, has in itself a *power*, a *faculty*, a *nature* by which it can move itself or give itself a figure that is now round, now square, so it is false that the soul, although naturally and essentially capable of knowledge and volition, has any *faculties* by which it can produce in itself its own ideas or its own impulse toward the good, for it necessarily wishes to be happy. (*Elucidation* X, OCM III, 145; LO 622)

Note, however, that by denying that there are primitive mental faculties, Malebranche does not want to imply that we do not have any capacities. He even stresses that the soul is "essentially capable of knowledge and volition." The crucial point simply is that we do not have these capabilities in virtue of being endowed with primitive powers, but because God takes our mental states as occasions to become causally active. On account of his occasionalism, then, Malebranche could treat bodily and mental capacities in basically the same way and even suggested conceiving of mental faculties in analogy to bodily properties:

> Matter or extension contains two properties or faculties. The first faculty is that of receiving different figures, the second, the capacity for being moved. The mind of man likewise contains two faculties; the first, which is the *understanding*, is that of receiving various *ideas*, that is, of perceiving various things; the second, which is the *will*, is that of receiving various *inclinations* or of willing different things. (*RV* I.1.1, OCM I, 40.; LO 2)

This analogy between mental and bodily powers is telling in at least two respects. First, it makes clear that powers should not be conceived in a reified way, as scholastic authors were often wont to. Second, the comparison shows that attributing faculties to the soul should not bring us to conceive of them as being endowed with irreducible causal powers to bring about certain effects quite similar to the way that we would not think that bodies possess such powers just because they are malleable and movable. Talking about cognitive faculties is just as inconsistent with occasionalism as talking about the dispositions of bodies.

Malebranche provided a more unified view of nature than Descartes, who accepted a primitive bifurcation between the analysis of mental and material powers. Nonetheless, his suggestion to conceive of mental faculties in the same way as bodily powers would surely not have convinced a staunch Cartesian.[52] As seen, Descartes thought that his immediate self-consciousness assured him of the existence of primitive mental faculties. Already in his *Second Meditation*, Descartes took himself to be entitled on purely a priori grounds to assert that he is a "thing that doubts, understands, affirms, denies, is willing, is unwilling, and also imagines and has sensory perceptions."[53] Consequently, it seems only legitimate to assume that mental faculties are irreducible powers. Does Descartes's reasoning not show that we can a priori know that we have irreducible cognitive powers or faculties and hence defeat occasionalism?

Even though Malebranche shares Descartes's view that—by being permanently conscious of our acts of thought—we are constantly aware of our soul,[54] he has a powerful argument against the Cartesian conclusion that this awareness assures us of possessing irreducible faculties. As he points out, this introspective awareness of our soul or this inner feeling (*sentiment intérieur*), as he calls it, can at most assure us of the

---

52 In fact, Malebranche's Cartesian archenemy, Antoine Arnauld, held that the metaphysical distinction between mind and body comes along with a methodological distinction, and rejected (in his *Des Vrayes et des fausses idées*, I, rule 6, 19) all analogies between bodies as minds.
53 *Second Meditation* (AT VII, 28; CSM II, 19).
54 See *RV* III.1.1.1 (OCM, I 382; LO 198).

soul's existence, but it cannot reveal its nature to us.[55] Knowledge of a thing's nature can only be gained by perceiving a thing's idea, and there is strong evidence for Malebranche that we have no access to the idea of the soul. We can see this, Malebranche argues,[56] if we reflect on our ideas of bodies in comparison to our notion of our mind. Since we grasp the idea of extension we can know a priori that a body—whose nature it is to be extended—can acquire all and only geometrical properties. This is different with regard to the soul. We can only know a posteriori what kinds of modification it is capable of. Or to put Malebranche's reasoning in more contemporary terms: based on our grasp of their essences we can have an objective knowledge of bodies, while the knowledge we can have of our soul is irreducibly subjective and phenomenal.[57] On account of this, we can at best have an experiential knowledge of our soul that is mediated by our inner feeling of its acts, and must lack a clear and distinct knowledge of its underlying faculties.

There is still another objection that a Cartesian defender of primitive faculties could launch against Malebranche's theory of the mind. By treating mental faculties, like bodily states, merely as occasional causes, she could argue, Malebranche unduly assimilates minds to bodies and thereby loses the ability to account for the mind's distinctive activity. Since only God can be a real cause according to Malebranche's occasionalism, finite things are doomed to uniform passivity. Yet, this runs against our general self-understanding as rational subjects who actively think and draw decisions as well as against Christian theology that requires that we can make free choices and are thus able to sin.[58]

Is Malebranche liable to this objection? Although it has been claimed that Malebranche held that everything except God is merely

---

55 See *RV* III.2.7.4 (OCM I, 451; LO 237–38).

56 In *RV* III.2.7.4 (OCM I, 451; LO 237–38).

57 This is nicely pointed out by Tad Schmaltz, *Malebranche's Theory of the Soul* (Oxford: Oxford University Press, 1996), 41–3.

58 This is even conceded by Malebranche: "If we had no freedom, there would be no punishment or future reward, for without freedom there are no good or bad actions. As a result, religion would be an illusion and a phantom" (*Elucidation* XV, OCM III, 225; LO 669).

passive,[59] Malebranche himself was eager to avoid this conclusion. Right before introducing the analogy between intellect and will and the bodily powers of receiving figures and being moved, he states that "these comparisons between mind and matter are not entirely appropriate."[60] Their inappropriateness consists in the fact that they suggest that the mind is—like bodies—entirely passive. But while the intellect is indeed a passive faculty for Malebranche, this does not hold for the will. Let me explain.

The intellect is the mind's capacity for receiving ideas and modifications. In this respect it is perfectly analogous to matter that has the capacity to receive figures and configurations. So, "just as the faculty of receiving different figures and configurations in bodies is entirely passive and contains no action, so the faculty of receiving different ideas and modifications in the mind is entirely passive and contains no action."[61] Given Malebranche's occasionalism and illuminativist theory of cognition this passive conception of the intellect makes perfect sense. Our intellect is only modified when God causes these modifications on appropriate occasions, and similarly we can only perceive ideas when God reveals them to us on occasion of our will to perceive them. Thus, like matter's capacity to receive certain figures and configurations, our mind's intellective capacity to receive ideas and modifications is entirely passive.

However, this does not hold for the will, which is an active faculty for Malebranche. Here the comparison between mental faculties and bodily powers involves a crucial disanology.

> It must be realized that there is a very significant difference between the impression or motion that the Author of nature produces in matter, and the impression or impulse [*mouvement*] toward the good

---

59 Most prominently by Nicholas Jolley, "Intellect and Illumination in Malebranche," *Journal of the History of Philosophy* 32 (1994): 209–24.
60 *RV* I.1.1 (OCM I, 41; LO 2).
61 *RV* I.1.1 (OCM I, 43; LO 3).

in general that the same Author continuously impresses in the mind. For matter is altogether without action; it has no force to arrest its motion or to direct it and turn it in one direction rather than another.... But such is not the case with the will, which in a sense can be said to be active, because our soul can direct in various ways the inclination or impression that God gives it. For although it cannot arrest this impression, it can in a sense turn it in the direction that pleases it, and thus cause all the disorder found in its inclinations, and all the miseries that are the certain and necessary result of sin. (*RV* I.1.2, OCM I, 46; LO 4–5)

Although the will is in its operation like the intellect completely dependent on God, who directs it toward the good in general, it still can autonomously turn this general inclination toward a particular good, and thus be active in a certain sense. It is in this peculiar "*force that the mind has of turning this impression toward objects that please us so that our natural inclinations are made to settle upon some particular object, which inclinations were hitherto vaguely and indeterminately directed toward universal or general good,*"[62] that our freedom consists.

According to Malebranche then, our faculty of will involves two components. We have the capacity to will something only to the extent that (1) God constantly sustains in us an inclination toward the good in general, which we (2) can direct toward a good in particular. In line with Malebranche's supernaturalist conviction, the first component of our willing ensures that our capacity to will is fundamentally owed to God. The second component of our faculty of will on the other hand allows us to determine our permanent inclination toward the good in general to a certain good in particular, such that we qualify as active and morally responsible agents.

There can be no question therefore that Malebranche conceived of the will as active to some extent in order to account for our freedom.

---

62 *RV* I.1.2 (OCM I, 46–47; LO 5, translation slightly modified).

The crucial question, however, is whether he was entitled to do so. Is our capacity to direct our continuous inclinations toward any good in particular not just a primitive causal power? Malebranche at least tried everything to avoid this conclusion. As he saw it, the power to determine our God-infused inclination toward the good in general to a particular good was no genuine causal power, since by determining our general inclination toward the good we do not *necessitate* anything. And Malebranche thought it essential for a cause to necessitate its effect. Indeed, it is this very thought that leads Malebranche to conclude that God is the only real cause, "because it is a contradiction that He should will and that what He wills should not happen."[63] Thus, the will, for Malebranche, is associated with the wrong kind of modality to count as a genuine causal power: causal powers necessitate their effects and hence are associated with the modality of necessity, while the will is only associated with the modality of possibility: our choices do not necessitate anything. Our "volitions are impotent in themselves; they produce nothing."[64] Their efficacy completely depends on God's grace, by which God takes our volitions as occasions to produce our wanted actions. Consequently, our capacity of forming free volitions requires no causal power on our part,[65] and so our freedom does not contradict Malebranche's occasionalism.[66]

We can now determine Malebranche's view on the metaphysics of cognitive faculties on the basis of the taxonomy I have suggested. Like Descartes, Malebranche accepts that rational beings are endowed with

---

63 *RV* VI.2.3 (OCM II, 316; LO 450).

64 *Elucidation* XV (OCM III, 225; LO 669).

65 This is to be objected to Susan Peppers-Bates, *Nicolas Malebranche: Freedom in an Occasionalist World* (New York: Continuum, 2009), 103–12, who thinks that Malebranche accepts a primitive, immanent-causal power of determining one's will. We indeed have a *power* to determine our will. But it is crucial for Malebranche's reconciliation of freedom and occasionalism that this power is not *causal*.

66 For a similar account of Malebranche's conception of freedom and further literature on this topic see Robert M. Adams, "Malebranche's Causal Concepts," in *The Divine Order, the Human Order, and the Order of Nature: Historical Perspectives*, edited by Eric Watkins (Oxford: Oxford University Press, 2013), 67–104.

faculties. However, unlike Descartes, he tries to defend a unified *reductive account* for as many kinds of dispositional properties as possible, regardless of whether they belong to bodies or to minds. He does so by showing their fundamental dependence on God's operations. Accordingly, our fundamental cognitive faculties of intellect and will consist in our capacity to apprehend ideas and to choose and to elicit voluntary actions, which are both owed to God. We can grasp ideas because God reveals them to us on occasion of certain perceptual circumstances, or on occasion of our will to contemplate them. We can will various things and initiate different actions because we can direct our God-induced inclination toward the good in general toward a determinate good in particular. This resolution in turn (usually) serves as an occasion for God to bring about a corresponding event according to the laws of nature.

By means of his occasionalism and illuminativism then Malebranche succeeded in giving a fully reductive account of all causal and cognitive roles of powers and faculties. According to occasionalism the causal roles of powers are reduced to the fact that God takes certain configurations of bodily and mental states as occasions to cause particular succeeding states; and according to illuminativism the cognitive role of our faculties is due to the fact the God shares certain ideas with us. The only role of faculties that escapes Malebranche's reductive account is the moral role of the will. And for good reason: God cannot perform our will's moral role without thereby dispensing us from the very responsibility that its possession and exercise is supposed to bestow on us. In order to count as responsible agents, our decisions must ultimately be up to us, and this requires us to have an irreducible power of choice.

Malebranche provided an undeniably impressive account of cognitive faculties, which is to a great extent reductive in nature. Whether this account is philosophically convincing is another question. With its strong theistic assumptions, it is not likely to be well received by secular readers. Let us turn to Hume, therefore, who avoided theistic assumptions as no other early modern philosopher did, and see whether his theory of faculties is more attractive.

5. David Hume: Faculties as Regularities of Thought

For Descartes and Malebranche, the investigation of cognitive faculties plays a decisive epistemological role: In their view, we can only assess the reliability of our thoughts when we get clear about their genesis. Therefore, we have to inquire into the various faculties of mind that our thoughts arise from, and understand their operations and mutual interplay. It is only by carefully distinguishing the cognitive outputs produced by sensation and imagination from those produced by the intellect that we can avoid the kind of errors and confusions that Aristotelians fell prey to in assigning sensible qualities to external bodies. For David Hume, the enquiry into the human mind and its operations became the main task of philosophy. As he saw it, philosophy was threatened not only by the morass of scholastic metaphysics but also by unrestricted rationalist speculations.[67] For there to be any hope of philosophy providing knowledge that was as firm as that achieved in the new science, it would have to be acquired, or so was Hume's basic empiricist assumption, by the same method of building one's theory on empirically amenable phenomena.[68] Thus, to avoid the "obscurity in the profound and abstract philosophy" (*EHU* 1.11), we have "to enquire seriously into the nature of human understanding" (*EHU* 1.12) and first and foremost provide a "mental geography, or delineation of the distinct parts and powers of the mind" (*EHU* 1.13).

What then are the parts and powers of the mind that philosophy is supposed to chart? According to Hume, the fundamental elements of

---

67 See also Stephen Buckle, *Hume's Enlightenment Tract: The Unity and Purpose of an Enquiry concerning Human Understanding* (Oxford: Clarendon Press, 2001), 33–43.

68 This is already reflected in the subtitle to Hume's *Treatise*: "An Attempt to introduce the experimental Method of Reasoning into Moral Subjects." I will indicate passages from Hume's *Treatise of Human Nature* by using the citation key *T* (followed by the number of the book, part, section, and paragraph of the indicated passage), and *EHU* for referring to his *Enquiry concerning Human Understanding*.

our mental lives consist in perceptions, which differ with regard to their force and vivacity. *Impressions* are those perceptions "which enter with most force and violence" (*T* 1.1.1.1). They comprise "all our more lively perceptions, when we hear, or see, or feel, or love, or hate, or desire, or will" (*EHU* 2.3). The other fundamental class of perceptions consists of *ideas*, which are but "faint images" of our impressions (*T* 1.1.1.1). Paradigmatic ideas are the memory of a tiramisu you had the other day or your thought of a tiramisu that has been evoked by my example. In these cases we form a perception of a tiramisu, which is much less vivid and forceful, though, than the impression you would have if you were to actually see and taste a tiramisu yourself. Your idea of the tiramisu is but a faint image of an impression of a tiramisu.

These two kinds of perceptions come along with two pairs of faculties: Our impressions are due to *sensation* and *reflection*. Impressions of sensations are perceptions of sensory qualities such as colors, odors, and tastes, while impressions of reflection consist of the vivid perceptions we experience in reaction to our perceptions, such as our desires, aversions, or emotions of hope and fear. Our ideas, by contrast, are mainly due to the faculties of *memory* and *imagination*. Both of them are faculties "by which we repeat our impressions." Yet, while our memory repeats our impressions in such a way that they retain a considerable degree of their first vivacity and (optimally) preserve their original order of appearance (*T* 1.1.3.1–2), the imagination is not bound to respect the original order of our impressions and thus "has unlimited power of mixing, compounding, separating, and dividing these ideas, in all the varieties of fiction and vision" (*EHU* 5.10). In return, ideas of imagination are usually less vivid than ideas of memory (*T* 1.1.3.1 and 1.3.5.3).[69]

---

69 It is subject of debate how the force and vivacity of perceptions is precisely to be understood. For a discussion and further references see Markus Wild, "Hume on Force and Vivacity: A Teleological-Historical Interpretation," *Logical Analysis and the History of Philosophy* 14 (2011): 71–88.

Apart from these explicitly introduced faculties, Hume was also happy to appeal to other traditional faculties,[70] by talking of *reason* (e.g. *T* 1.3.6.4 and 1.4.1), *understanding* (e.g. *T* 1.4.2.14), *judgment* (e.g. *T* 1.4.1.6), and the *will* (e.g. *T* 2.3.3.4). In addition, Hume also allotted the mind a range of further capacities such as those of *comparing* (e.g. *T* 1.1.5.1), *abstracting* (e.g. *T* 1.1.7.2) and of *separating* and *conjoining* ideas (e.g. *T* 1.2.3.10). Unfortunately, Hume often simply used the traditional language of faculties without clarifying its meaning. It has therefore become a highly debated issue whether Hume accepted the mentioned traditional faculties as genuine faculties at all and how they are distinguished from one another.[71] This is not the place where these problems can be solved. Neither is it necessary for the purpose of this chapter. Nonetheless, it will be worthwhile to start our inquiry into Hume's conception of faculties with a general observation about his use of faculty-language.

As already mentioned, Hume resorted to traditional faculty-language in order to provide a mental geography. Yet this was not the only project Hume pursued. In addition to his descriptive project, he also engaged in an explanatory project of devising a comprehensive theory of the mind. Indeed, Hume's theory of mind became so revisionary that it increasingly undermined the traditional faculty-language, by which he first accessed the phenomena that he set out to analyze. This can best

---

70 Unlike in the case of imagination and memory Hume never talked about our ideas of reason or understanding. This brought David Owen, *Hume's Reason* (Oxford: Oxford University Press, 1999), 74, to maintain that memory and imagination are the only genuine faculties for Hume, whereas other "faculties" such as reason and understanding are but characteristic activities. As I will argue, this ultimately holds for all faculties.

71 In particular, scholars quarrel about the role of reason in Hume and its relation to the understanding. While Don Garrett, *Cognition and Commitment in Hume's Philosophy* (Oxford: Oxford University Press, 1997), 84–85, holds that reason is simply the faculty of drawing inferences, which is subordinated to the general cognitive faculty of understanding, Peter Millican, in "Hume on Reason and Induction," *Hume Studies* 24 (1998): 141–59, argues that Hume used the expressions "reason" and "understanding" interchangeably. Moreover, there is a debate about whether reason is a normative notion that refers to a faculty, which is supposed to yield to the conception of truths (as defended by Kenneth P. Winkler in "Hume's Inductive Skepticism," in *The Empiricists: Critical Essays on Locke, Berkeley, and Hume*, edited by Margaret Atherton [Lanham, MD: Rowman and Littlefield, 1999], 183–212), or whether reason is a descriptive notion of our faculty to process ideas as it is manifested in our actual reasoning (as defended by Don Garrett, *Cognition and Commitment in Hume's Philosophy* (Oxford: Oxford University Press, 1997), 91–92).

be illustrated by Hume's statements about the relation between the faculties of understanding and reason on the one hand and imagination on the other. Especially in his discussion about the source of our idea of causation, Hume often opposes these faculties to one another. He asks whether our idea of a causal connection is produced "by means of the understanding or of the imagination; whether we are determin'd by reason to make the transition, or by a certain association and relation of perceptions" (T 1.3.6.4). He famously answers this question by maintaining that "when the mind... passes from the idea or impression of one object to the idea or belief of another, it is not determin'd by reason, but by certain principles, which associate together the ideas of these objects, and unite them in the imagination" (T 1.3.6.12). In doing so, Hume clearly contrasts the faculties of understanding and reason with the faculty of imagination. In the conclusion of his first book of the *Treatise,* however, he throws this opposition overboard: "The memory, senses, and understanding are, therefore, all of them founded on the imagination, or the vivacity of our ideas" (T 1.4.7.3). In line with this, Hume no longer contrasts the understanding with the imagination, but goes on to identify the understanding with "the general and more establish'd properties of the imagination" (T 1.4.7.7).

Hume obviously uses the language of faculties ambiguously.[72] He starts off by employing it rather traditionally in order to classify the distinct kinds of mental operations that he wants to account for, while he ends up with the revisionary view that most of our cognitive faculties are ultimately "founded on the imagination or the vivacity of our ideas." While Descartes and Malebranche tried to reduce all cognitive faculties to the operations of the will and intellect, Hume tried to

---

[72] And he did so intentionally, as he makes clear in footnote T 1.3.9, n. 22, with regard to the imagination. See also Barbara Winters, "Hume on Reason," *Hume Studies* 5 (1979): 20–35; Annette C. Baier, *A Progress of Sentiments: Reflections on Hume's "Treatise,"* (Cambridge, MA: Harvard University Press, 1991), 9–15, and Michael Ridge, "Epistemology Moralized: David Hume's Practical Epistemology," *Hume Studies* 29 (2003): 165–204. For a defense of a univocal reading see Don Garrett, "Hume's Conclusions in 'Conclusion of This Book,'" in *The Blackwell Guide to Hume's Treatise,* edited by Saul Traiger (Oxford: Blackwell, 2008), 154–6.

account for their operations in terms of sensation, the passions, and imagination and the vivacity of ideas. But why should it be the case that the traditional cognitive faculties such as the "memory, senses, and understanding are...all of them founded on the imagination, or the vivacity of our ideas"? (*T* 1.4.7.3)

Let us start with memory. At first sight, the claim that our memory ultimately relies on our imagination seems to fly in the face of Hume's initial distinction between the memory as a faculty that reproduces vivid ideas by preserving their original order and imagination as an unrestricted power of recombining and mixing ideas. Yet, as Hume hastens to add, this distinction is not so clear-cut: we all have been uncertain about our memories of times long past, and we also know "of liars; who by frequent repetition of their lies, come at least to believe and remember them, as realities" (*T* 1.3.5.6). Although memory, which reproduces our past impressions in a vivid way, ought to do so by preserving their original order, this is not always the case. From a subjective point of view ideas of memory are often indistinguishable from ideas of imagination, and the perceived difference (if there is one) "betwixt it and the imagination lies in its superior force and vivacity" (*T* 1.3.5.3).

The enhanced vivacity of ideas of memory also explains why ideas of memory typically come in the form of beliefs, while we do not endorse fictional or other more faint ideas of imagination. In contrast to Descartes and Malebranche, who analyzed our beliefs as resulting from the interplay of our understanding or intellect that perceives a certain idea and the will that gives its assent to it, Hume argues that "the difference between *fiction* and *belief* lies in some sentiment or feeling, which is annexed to the latter, not to the former, and which depends not on the will, nor can be commanded at pleasure" (*EHU* 5.11). Thus, for Hume, "the *belief* or *assent*, which always attends the memory and senses is nothing but the vivacity of those perceptions they present; and...this alone distinguishes them from the imagination" (*T* 1.3.5.7).

This shows that ideas of memory are only gradually distinct from mere ideas of imagination due to their increased degree of force and vivacity. However, to what extent are they *founded* on imagination? Given that Hume thought it to be "impossible, that this faculty of imagination, can ever, of itself, reach belief" (*EHU* 5.12), this question seems to even amount to a challenge.

Fortunately this challenge can be met by noticing an important qualification of Hume's notion of imagination. Although Hume describes the imagination as an "unlimited power of mixing, compounding, separating and dividing" ideas (*EHU* 5.10), he stresses that its operations are typically governed by some universal principles, which he identifies as the three principles of *association*, namely "RESEMBLANCE, CONTIGUITY in time or place, and CAUSE and EFFECT" (*T* 1.1.4.1). Though by nature unrestricted, the imagination tends to reproduce ideas according to these principles. This is confirmed by our everyday experience: "A picture naturally leads our thoughts to the original [so we are led by *resemblance*]: the mention of one apartment in a building naturally introduces an enquiry or discourse concerning the others [so we are determined by *contiguity*]: and if we think of a wound, we can scarcely forbear reflecting on the pain which follows it [and are thus governed by *causation*]" (*EHU* 3.3). Now, our imagination's tendency to reproduce its ideas in this associative order is, at least with regard to contiguity and causality, due to the fact that the imagination grows *accustomed* to such an order by the common sequence of our sense impressions: my senses usually provide me with impressions of contiguous things (when I move my eyes, my impressions change in accordance with the contiguity of seen objects), and likewise impressions of causes are usually followed by the impressions of their effects.[73] These principles of association, which govern the operation of imagination, are at the same time the principles according to which impressions transfer their share of force and vivacity to their related ideas. This general rule is again

---

73 Things are more complicated for Hume since according to him we have no impressions of causes and effects, literally speaking. We are only aware of regular sequences of impressions, which we then begin to conceive as causal sequences once we have experienced these sequences often enough to acquire the habit to expect the one impression on perceiving the other (see *T* 1.3.14 and *EHU* 7).

warranted by our ordinary experience: Seeing "the picture of an absent friend, our idea of him is evidently enlivened by the *resemblance*" (*EHU* 5.15); seeing the front of a car behind the corner immediately brings me to think about the rest of the car, *contiguous* to its front, and seeing a broken window leads me to think about its *cause* of it being smashed, and seeing a stone thrown toward a window prompts me to think about its *effect* and to expect the window to shatter. This finally solves our problem: To the extent that imagination produces its ideas in accordance with the general principles of association it has grown accustomed to, it automatically confers to its ideas a share of the vivacity and force of the impressions, which they are caused by. Thus, although it is impossible for the imagination to form genuine beliefs if it is conceived with respect to its unrestricted power to combine and transpose ideas, this is not the case if we conceive of the imagination more narrowly as being governed by the vivacity-transferring principles of association.[74] In this case, the imagination can indeed be taken as the faculty responsible for the transfer of vivacity, because it produces its ideas exactly in accordance with the principles of association, which govern the transfer of vivacity.[75]

We can now see why Hume took our memory to be founded on the imagination or the vivacity of our ideas. Our memory is the faculty of reproducing lively ideas that are supposed to conform to the order of our original impressions. From our subjective perspective, however, ideas of memory are just characterized by being particularly vivid so that we assent to them and entertain them in the form of beliefs. Their enhanced vivacity in turn is due to their high resemblance to previous impressions and can hence be attributed to the operation of the imagination, which is responsible for the transfer of vivacity when it operates pursuant to its accustomed principles of association.

---

74 In *T* 1.3.9.2 Hume even claims that "belief arises only from causation" such that the principles of contiguity and resemblance do not suffice to establish beliefs.

75 Hence Hume distinguished "in the imagination betwixt the principles which are permanent, irresistible, and universal; such as the customary transition from causes to effects, and from effects to causes: And the principles, which are changeable, weak, and irregular" (*T* 1.4.4.1). Only the latter principles characterize the imagination as a faculty of fictions, which can never, of itself, reach belief.

The fact that the imagination habitually operates according to associative principles also explains why the senses and the understanding are founded on the imagination. The operations of the senses rely on the associative principles of the imagination insofar as they seem to inform us about mind-independent persisting objects. However, such perceptual beliefs about persisting external objects cannot be due to our senses, Hume argued, since our senses only provide us with ever-changing sensory impressions. They can only result from our imagination, which becomes accustomed to the regular occurrence of qualitatively similar or constant or coherent impressions so that we begin to form beliefs about enduring objects in our surroundings, to which we then attribute our sensory impressions (by taking them as corresponding to qualities *of* these objects).[76]

Similarly, our understanding or reason is founded on the imagination—at least when it is engaged with truths about matters of facts, which we can only infer by reasoning from causes to effects (or vice versa).[77] This reasoning is founded on the imagination because the transition from impressions or vivid ideas of one type of event to the vivid idea or belief of another type of event is due to custom and habit, that is, due to certain associative connections we have acquired by having experienced these events succeeding each other many times.[78]

In light of these considerations Hume takes our cognitive faculties to be founded on the imagination or the vivacity of ideas such that all their operations can be explained in terms of the operation of imagination and its transfer of vivacity. This immediately raises the question about imagination itself. Did Hume accept imagination as a genuine

---

76 This is the upshot of Hume's investigation of the reliability of the senses in *T* 1.4.2.

77 As Hume explains (*EHU* 4.1–3), reason can also be engaged with truths founded in relations of ideas, which give rise to certain knowledge. Reason arrives at this certain knowledge by coming to see that the negations of truths about relation of ideas "would imply a contradiction, and could never be distinctly conceived by the mind" (*EHU* 4.2). Thus, the operation of demonstrative reason rests on there being incompatible ideas, which cannot be distinctly coconceived.

78 This is the upshot of Hume's investigation of the reliability of probable reasoning in *T* 1.3.

entity or did he rather join Spinoza in holding that faculties are nothing over and above certain patterns of mental processing?

There are several considerations that jointly speak in favor of seeing Hume, like Spinoza, as a defender of an eliminativist conception of faculties. The first of these is provided from the way in which Hume spells out the distinction between impressions of sensation on the one hand and impressions of reflection on the other.

> The first kind arises in the soul originally, from unknown causes. The second is derived in a great measure from our ideas, and that in the following order. An impression first strikes upon the senses, and makes us perceive heat or cold, thirst or hunger, pleasure or pain of some kind or other. Of this impression there is a copy taken by the mind, which remains after the impression ceases; and this we call an idea. This idea of pleasure or pain, when it returns upon the soul, produces the new impressions of desire and aversion, hope and fear, which may properly be called impressions of reflexion, because derived from it. (T 1.1.2.1)

As is evident here, the appeal to the faculties of sensation and reflection only serves a classificatory purpose. Characterizing an impression as an impression of sensation tells us nothing about its causes. Hume even frankly admits (in T 1.3.5.2) that impressions of sensations arise from causes that cannot be known with certainty. Likewise, passions and emotions are not called impressions of reflection because they are somehow caused or produced by a reflective faculty. These impressions arise from a rather complex process involving the intermediary production of ideas. Thus, neither sensation nor reflection should be understood as causes of their impressions, which somehow explain their occurrence. It is clear therefore that Hume took a nominalist stance toward our talk of sensation and reflection: the terms "reflection" and "sensation" can be used to classify certain mental states, but not as designating certain entities, which figure as causes of these states.

In introducing and describing our fundamental faculties of ideas, that is, memory and imagination, Hume is less clear. He describes them as faculties "by which we repeat our impressions" and thus produce ideas with content derived from them (*T* 1.1.3.1). This suggests that he conceived of them in a genuinely explanatory way, such that the appeal to these faculties substantially accounts for our ability to make copies of our impressions, and to combine and separate them. Nonetheless, there are at least three good reasons to not attribute a realist conception of these faculties to Hume.

The first reason is intimately linked with Hume's analysis of causation. As already seen in discussing the associative principle of causality, Hume thought that we grow accustomed to associate ideas of events that we have often observed to follow each other. Hume also thought that there is nothing more to causality that we can conceive of. He thus famously defined a cause as "*an object followed by another, and whose appearance always conveys the thought to that other*" (*EHU* 7.29) and he dispelled our (modal) intuition that a cause necessitates its effect by tracing it back to a mere feeling that attends the "customary transition of the imagination from one object to its usual attendant" (*EHU* 7.28). On Hume's view, our ideas of power and causation are simply derived from observing a great number of succeeding events by which we acquire the habit of associating the ideas of these events. "These ideas, therefore, represent not any thing, that does or can belong to the objects, which are constantly conjoin'd" (*T* 1.3.14.19). Rather, by describing something as a power or a cause, we project our accustomed propensity of associating ideas to their corresponding objects and consequently take these objects to be inherently connected. In light of this reductive theory of causality, it would be highly dubious to impute a substantial conception of cognitive faculties to Hume.[79] After all, Hume clearly

---

79 This conclusion runs contrary to the "New Hume readings," which construe Hume as a causal realist after all. (See the essays collected by Rupert J. Read, ed., *The New Hume Debate* [London: Routledge, 2000]). I take these readings to be convincingly refuted by Kenneth P. Winkler, "The New Hume," *Philosophical Review* 100 (1991): 541–79.

says that our causal concepts do not represent anything belonging to the objects they are applied to. (They rather express our accustomed propensity to associate their ideas in a certain way.) Strictly speaking then we do not even have an idea of a power of an object since there is no impression of the object from which we could derive this idea.[80] For this reason a substantial realist conception of faculties is literally inconceivable for Hume.

The second reason against attributing a substantial conception of faculties to Hume arises from his conviction that the "distinction we sometimes make betwixt a *power* and the *exercise* of it, is entirely frivolous, and that neither man nor any other being ought ever to be thought possest of any ability, unless it be exerted and put in action" (*T* 2.1.10.4). For Hume the untenability of a distinction between a power and its exercise is an immediate consequence of his analysis of causation and necessary connections. Given that it is impossible to conceive of modal relations among things, it is equally impossible to conceive of a distinction between a power and its exercise: the exercise of a power is something the power *can* engage in but actually never has to. Just as experience cannot show us any necessary connections it cannot reveal any other modal connections. On the basis of our past experience, we of course do expect things of a certain kind to behave in certain ways under certain circumstances and thus *say* that it has a range of unexercised powers. Yet, just as in the case of causation, this way of talking is but an expression of the fact that we have grown accustomed to associate the idea of such kinds of objects with certain other ideas when we imagine these objects to be in certain circumstances.[81] Now, Hume's

---

80 Things are more complex because Hume employed the term "idea of *X*" ambiguously. In one sense, call it the representational sense, we only have an idea of *X* if this idea represents *X* and is thus a copy of an impression of *X*. (In this sense, we have no idea of a power). In another sense, however, call it the attributive sense, no representation is required for having an idea of *X*. For having an idea of *X* in this sense it is enough if we have this idea from elsewhere, but attribute it to *X*. (In this attributive sense we have an idea of a power of an object since we project our internal impression of being inclined to perform certain associations to objects outside of us).

81 Hume himself provides this expressivist reconstruction of our talk about power when he explains the power involved in property and riches in *T*. 2.1.10.4–9.

rejection of any distinction between the existence and the exercise of a power gives us a strong reason not to impute a substantial conception of faculties to him. For, unlike Descartes, Hume denied that we think permanently, and took it for granted that in phases of sound sleep our perceptions are completely removed for some time (see T 1.4.6.3). If we therefore assumed that Hume held a substantial conception of faculties and hence conceived of perceptions as products of the exercise of these faculties, we would be bound to additionally assign him the view that there are unexercised mental powers. But this, as just seen, was a view he explicitly denied.

The third reason against attributing a realist conception of faculties to Hume is related to his conception of the mind. As seen in our discussion of Spinoza, a realist conception of cognitive faculties is only intelligible on the assumption that these faculties belong to a substance that can perform certain acts. Indeed, a faculty, on a realist conception, is a property that explains why a substance can perform those acts. Thus, if one denies that the mind is a substance and holds instead that the mind is a mere bundle of mental episodes, it is absurd to conceive of faculties in a substantial way: a bundle of ideas or perceptions cannot literally *form* ideas or perception. Such a mind can only contain them or not.[82] Consequently, it also cannot have a property in virtue of which it can perform certain acts. Now, like Spinoza, Hume also held a bundle theory of mind, arguing that a mind is "nothing but a bundle or collection of different perceptions, which succeed each other with an inconceivable rapidity, and are in a perpetual flux and movement" (T 1.4.6.4). And this rules out the possibility of it being endowed with genuine faculties.

These observations make plain that Hume could not consistently conceive of faculties in a realist way. Instead, he conceived of them as distinctive patterns of regularly succeeding perceptions, on account of

---

82 See Nelson Pike, "Hume's Bundle Theory of the Self: A Limited Defense," *American Philosophical Quarterly* 4 (1967): 164–65.

which perceptions can be classified. Pretty much like Spinoza, then, Hume held an eliminativist conception of faculties. Unlike Spinoza, however, he arrived at this view not only by taking a nominalist stance concerning our talk of faculties but also by giving it, at least in parts, an expressivist interpretation. On Hume's view, our talk about faculties contains cognitive and noncognitive aspects alike. It is cognitive insofar as by assigning faculties to persons, we refer to patterns of regularly succeeding perceptions, such that our ascription can be true or false depending on whether these patterns of regularly succeeding perceptions obtain or not. At the same time our talk about faculty contains a noncognitive element that is due to its modal aspect, of which Hume gave an expressivist analysis. So, by assigning a certain faculty to someone, I not only describe an actual state of her but also say something about what is *possible* for her: saying that she has a faculty of reason, for instance, means that she *can* draw valid inferences—even if she may not actually do so. As Hume makes clear in his analysis of powers in general, this modal aspect cannot be understood referentially (for we simply have no impressions of possibilities). Rather, this modal aspect of our talk of faculties is to be understood as an expression of an immediate expectation we form on the basis of our experiences of succeeding perceptions and our consequential habit of classifying these perceptions according to their regular patterns of occurrence.

6. Conclusion

If there is a general lesson to be drawn from my survey of early modern accounts of faculties, it is surely that talk of mental faculties is surprisingly metaphysically neutral. By saying that humans have an intellect, a will, imagination, memory, and perceptions one has not yet entered into a very specific ontological commitment, for such claims can be interpreted and metaphysically substantiated in various ways, as has become plain regarding the accounts of Descartes, Spinoza, Malebranche, and Hume.

On the one hand one can take this talk seriously as a way of attributing certain powers or dispositions to the mind, with regard to which one can explain why rational beings can engage in different sorts of cognitive activities. This is to adopt a realist view of faculties. Descartes and Malebranche were both realists about faculties in this sense, although they ultimately fleshed out their realist stance in different ways. Descartes held a substantial realist conception of faculties (in an actualist version), according to which the mind was endowed with two primitive mental powers, the will and the intellect, which he conceived of as permanently actualized forces that, depending on the ideas or bodily sense impressions they were directed at, gave rise to our different mental acts. Malebranche, by contrast, took the assumption of primitive causal powers in creatures to be unacceptable and thus argued for a reductive conception of faculties according to which the cognitive and causal roles of our basic mental faculties of will and intellect were to be given a reductive analysis. In Malebranche's view, the causal and cognitive roles of mental faculties are owed to God's decree to give rise to certain mental states and to reveal us his ideas whenever there is an appropriate occasion. Even our noncausal but irreducible power of will, on account of which we are morally responsible agents, depends on God insofar as we could never exercise it if God did not incessantly endow us with an inclination toward the good in general.

Spinoza and Hume on the other hand adopted what I have called a nominalist stance toward talk about faculties. In their view, saying that we have imagination or intellect or reason, for example, simply means that we have different kinds of mental states. Hence, when employing faculty-language we do not refer to certain entities over and above mental states, which could figure as their causes. In accordance with this, faculty terminology does not have a primarily explanatory role, but rather a classificatory function: calling an idea an idea of imagination, for example, informs us of the mental kind the given idea belongs to, and not (directly) about its cause. Given their bundle theories of the mind, it is not surprising that Spinoza and Hume adopted a nominalist stance toward faculty-language and consequently endorsed an eliminativist conception of faculties. After all, a bundle of ideas is nothing that could possibly perform any act and consequently nothing that could have any genuine faculty in virtue of which it can perform such acts. A Spinozist or Humean mind can only contain certain series and patterns of mental states. Despite their agreement on the eliminativist view that there were no faculties, Hume and Spinoza disagreed about the way one should accommodate the modal aspect of faculty-language. Spinoza accounts for this aspect by conceiving of ideas as intrinsically powerful entities, which have a capacity to produce further ideas. For this reason, faculties provide a modally robust classification of mental states for Spinoza. An idea of intellect, for example, is an idea that *would* bring us to only think true ideas, if its influence *were* not to be prevented by other more powerful ideas. Hume, by contrast, accounts for the modal aspect of faculty terminology by giving it an expressivist analysis, according to which this aspect is simply due to the expectations we have concerning the types and patterns of mental states classified by a certain faculty-term because we have experienced such states to behave in a certain way in the past.

In light of the wide use of faculty-language in early modern theories of mind and its amenability to various interpretations, it is not surprising that Immanuel Kant, who was primarily interested in the conditions of our ability to ask metaphysical questions rather than in answering these metaphysical questions themselves, found it unproblematic to examine these conditions by analyzing our cognitive faculties and their limits. As the various accounts of his early modern forerunners show, one can perfectly well theorize about our faculties and their interrelations without being committed thereby to any specific metaphysical view about what faculties really are.[83]

---

83 I am grateful to Donald Ainslie, Christian Barth, Julia Borcherding, Rebekka Hufendiek, Yitzhak Melamed, Dominik Perler, Marleen Rozemond, Kelley Schiffman, Barbara Vetter, and Valtteri Viljanen for their helpful comments on earlier versions of this chapter. It has profited a lot from them.

CHAPTER FIVE

Faculties in Kant and German Idealism

Johannes Haag

1. Introduction: The Transcendental Approach to Faculties

Although at first glance it might seem as if Kant and his successors engaged in the *metaphysical* project of their predecessors, nothing could be further from the truth. In the wake of the methodological turn instigated by the newly discovered *transcendental* approach to important philosophical questions, the cluster of problems surrounding metaphysical categorization in general—and the metaphysical status of faculties in particular—lost its predominance. This development will form the focus of this chapter.

According to the methodological standards set by transcendental philosophy, philosophical theorizing—at least in the realm of theoretical philosophy—is only appropriate in the context of an inquiry into the conditions of the possibility of knowledge or, more generally, the possibility of our conscious intentional reference to objects. Such an inquiry stands in stark contrast to one that only describes the psychological constitution the human mind is presumed to have. The modality involved is, consequently, quite strong: for something to be established by the method of transcendental philosophy it has to be part of the only possible explanation of our capacity for intentional reference.

The existence of faculties, from the perspective of transcendental philosophy, likewise has to be established by reflecting on the conditions of the possibility of conscious experience. Thus, the metaphysical status of the faculties invoked in this type of reasoning no longer carries any importance. For it is the function that transcendental reflection reveals as needing to be fulfilled that justifies the introduction of a particular faculty.

The first part of this chapter will examine Kant's critical work and show that the concept of a faculty served a twofold methodological purpose. On the one hand it was frequently used to delineate our own epistemic capacities *from within*, insofar as faculties were postulated as conditions of the possibility of these very capacities. On the other hand the concept of faculties proved useful in delineating our epistemic capacities *from outside*, as it were, insofar as they helped to articulate the conceptual possibility of capacities we—as human beings or, broadening the scope of the investigation, as finite rational beings—do not and cannot have for principled reasons.

A case in point for the first methodological strategy is, of course, the passive or receptive capacity to receive sensory impressions, a receptivity that is ascribed to the faculty of sensibility, and the active or spontaneous capacity to synthesize those impressions into conceptually structured representations—a spontaneity that is, in turn, ascribed to the discursive intellectual faculty or the understanding. Since neither of these two faculties, as Kant famously argued, can provide us with knowledge on its own, it is only in their interplay that these faculties become genuinely *epistemic* faculties—an interplay we consequently have to accept as delineating the epistemic scope of each of them from within.

The second methodological strategy is on display, for instance, in Kant's conception of an intellectual intuition, which specifically contrasts with our own nonproductive epistemic faculties. This limiting faculty, which already figures in the first *Critique*, is complemented in the third *Critique* by the faculty of an intuitive understanding, that is, a faculty of understanding that operates nondiscursively (as opposed to our discursive intellect).

The latter, contrastive approach to faculties was of the utmost importance for the development of post-Kantian, idealist philosophical systems. For in one form or another, all of the German idealists took up one of these contrasting capacities and put it to a quite different use, claiming that we finite beings do, in fact, possess the faculty in question, albeit in a carefully modified sense. In so doing, they transcended the Kantian framework in different, but equally radical ways. Fichte and Schelling took up the faculty of intellectual intuition, while the concept of an intuitive understanding—via Goethe's mediating influence—paved the way for the system presented in Hegel's *Phenomenology of Spirit*. The second part of the chapter will be dedicated to sketching this development, focusing largely on Fichte.

The overall lesson to be learned from the use Kant and his idealist successors made of the concept of epistemic faculties is the following: epistemic faculties should only be admitted insofar as they can be justified by way of transcendental reasoning. For Kant, such reasoning can either take the positive form of reflection on the conditions of the possibility of our empirical knowledge or the negative form of delineating this kind of knowledge from without by introducing limiting faculties as contrastive concepts. The idealists—and Fichte in particular—can then be seen as transforming the contrastive concepts Kant had introduced into concepts of faculties that play an important role in establishing the possibility of empirical knowledge. In so doing, the idealists go beyond the purely negative use Kant made of those concepts and instead employ them in a positive, constructive argument concerning the possibility of knowledge itself.

2. KANT ON EPISTEMIC FACULTIES

Let us begin with the first task faculties fulfill in Kant's critical philosophy, that is, their delineation of our own epistemic capacities *from within*. To introduce a faculty in this context, one has to show that the faculty in question is indeed a condition of the possibility of knowledge. So which faculties prove to be conditions of the possibility of knowledge in the required sense? For Kant it is, first and foremost, two "stems" of our knowledge that have to be brought into play here: *sensibility* and *understanding*.

2.1. Intentionality and the Two Stems of Knowledge

Why those two stems? Here a variety of reasons might be given, some of them more historical, others more systematic. Since our aim is to prove that these faculties are conditions of the possibility of our epistemic access to the world (of which we ourselves are part), the systematic arguments are of greater interest. There are, generally speaking, two different dimensions to the relevant systematic questions: one concerns the possibility of a priori knowledge, the other the possibility of a posteriori knowledge. Both dimensions are, I would like to argue, of pivotal importance for an adequate appreciation of Kant's systematic reasons for introducing the two stems—and, accordingly, the two *distinct* faculties of sensibility and understanding. Both dimensions are, however, ultimately related to questions we would today subsume under the broader heading of the problem of intentionality. I take the core problem of intentionality to be the question of how we are able to refer to something *as* something—even if this reference should, on certain occasions, turn out to be only *ostensible* reference.[1]

---

[1] For this use of the term "ostensible" see Wilfrid Sellars, "Kant's Transcendental Idealism," *Collections of Philosophy* 6 (1976): 16.

At least initially, Kant thought this question to be problematic only in connection with a priori knowledge. Thus, when he asks in a famous letter to Marcus Herz of 1772: "What is the ground (*Grund*) of the relation of that in us which we call 'representation' to the object?" (AA. 10:130) he is only addressing the a priori dimension of this question.[2] The corresponding question for our a posteriori ideas or representations did not seem to him to pose particular difficulties at that time. He thought, roughly, that a causal story should suffice to explain how empirical representations (purport to) refer to their objects as such. It is only in the course of answering the question about a priori reference—the official topic of the 1781 *Critique of Pure Reason* (CPR)—that empirical knowledge turns out to be problematic in a way that cannot be solved simply by invoking the causal connection between objects and our experience of them.

This becomes obvious once one realizes that this causal relation itself is not something *given* to us, but is rather the way in which we as finite rational beings have to conceive of any reality that is to be thought of as existing independently of our conception. We have to take causality for granted in order to arrive at a concept of a world that exists independently of our representing it. Causality, in other words, turns out to be one of the concepts necessary for a conception of an object as something existing independently of our mental activity. Such concepts, which are necessary in order for us to refer to objects in general, are Kant's forms of thought or forms of the understanding, that is, the categories. One thing these forms have in common is that they cannot be abstracted from experience, though they enable every abstraction from experience in making experience itself possible.

Kant arrived at this view in the course of working on what later became the *CPR* as he came to realize that it is not only our sensible access to the world that is endowed with a priori forms—as he had already argued in his 1770 *Inaugural Dissertation*—but our conceptual

---

[2] All citations of Kant's works refer to the Akademie-Ausgabe (AA).

access to reality as well. If our knowledge does indeed have two stems, as he postulates in the first *Critique*, they must each be equipped with their own a priori forms: sensibility with the forms of intuition (in the case of human beings: space and time) and the understanding with its forms of thought or categories.

2.2. Sensibility and Understanding

That sensibility and understanding have different sets of a priori forms constitutes an important reason to distinguish between them as faculties.[3] Another reason is connected with an ultimately phenomenological distinction that becomes apparent in our way of experiencing the world (and, indeed, ourselves as part of that world). Kant is impressed by the fact that our epistemic access to reality is characterized by a strongly *passive* element. Put simply: we cannot choose what we perceive. Something, at least, is *given* to us in experience. Still, we need a capacity to receive what is given to us. This "receptivity of our mind," indeed a "receptivity of impressions" (CPR A50/B74), is called sensibility. Sensibility is, consequently an essentially receptive faculty.

That something is *given* to us in experience, however, does not preclude what is thus given from being at the same time *taken* in a certain way, to borrow Wilfrid Sellars's helpful expression. The taking has to be something essentially spontaneous—and for Kant the faculty of such spontaneity is the understanding.

Consequently, in his introductory remarks to the *Transcendental Analytic* of the first *Critique*, Kant writes that while objects are given to us in sensibility, it is through the understanding that they are thought.[4] This is, as I will shortly show, a claim that must be read with some

---

3 While we have considered a tentative argument in relation to the forms of thought (insofar as those forms are impossible to abstract from experience), no such grounds have so far been given with regard to the forms of sensibility, i.e. space and time. Kant presents these arguments in the *Transcendental Aesthetics* of the CPR. The argument for the necessity of spatial representation may serve as an example: In order to represent something as different from us, Kant argues we have to represent it as external to us. And to represent it, in turn, as external we have to represent it in space. See CPR A23/B38.

4 See CPR A51/B75.

caution. Its interpretation depends on how we understand the remark that objects are thought. Given the overall picture Kant develops in the *CPR*, it could mean either that in so much as representing an object we actively *judge* it as being a certain way or it could mean that taking up an object in conscious experience we already represent it *as* being a certain way, that is, our experience of the object already involves some conceptual characterization on its own.

On the first reading, the understanding would be *essentially* a capacity to judge. But Kant introduces the faculty of judgment as an additional faculty for subsuming one representation under another in a judgment. The representations thus related do not have to be of the same kind. They may be of the same kind, that is, both may be concepts; but more important for our epistemic access to reality are judgments that relate concepts to a very different kind of representation, namely intuitions.

Intuitions are representations that, unlike concepts, relate to their objects immediately. Judgments that subsume intuitions under concepts are, accordingly, the only judgments that contain elements relating directly to the objects given to us in experience. This unique feature allows them to play a foundational role in our picture of empirical reality.

That there are judgments that can play such a foundational role does not, however, mean that the Kantian picture of empirical knowledge as a whole is a foundationalist one. To see how Kant can avoid what Wilfrid Sellars famously castigated as the "Myth of the Given,"[5] we have to turn to intuitions—and to the second way of reading the Kantian remark that objects given by sensibility are thought through the understanding.

Thinking, according to this reading, should be interpreted simply as having conscious representations of objects *as* objects (of a certain

---

[5] Wilfrid Sellars, "Empiricism and the Philosophy of Mind," *Minnesota Studies in the Philosophy of Science* 1 (1956): 253–329.

kind, even if this "kind" is sometimes very general).[6] Intuitions would be just such representations. While concepts are general and only indirectly related to their objects (via judgments), intuitions are singular and relate directly to their objects.[7] We can think of intuitions as representations that refer to their objects demonstratively. As such, they are the joint products of sensibility and understanding. It is the understanding that guarantees that intuitive representations refer to objects. Yet the activity of the understanding that secures this object-reference cannot be judgment, since judgment already presupposes conscious representations to subsume under concepts. Kant calls the joint activity of sensibility and understanding that generates such conscious representations in the first place *synthesis*.

Synthesis unites, in a manner yet to be explained, the sensory impressions of receptivity and concepts—and it does so either a priori or a posteriori. The primary a priori concepts that synthesis unites with the deliverances of sensibility are, of course, the categories or forms of the understanding. Accordingly, Kant writes in a notorious passage:

> The same function that gives unity to the different representations in a judgment also gives unity to the mere synthesis of different representations in an intuition, which, expressed generally, is called the pure concept of understanding. (CPR A79/B104/5)

This complicated passage at the very least makes clear that the empirical judgments that directly relate concepts to intuitions do not thereby relate concepts to something given to us independently of any conceptual structuring. Concepts are essentially involved in the process of synthesis itself—and empirical judgments, if correct, only make explicit what is already implicit in the conceptual structuring of an intuition.

---

[6] Just how general or specific this classification will be in a given case depends on our individual abilities to classify the object given—which depend, in turn, on our background knowledge and the information available to us.

[7] CPR A320/B376.

One worry that might arise about the Kantian picture of the interplay of faculties developed so far would be that it somehow fails to capture the sense in which our knowledge is *objective*.

Things are further complicated by the fact that, on Kant's account of the a priori forms of sensibility and understanding, experience only presents us with *appearances* (empirical reality), which are to be contrasted with something that appears, that is, things-in-themselves (noumenal reality). We have no epistemic access to things-in-themselves, due to our a priori way of forming the sensory manifold or manifold of sensibility given to us in experience. Reality as it is in itself is consequently inaccessible to us.

Nevertheless, those things-in-themselves are a condition of the possibility of the appearances: by affecting our receptivity a certain way (an affection we can only conceive of as causal, but of whose "true nature" we can likewise know nothing), things-in-themselves provide us with the sensory manifold we take up in the complex synthetic activity that generates conscious experience of objects, that is, appearances.

Appearances are representations that are perceived (in the prevailing technical sense, i.e., cognized) *as* objects existing independently from our perception of them.[8] To perceive appearances in that way, we have to abstract from all properties that the purported objects can only have in relation to perceiving subjects. Those properties cannot be objective properties in the required sense, since they belong to the objects only *as perceived*. Thus, for instance, no perspectival property can be a property in the required sense: objects are perceived from a certain perspective, but they do not themselves have perspectival properties.

In order to develop a conception of objectivity Kant has to elaborate this distinction. This elaboration is one of the central aims of the *Transcendental Analytic* in the first *Critique*. Its result is the doctrine of the categories or forms of the understanding I introduced earlier. Ultimately, categories are the necessary characteristics of our intentional

---

8 See, for example, CPR A190/B235.

reference to objects as objectively existing. They serve as the conceptual scaffolding, as it were, of our construction of objectivity.

Kant's deployment of the concept of things-in-themselves, then, ensures that this conception of objectivity is not grounded in the reality-constructing activity of the perceiving subject alone. Thus, Sellars observes,

> the [manifold of intuitions] has the interesting feature that its existence is postulated on general epistemological or, as Kant would say, transcendental grounds, after *reflection on the concept of human knowledge as based on, though not constituted by, the impact of independent reality.*[9]

This impact of an independent reality corresponds to the "guidedness" of our perceptual content.[10] This guidedness, for Sellars, is ultimately phenomenologically grounded in the *passivity* of our experience.[11] Kant, as I already indicated, throughout his critical writings emphasizes this passivity with respect to the content of our experience.[12] There has to be something that explains the basic phenomenological fact that we are passive with respect to the actual content of our experience. For Kant (as for Sellars) this guidance has to be strictly "from without" the conceptual order. Independent reality, the Kantian thing-in-itself, guides us from without via the impressions of sheer receptivity.[13] Only the

---

9 Wilfrid Sellars, *Science and Metaphysics: Variations on Kantian Themes* (London: Routledge, 1968), 9 (emphasis added).
10 This is Sellars's concept; see Wilfrid Sellars, *Science and Metaphysics: Variations on Kantian Themes* (London: Routledge, 1968), 9. It is taken up in the literature on Kant for instance by Robert Pippin, *Kant's Theory of Form* (New Haven: Yale University Press, p. 46ff., and John McDowell, "Sellars on Perceptual Experience," 22, and "The Logical Form of an Intuition," 38-40, both in McDowell, *Having the World in View: Essays on Kant, Hegel, and Sellars* (Cambridge, MA: Harvard University Press, 2009).
11 See Wilfrid Sellars, *Science and Metaphysics: Variations on Kantian Themes* (London: Routledge, 1968), 16.
12 See e.g. A 50/B 74; Kant, *Grundlegung zur Metaphysik der Sitten*, 4:452; Kant, *Anthropologie in pragmatischer Hinsicht*, 7:141.
13 For more on this notion, see Wilfrid Sellars, *Science and Metaphysics: Variations on Kantian Themes* (London: Routledge, 1968), 16. Sellars thinks that Kant unduly neglects the idea of sheer receptivity; but see 2.3 here.

latter are immediately accessible to the workings of our spontaneity. Yet even this immediate contact with sense-impressions is nevertheless guidance "from *without*" the conceptual order in the sense that these impressions are not given as what they are in themselves, but are always synthesized.

These all too sketchy remarks give us a first clue as to how the conception of an objectively existing reality can be spelled out in a Kantian framework even though the idea of such a reality cannot be equated with the notion of things-in-themselves, which, according to the doctrine of a priori forms, are not possible objects of our knowledge. To give a more detailed account, we have to considerably extend the theoretical framework presented so far. In particular, we have to invoke two further faculties, the faculty of *imagination* and the faculty of *apperception*; and we have to further differentiate our conception of a faculty already at hand, that is, the receptive faculty of *sensibility* or *sense*. At one point at the very beginning of the *Transcendental Deduction* of the first edition of the CPR (the so-called A Deduction), Kant brings the three ultimately required elements together in the following way:

> There are three original sources (capacities or faculties of the soul) which contain the conditions of the possibility of all experience, and cannot themselves be derived from any other faculty of mind, namely, *sense, imagination*, and *apperception*. Upon them are grounded (1) the *synopsis* of the manifold *a priori* through sense; (2) the *synthesis* of this manifold through imagination; finally (3) the *unity* of this synthesis through original apperception. All these *faculties* have a transcendental (as well as an empirical) employment which concerns the form alone, and is possible a priori. (CPR A94)

2.3. Synopsis of Sense and Sheer Receptivity

In the foregoing quotation, synopsis of sense is sharply distinguished from any kind of synthesis: synthesis is always a function of the imagination that in turn, as I shall show shortly, is a function of the understanding in a certain application. Synopsis on the other hand, being a function of sense, does not require any synthetic activity. The synopsis of sense functions as sheer receptivity.

Synopsis nevertheless seems to involve *some* kind of structuring of the given sensory material, as Kant makes clear in his sole subsequent appeal to synopsis shortly afterward:

> If each representation were completely foreign to every other, standing apart in isolation, no such thing as knowledge would ever arise. For knowledge is a whole in which representations stand compared and connected. As sense contains a manifold in its intuition, I ascribe to it a synopsis. But to such synopsis a synthesis must always correspond; receptivity can make knowledge possible only when combined with spontaneity. (CPR A97)

If synopsis *corresponds* to synthesis insofar as it is a faculty that unites otherwise distinct and isolated representations, we can certainly conclude that some order is already imposed on the manifold "in intuition" by the synopsis of sense.

This characteristic receptivity of the mind is an innate disposition to receive sensory affection.[14] It is a mere disposition to react (in no sense spontaneously but merely passively) when acted on by things-in-themselves. This is the sense in which even the synopsis of sense can be a priori.

One might be tempted to think about this passive reaction as the shaping of a piece of wax when a seal is pressed on it. But this picture would be misleading, because it would neglect the important fact that this disposition to react to a given affection of sense with the forming of the material thus given has "a strong voice in the outcome,"[15] in

---

14 See AA. 8:222.
15 Wilfrid Sellars, *Science and Metaphysics: Variations on Kantian Themes* (London: Routledge, 1968), 16.

terms of giving this input some structure. A better picture might be light falling through a grate:[16] in this case the shape of the grate would stand for the specific form of the receptivity—in our case (something that systematically relates to) spatial and temporal order. The light would represent the sensory input that, by falling through the grate, is structured in a certain way. Resulting from this process are synoptically structured impressions that present a manifold "for intuition," that is, a manifold that has to be "given *prior to* any synthesis of understanding and *independent* of it" (CPR B145; my emphasis). Kant's picture of sensory consciousness, on this reading, implies the existence of sensory structured *material* that is completely non-synthetic, nonspontaneous, and, a fortiori, neither conceptual nor intentional—namely, the synoptically structured impressions of sheer receptivity.[17]

The products of synopsis would be sensations completely located below the line that separates the realm of spontaneity from sheer receptivity.[18] As such, they cannot be structured by space and time as forms of intuition in the sense elucidated above. For such forms of intuition are themselves products of an a priori synthesis, whereas sheer receptivity is precisely marked by the absence of spontaneity.

But what, then, is the nature of those forms of *sheer* receptivity? Again, we have to restrict our claims to what can be founded on transcendental reasoning. We consequently have to explain why we need to invoke a synopsis of sense in the first place. Now, the passivity of experience and reflection on the need for guidance from without only give us reason to assume affection by things-in-themselves, so these considerations cannot furnish us with an argument for the

---

16 I owe this metaphor to Eckart Förster.
17 On this reading (contra both Wilfrid Sellars, *Science and Metaphysics: Variations on Kantian Themes* (London: Routledge, 1968), and John McDowell, "The Logical Form of an Intuition," in McDowell, *Having the World in View: Essays on Kant, Hegel, and Sellars* (Cambridge, MA: Harvard University Press, 2009)), Kant not only has the conceptual resources to account for Sellars's "sheer receptivity," but explicitly invokes such sheer receptivity in his notion of the synopsis of sense.
18 For this metaphor see John McDowell, "Sellars on Perceptual Experience," in McDowell, *Having the World in View: Essays on Kant, Hegel, and Sellars* (Cambridge, MA: Harvard University Press, 2009), 5.

necessity of a synopsis of sense. Why do we need to posit sensory input that is *structured* by our receptivity alone? Why, in other words, is this synopsis a "condition of the possibility of all experience" (CPR A94)?

The reason, ultimately, is that our forms of intuition are, as John McDowell puts it, a "brute fact about our subjectivity."[19] For we are aware that other forms of intuition are at least logically possible. We can conceive of finite, rational beings whose sensibility is distinct from our own but is, on account of their finitude and rationality, nevertheless subject to the same categories of the understanding, albeit in an alternatively schematized form.[20]

Our own spatiotemporally structured empirical reality is therefore only one particular way to structure reality (by schematizing categories accordingly and providing a framework for the spatial and temporal location of objects of experience). Indefinitely many other structures are logically possible. This *specificity* of the way objects are given to us in experience therefore stands in need of explanation. And the forms of sheer receptivity are introduced by a purely functional characterization to explain just this specificity.[21]

Before moving on, let me note by way of conclusion that the introduction of this faculty of sheer receptivity is motivated solely by reference to the conditions of the possibility of our experience. It can therefore serve as a particularly convincing example of the methodological turn the treatment of the faculties underwent under Kant's hands: sense as sheer receptivity serves to delineate our own epistemic capacities *from within* insofar as it is introduced as a condition of the possibility of our

---

19 John McDowell, "Hegel's Idealism as a Radicalization of Kant," in McDowell, *Having the World in View: Essays on Kant, Hegel, and Sellars* (Cambridge, MA: Harvard University Press, 2009), 85.
20 See CPR B 148.
21 Paul Franks refers to the importance of that distinction between "receptivity in general and specifically human sensibility," and he points out that the specific features of our sensibility cannot be "derived from the forms of the understanding" (Paul Franks, *All or Nothing: Systematicity, Transcendental Arguments, and Skepticism in German Idealism* [Cambridge, MA: Harvard University Press, 2005], 58). Thus he is aware of this "brute fact" about our sensibility without, however, attempting to give a further explanation in Kantian terms of how exactly it is to be understood.

knowing relationship to a reality of which we conceive ourselves to be a part.[22]

2.4. Synthesis of Imagination

I can now turn to the next element Kant mentions in his list of the "three original sources (capacities or faculties of the soul)" (CPR A94). Imagination was introduced in the list of the three original sources or faculties as a capacity to subject the synoptically given sensory manifold to a process Kant calls *synthesis*. In this section I would like to throw some light on how this process (and the faculty responsible for it) is to be understood.

We already know not only that the imagination takes up the sensory manifold into consciousness, but that it furthermore fulfills this task by subjecting this manifold to the conceptual forms of the understanding, thereby guaranteeing that the synthetic process itself is structured by concepts. In short, it connects sensibility and understanding in a way that applies conceptual representations to sensible representations already in the construction of intuition—and thus ultimately allows for intuitions to be explicitly subsumed under concepts in judgments.

Kant in one place calls imagination a "forming [bildendes] faculty of intuition" (AA. 28.1:235). It is the task of the imagination "to bring the manifold of intuition into a picture" (CPR A120). It does so by synthesizing this manifold into complex representations. Synthesis is introduced as "the action of putting different representations together with each other and comprehending their manifoldness in one cognition" (CPR A77/B102/3). The concept of synthesis is, accordingly, the idea of merging different given representations (which are, as such, unconscious) into a conscious representation of something (as something). Note that it is the resulting complex representation that is sup-

---

22 For a more detailed account of synopsis of sense see Johannes Haag, "Kant on Imagination and the Natural Sources of the Conceptual," in *Contemporary Perspectives on Early Modern Philosophy: Nature and Norms in Thought*, edited by Martin Lenz and Anik Waldow (Dordrecht: Springer, 2013), 76–82.

posed to be conscious and not the process of synthesis itself. Through this process, the representations engraved, as it were, by the forms of sheer receptivity become spontaneously (as opposed to receptively) united in such a way that the resulting representations are epistemically accessible to us. Without this capacity it would still be possible for the passive senses to produce sensory impressions—but those impressions would be a merely unconnected manifold and, as such, "nothing to us" (CPR A120), "less than a dream" (CPR A112).

The products of the synthesis of the imagination are, in the first place, mere images of objects, but not yet the intuitions they will eventually become. In his description of what he calls the "threefold synthesis," Kant explains how to understand the details of this synthetic process.[23] The individual aspects of this "threefold" synthetic process are, however, best understood as three aspects of one complex synthesis neither of which would be possible without the others.

Kant himself illustrates what those aspects are supposed to contribute to this complex synthesis by way of example. He asks the reader to think of intuiting a line in space. This intuition, he argues, presupposes a "running through" (or, alternatively, a "taking up" of) a set of impressions into the mind. This running through is what he calls the *synthesis of apprehension*. However, it is not enough to run through the impressions: every impression taken up into this process has to be constantly reproduced—otherwise only the actual impression would be present to the mind and consequently no complex representation would be possible. In other words, we would not have a representation of a line, but only impressions of points, as it were, one following the other. This is the reproductive aspect of the threefold synthesis or the *synthesis of reproduction*. And finally, the representing subject needs to be able to separate the represented object from the objects in its vicinity. For the process of synthesis to involve such differentiation, the representing

---

23 See CPR A97/8. This description is to be found in the *Transcendental Deduction* in the first edition, CPR A98–110.

subject has to conceptually classify the object in question. She must, therefore, synthesize the manifold in accordance with a concept under which the object could, in principle, be explicitly subsumed in a judgment. This is what Kant calls the *synthesis of recognition in a concept*.

This complex threefold synthesis results in spatiotemporally structured, perspectival images of empirically real objects. As such they are not only subject to spatiotemporal structuring, but are already informed by the categories of the understanding (and, at least in the ordinary case, empirical concepts as well). The pivotal role of the faculty of imagination in this context should by now be obvious: even in its empirical activity it is not merely reproductive but essentially productive.

The categorical structuring at this point of the synthetic process is, however, restricted to the so-called mathematical categories of quantity and quality, that is, categories that concern solely the sensible properties of empirical objects and their ordering. At this point the dynamical categories of relation (for instance causality and substance-attribute) and modality (possibility, actuality, necessity) are not yet part of the picture. The latter rather relate to the dispositional and causal properties associated with our conception of objects. It is only once these properties come into play that we are able to take the images produced by the synthetic process up to this point *as* images of *objects existing independently from our picturing them*.

To get to this level of truly objective reference we have to invoke another objectifying step that can be contributed solely by the conceptual faculty we have called understanding. Yet this particular contribution is, once again, a process that applies representations to other representations in a way that is distinct both from the work of the imagination in its threefold synthesis and from the subsuming activity of judgment.

It is important to remember that, from a transcendental perspective, appearances are just representations and do not really have this independent existence: it is only in *taking them to be* objects that we can construe them as proper objects of reference that constitute an empirical reality.[24] In intuition, it follows, sensible and conceptual properties are inextricably intertwined.

To make this possible, we have to restrict the categories to the specific forms characteristic for our human sensibility, that is, space and time. For instance, we can represent the mathematical category of quantity only as magnitude in space and time, that is, as a juxtaposition of impressions in space. And we can represent the mathematical category of causality only in space and time, that is, as a temporal ordering of cause and effect. (The categories thus restricted to our forms of sensibility are the *schematized* categories.) Only when they are thus restricted can the categories play a determining role in the synthetic process that produces images of objects *and* (by way of the schematized dynamical categories) in the further objectification of those images as transcendental objects of our intuitive representations. In order to account for this further objectification Kant introduces another faculty, which I have yet to consider: the faculty of apperception.

2.5. A Priori Synthesis and the Question of Objectivity

Recall that the passage from the beginning of the A Deduction, which structured our deeper investigation of the interaction between sensibility and understanding, indicates that imagination (like the other faculties it mentions) "has a transcendental (as well as an empirical) employment which concerns the form alone, and is possible a priori" (CPR A94). To this transcendental employment I now turn, in order to further our understanding of the complex interplay of those faculties as well as to flesh out our assessment of Kant's conception of objectivity. In some of the central passages of the A Deduction, Kant points out that the empirical employment of the imagination would be impossible without its a priori employment.[25]

---

24 See for example A190/B235–A191/B236.
25 See e.g. CPR A123.

Kant is very clear that it is one and the same faculty that has both an empirical and an a priori function. In its a priori function, the imagination is truly transcendental in that it makes its empirical activity possible. In this a priori function, it has "as its aim in regard to all the manifold of appearance... nothing further than the necessary unity in their synthesis" (CPR A123). Kant's insistence on the *necessity* of the unity in question serves his aim of securing a substantial conception of objectivity. Without such objectivity, Kant argues in the *Transcendental Deduction*, neither consciousness of objects nor self-consciousness would be possible. In effect, Kant argues that the self-ascription of mental states is a condition of the possibility of *having* conscious mental states.[26]

The categories, Kant emphasizes, are applied by a thinking (i.e. representing) subject—a subject that consciously refers to objects precisely by applying the categories. It is therefore the subject herself who, by means of this structuring of the empirically given, lends *objectivity* to her reference to objects, namely by making it *lawful*.

This structuring of representations, Kant argues, is only possible if our representations can be treated as belonging to *one and the same* consciousness. We can unite unconnected mental states into representations *of objects* by means of the categories only if we can conceive of them as the mental states of a unitary subject, that is, only if we are able to ascribe them to one and the same subject.[27]

In this way, the conscious subject constitutes a point of reference for the self-ascription of representations and thereby furnishes the empirically given with the unity necessary for consciousness of objects: the standing possibility of consciously referring to a unitary subject of representations, which is characteristic of all our conscious life, is, as it

---

26 There are countless attempts to reconstruct the argument in the *Transcendental Deduction*. Among the most influential are the discussions of Peter Strawson, *The Bounds of Sense: An Essay on Kant's Critique of Pure Reason* (London: Methuen, 1966), and Dieter Henrich, *Identität und Objektivität: Eine Untersuchung über Kants transzendentale Deduktion* (Heidelberg: C. Winter, 1976). For my own take of the details of this argument see Johannes Haag, *Erfahrung und Gegenstand: Das Verhältnis von Sinnlichkeit und Verstand* (Frankfurt: Klostermann, 2007), ch. 6.

27 CPR A123/4 and similarly, B137.

were, the constant element in the steady flow of the empirically given. And that is the sense in which consciousness of objects is not possible without self-consciousness—the unity of apperception.

On the other hand we can become conscious of ourselves only in this spontaneous activity of structuring: in bringing the empirically given to a unity that we can consciously experience, we simultaneously experience ourselves as subjects of those conscious experiences of unity.[28] We experience ourselves as that which applies the same structuring principles over and over again in connecting its representations. Without consciousness of objects, therefore, there could be no self-consciousness.

If it is possible to fill in the details of the argument thus outlined, then consciousness of objects and self-consciousness will, indeed, be mutually dependent: one must ascribe states of consciousness to a subject *because* one would not be conscious of objects if one were unable to thus ascribe them.

In this sketch of an argument, one can discern the activities of at least two of the faculties I am concerned with: imagination and apperception. Apperception is the faculty that gives unity to the activity of the imagination in subjecting this activity to a priori rules. The faculty of apperception gives "the *unity* of this [i.e. the imagination's] synthesis" (CPR A94). Apperception is the faculty that applies the forms of the understanding (the categories) to the activity of the imagination in synthesizing the a priori manifold of sensibility. While it is the productive imagination without which "no concepts of objects would converge into an experience," it is the "the standing and lasting I (of pure apperception)," the "all-embracing pure apperception" (CPR A 123), that enables the productive imagination to accomplish this, by making "its function intellectual" (CPR A124).[29]

It is therefore the function of concepts of objects to transform an appearance into an (as yet undetermined) object of intuition. To this

---

28 CPR A108 and similar: B135/6.
29 See the similar reasoning in sec. 24 of the B Deduction.

end, such concepts have to be related by the activity of the imagination to the sensible manifold both a posteriori and a priori. The imagination, therefore, is able to relate sensibility and understanding empirically, in its productive role of generating images of objects, only because it also operates a priori in relating the a priori representations of sense (i.e. the formal intuitions of time and space) and the a priori representations of the understanding (i.e. the categories).[30]

With transcendental apperception I have now covered all the faculties operative in our epistemic access to empirical reality. The two stems of knowledge, sensibility and understanding, produce knowledge (both empirical and a priori) only in working on synoptically given sensible material (*faculty of sense*), which guides "from without" the synthetic activity that transforms the sensible manifold thus given into a conscious, complex image of an object (*faculty of imagination*). The latter faculty, in turn, takes its unity from an intellectualizing activity that guides the process of synthesis "from within" by subjecting it to mathematical categories and that, by means of rather different, purely conceptual activity, turns the images of objects into representations of transcendental objects that are taken to be subject as well to the dynamical categories, like causality and substance (*faculty of apperception*).

The objectivity of representation, and hence the possibility of knowledge, is thus secured by a complex process "from without" *and* "from within." The result is the concept of an empirical reality that exists independently of being represented, and hence objectively, without existing in-itself.[31] Knowledge is consequently restricted to the reality thus conceived. In other words, I have shown how the faculties that Kant introduced simply in order to outline the conditions of the possibility of knowledge serve, at the same time, to delineate our own epistemic capacities *from within*. The limitations of the faculties thus introduced are the limitations of our epistemic access to reality.

---

30 The special form of transcendental synthesis required to achieve this aim is called *synthesis speciosa* or figurative synthesis. See B 151/2.
31 See Wilfrid Sellars, "Kant's Transcendental Idealism," *Collections of Philosophy* 6 (1976), sec. 24.

3. Strange (Forms of) Faculties as Limiting Concepts

As indicated at the outset of this investigation, Kant finds the concept of a faculty useful in delineating our epistemic capacities not only from within but also *from without* by outlining the conceptual possibility of capacities we—as human beings or, broadening the scope of the investigation, as finite rational beings—do not and cannot, in principle, possess.

This methodological strategy is employed throughout Kant's critical philosophy, but usefully pinned down in a section of the *Critique of the Power of Judgment* (CPJ) from 1790 that proved of immense importance for the post-Kantian idealists. In section 76 of this work, Kant outlines how such *concepts of limiting faculties* serve the important purpose of making apparent the limitations of three of our faculties: understanding, reason, and the power of judgment. In contrasting our own faculties with each of those limiting faculties, we become aware of our limitations as the finite rational beings we happen to be.

Kant has already employed a similar strategy in his argument for the specificity of our human forms of sensibility. It is, he argued, at least conceivable that other forms of sensibility, different from our own spatiotemporal forms of intuition, might serve the purpose of intuitively forming the manifold of sensibility in other finite, rational creatures. Accordingly, we must not take our own forms of sensibility to correspond to the order of things-in-themselves.

In section 76 of the *CPJ*, Kant addresses this methodological approach explicitly, for purposes of illustration, before applying it to our forms of understanding, reason, and, ultimately, our power of judgment. Let us consider the three limiting faculties and their human counterparts in the order Kant discusses them.[32]

(1) *Intellectual intuition.* In discussing our own, discursive understanding, I have already detailed the epistemic limits that are due to the dual

---

32 In what follows I am deeply indebted to Eckart Förster's discussion of sec. 76 in *The 25 Years of Philosophy: A Systematical Reconstruction*. (Cambridge, MA: Harvard University Press, 2012), ch. 6.

dependence of our knowledge on understanding *and* sensibility. I did not, however, explicitly invoke the categories of modality, that is, possibility, actuality, and necessity. It is harder to see how these categories contribute to our concept of an object as compared to, for instance, relational categories such as substance or causality. Consequently, concepts like possibility, actuality, and necessity may not seem to be specific to finite rational beings like us, but rather seem to pertain to distinctions among things-in-themselves.

It is exactly at this point, Kant tells us in section 76, that we have to invoke the limiting concept of an *intellectual intuition* in order to show that such modal distinctions likewise rest on our finitude as epistemic subjects who have to rely on something being sensibly given to them in order to apply the concepts of the understanding. This property of our mind makes our understanding *discursive*: it has to start from the particular—given in sensibility—and subsume it under the conceptually general or universal.[33] This necessary reliance on sensible intuition, Kant argues, accounts for the distinction between actuality and possibility:

> For if understanding thinks it [i.e. a thing] (it can think it as it will), then it is represented as merely possible. If understanding is conscious of it as given in intuition, then it is actual without understanding being able to conceive of its possibility. (CPJ 5:402)

But we can at least conceive of a being that does not likewise depend on something being sensibly given to it in order for it to apply its conceptual resources. This would be a being whose understanding was productive in its very act of thought: it would, as it were, generate the actuality of something merely by thinking it. What it would think would be actual in and through its mere act of thought. Kant calls this faculty

---

33 See CPJ 5:407.

intellectual intuition.³⁴ The conceivability of a being equipped with this faculty contrasts with our own intellect in a way that highlights the limitations of our epistemic capacities:

> The propositions, therefore, that things can be possible without being actual, and thus that there can be no inference at all from mere possibility to actuality, quite rightly hold for the human understanding without that proving that this distinction lies in the things themselves. (CPJ 5:402)

The concept of an intellectual intuition is merely problematic for our understanding, in the sense that it transcends the limits set by our discursivity. Yet it is nevertheless an "indispensable idea of reason" that we have "to assume some sort of thing (the original ground) as existing absolutely necessarily, in which possibility and actuality can no longer be distinguished at all" (CPJ 5:402).

(2) *Holy will*. Reason, however, as a faculty can be viewed not only from a theoretical perspective—as the faculty of both inferential reasoning and regulative principles (ideas)³⁵—but also from a practical perspective. From a practical perspective reason "presupposes its own unconditioned (in regard to nature) causality, i.e., freedom, because it is aware of its moral command" (CPJ 5:403). Here, we are confronted with the limitations of a *practical* faculty: we are aware of the moral command only as a command, and not as a law, because we perform the actions that are commanded of us under the condition that we are part of empirical reality (nature) and are thus constituted by understanding *and* sensibility. Under these conditions, the moral command does not and cannot express

---

34 See e.g. CPJ 5:409.
35 It should be noted that it is possible to distinguish and order the faculty of imagination (as informed by concepts), the power of judgment, and the faculty of reason by reference to the increasing complexity of their respective products: intuitions, judgments, and inferences.

what *is* (*ein Sein*), but only what *should* be (*ein Sollen*)—something that, even if it does turn out to be the case, obtains merely accidentally *sub specie* the laws of nature. But we can also conceive of an alternative faculty of practical reason:

> If reason without sensibility (as the subjective condition of its application to objects of nature) were considered, as far as its causality is concerned, as a cause in an intelligible world, corresponding completely with the moral law, where there would be no distinction between what should be done and what is done, between a practical law concerning that which is possible through us and the theoretical law concerning that which is actual through us. (CPJ 5:403/4)

Again, this conceivable practical faculty need not be real; we only need to conceive of it in order to recognize the limitations of our own practical faculty and, at the same time, to accept the guidance of our action by what, in our case as finite rational beings (i.e. as dependent on sensibility in general), can be only a moral command. Through the contrastive use of the concept of a limiting faculty, we can see that what is a command for beings like us would be a law for a "holy will" (AA. 18:469).

(3) *Intuitive understanding.* Kant's discussion in section 76 determines this contrastive limiting concept only negatively. We are to conceive of an understanding that does not "go from the universal to the particular" (CPJ 5:403). For our discursive understanding commits us to an attitude toward nature that distinguishes between a mechanistic explanation of natural phenomena and the teleological explanation we are forced to give of some very peculiar appearances, namely living beings or organisms. Both ways of explanation are—in a sense yet to be determined—necessary for us to adopt, and both are, ultimately, ways of completely explaining nature. Unfortunately, the two are not compatible with each other

and thus lead to an antinomy that can only be resolved by treating the maxims that employ them as merely reflective or regulative.[36]

But this distinction should not be dismissed as an ad hoc solution to the antinomy.[37] We should rather try to understand Kant's compelling reason for treating the maxims in question as regulative. And this reason can be found, once again, in the contrastive concept of a limiting faculty—this time in the concept of an understanding that is not discursive, but intuitive. The conclusion this concept should warrant is that teleological description of the phenomena in question, that is, organisms, "is necessary for the human power of judgment in regard to nature but does not pertain to the determination of the objects themselves, thus a subjective principle of reason for the power of judgment which, as regulative (not constitutive), is just as necessarily valid for our human power of judgment as if it were an objective principle" (CPJ 5:404).

In order to understand Kant's argument we need to further elaborate the relevant antinomy and the details of its solution as they are presented in section 77 of *CPJ*. The purpose of this section is to resolve the difficulties that arise from the fact that experience presents us with a class of objects that seem to defy the mechanical description of nature: namely, living beings or organisms.

Mechanical explanation is always an explanation in accordance with our discursive understanding, that is, an explanation that explains a given entity as the sum of its parts. But organisms are not mere sums of their parts:

> In such a product of nature each part is conceived as if it exists only through all the others, thus as if existing for the sake of the others and on account of the whole, i.e., as an instrument (organ), which is, however, not sufficient (for it could also be an instrument of art, and

---

36 See CPJ 5:387.
37 Similarly Henry Allison, "Kant's Antinomy of Teleological Judgment," in *Kant's Critique of the Power of Judgment*, edited by Paul Guyer (Oxford: Rowman and Littlefield, 2003), 226.

thus represented as possible at all only as an end); rather it must be thought of as an organ that produces the other parts (consequently each produces the others reciprocally), which cannot be the case in any instrument of art, but only of nature, which provides all the matter for instruments (even those of art): only then and on that account can such a product, as an organized and self-organizing being, be called a natural end. (CPJ 5:373–74)

This mutual causality of whole and part we encounter in organisms is, as Kant clarifies, "strictly speaking...not analogous with any causality that we know." It is "not thinkable and not explicable even through an exact analogy with human art" (CPJ 5:375).

It is, for a discursive understanding like ours, only graspable by a "remote analogy" (CPJ 5:375) with a causality we do know, that is, a final causality in which the representation of the end precedes the result of the process of production. In this case the representation of the whole does indeed precede the existence of the parts: we first form the idea and only afterwards work on its realization.

The analogy, however, is not only remote, but, "strictly speaking" (CPJ 5:375), not an analogy at all, since the object in question is at the same time represented as a natural object, that is, as an object that exhibits this causality in itself, and is not caused by a rational being external to it. (Otherwise it would indeed be an "analogue of art" (CPJ 5:374).) We conceptually struggle with this phenomenon, since the mutual causality of whole and part does not fit within the constraints of our conceptual system and we have to contend ourselves with the construction of an auxiliary, mongrel concept that does fit this framework—at least by analogy with the familiar concept of intentional final causation.[38]

---

38 This is the point at which Hannah Ginsborg's influential criticism of Peter McLaughlin's interpretation seems to go astray: "But for Kant there is no less of a need for teleology in understanding a machine such as a watch, than there is in understanding an organism." Ginsborg, "Two Kinds of Mechanical Inexplicability," *Journal of the History of Philosophy* 42 (2004), 37. This would only be

So, our use of the concept of a natural end is not only necessary, given our cognitive constitution, it is, at the same time, experienced as ultimately failing to fully do justice to the phenomena to be explained. The concept of a natural end is, in other words, inevitably formed by the understanding in reaction to certain phenomena intuitively given by the synthesizing activity of the imagination. In the special case of organisms, we find that the synthesized material cannot be understood through empirical concepts we already possess. Consequently, the understanding has to react with the formation of a new concept, that is, the concept of a *natural end*.

In doing so, however, we find that the resulting empirical concept aims to integrate two different kinds of dependence that cannot ordinarily be united in one and the same object: namely, a dependence of the whole on the parts *and* a dependence of the parts on the whole. This dependence can be conceived by a *discursive* understanding like our own only by analogy to the teleological dependence of the artefact on its creator's idea of it: organisms have to be conceived as *ends*. Yet since organisms, unlike artefacts, are *natural* objects, that is, products of nature and not products of thinking beings, we have to think of organisms as *natural* ends. The mongrel concept of a *natural end*—a concept that we cannot help but construct in the face of certain phenomena that the imagination presents us with in intuition—therefore "includes natural necessity and yet at the same time a contingency of the form of the object (in relation to mere laws of nature) in one and the same thing as an end" (CPJ 5:396).

But this concept of a natural end, even if it does not contain an outright contradiction, is still only a "problematic concept" (CPJ 5:397) in the Kantian sense, since it cannot be abstracted from experience.[39]

---

right if organisms did indeed exhibit a causality "analogous to a causality we know," i.e. the final causation of intentional action.

39 See CPJ 5:408.28/9. Since the concept could not simply be abstracted from the objects in question—organisms—it had to be formed, as we have seen, by an *analogical* transformation from the concept of an end.

Thus, due to the limited constitution of our understanding, our concept of a natural end must draw on another faculty for its construction, namely the faculty of *reason*. To conceive of something as an *end* involves an essential appeal to the faculty of reason as the faculty that is responsible for the explanation of intentional action, that is, the faculty responsible for teleological explanations. Thus the concept of an end, which has its home in the explanatory discourse surrounding intentional action, is here united with the principle of mechanical explanation in the concept of a *natural end*.

If that is right, then teleological judgment is by no means optional: we are compelled to introduce forms of teleological explanation not only in our overarching scientific pursuit of a unified empirical reality, but even in the synthetic construction of some of our intuitions of appearances (namely organisms).

Yet however necessary these teleological judgments are, in rendering explicit the contents of our intuitions of natural ends, they remain as problematic as the intuitive representations they describe and are based on, since we cannot understand how it is possible for natural ends to exist in empirical reality in the first place. We cannot, in other words, distance ourselves from the concept of a natural end: we are forced to synthesize objects in accordance with it, for natural objects are ineluctably represented in imagination as exhibiting (*darstellen*) the concept of objective purposiveness. The step from heuristic judgment (an exercise of our faculty of judgment) to intuitive presentation (an exercise of our faculty of imagination) consequently leads to an *antinomy*, that is, an "unavoidable illusion" forced upon us by a "natural dialectic."[40]

This antinomy is supposedly resolved by the distinction between reflective (or regulative) judgment and determining judgment. To ultimately understand why this move is not ad hoc, we have to recall Kant's distinction between *discursive* and *intuitive understanding*. Eckart Förster convincingly argues that we have to distinguish not only two, but four

---

40 See CPJ 5:386.

different concepts that are subsumed under that labels in different places of Kant's work:[41]

(1) intellectual intuition as a nonsensible intuition of things-in-themselves
(2) intellectual intuition as a productive unity of thought and reality
(3) intuitive understanding as the ground of all possibilities
(4) intuitive understanding as a synthetically universal intuition of a whole as such.

While the first concept does not play an important role in the argument of sections 76 and 77 of the *CPJ* (though it is important in the first *Critique* as a limiting concept for our own inability to perceive things-in-themselves),[42] the other three will turn out to be of considerable importance for the solution of the antinomy presented here.

How can a peculiarity of our mind contribute to this solution?[43] First, we have to recall that, at root, the concept of a natural end is a marriage, not of convenience, but of necessity—one predicated on a "remote analogy with a causality we know" whose sole purpose it is to unite mechanism and teleology in a single concept. But such a unification is only desirable or even necessary for discursive beings. Only discursive beings feel the conceptual pressure resulting from the two modes of explanation that are, *for them*, incompatible. And that compels them to think of a supernatural point of reference, an intuitive understanding qua cause of the world (3) or a being capable of intellectual intuition (2).

---

41 See Eckart Förster, *The 25 Years of Philosophy: A Systematical Reconstruction* (Cambridge, MA: Harvard University Press, 2012), ch. 6.
42 See Eckart Förster, *The 25 Years of Philosophy: A Systematical Reconstruction* (Cambridge, MA: Harvard University Press, 2012), 120.
43 While the following interpretation is inspired by Eckart Förster's reconstruction in ch. 6 of *The 25 Years of Philosophy: A Systematical Reconstruction* (Cambridge, MA: Harvard University Press, 2012), it differs in some important details, in particular with regard to the role of an intuitive understanding in the solution of the antinomy.

Such a concept of a supernatural substratum (*Urgrund*) dissolves the difference between mechanistic and teleological explanation by providing a common root for both in the unity of thought (idea) and being (reality):

> But since it is still at least possible to consider the material world as a mere appearance, and to conceive of something as a thing in itself (which is not an appearance) as substratum, and to correlate with this a corresponding *intellectual intuition* (even if it is not ours), there would then be a supersensible real ground for nature, although it is unknowable for us, to which we ourselves belong. (CPJ 5:409; emphasis added)

We would thus be justified in judging nature "in accordance with two kinds of principles, without the mechanical mode of explanation being excluded by the teleological mode, as if they contradicted each other" (CPJ 5:409). And this part of the conclusion would be justified by the conceivability of such a "real ground for nature," be it (infinite) intuitive understanding or intellectual intuition.

The second step in the solution would be to invoke the intuitive understanding in its finite guise, that is, (4) an understanding that is able to have an intuition of a synthetic universal (i.e. a whole) as such and to go "from the whole to the parts" that is, an understanding "in which, therefore, and in whose representation of the whole, there is no contingency in the combination of the parts, in order to make possible a determinate form of the whole" (CPJ 5:407).

For an intuitive understanding thus conceived there would not even be any tension between mechanism and teleology to dissolve. Such an understanding would be able to *explain* organisms as natural products. This contrastive concept of a limiting faculty has a merely problematic status: it is a concept we think up in order to illustrate the possibility of a *mechanistic* explanation (broadly conceived) of organisms *within nature*—an explanation that is impossible for a discursive understanding such as our own to deliver. An intuitive understanding is consequently characterized as one "in relation to which, and indeed *prior to any end attributed to it*, we can represent that agreement of natural laws with our power of judgment, which for our understanding is conceivable only through ends as the means of connection, as necessary" (CPJ 5:407; emphasis added). An understanding like that, I would like to suggest, should not be identified with an infinite intellect to which we attribute ends.[44]

Accordingly, Kant shortly thereafter identifies this understanding as one concerned with the synthetic-universal, that is, as the intuitive understanding that is to be contrasted with our discursive intellectual faculty. It is the intuitive understanding thus conceived that is able to "represent the possibility of the parts (as far as both their constitution and their combination is concerned) as depending upon the whole," while "given the very same special characteristic of our understanding, this cannot come about," for this would be "a contradiction in the discursive kind of cognition" (CPJ 5:407). That is why *we* need to construe this dependence in teleological terms, that is, "by the representation of a whole containing the ground of the possibility of its form and of the connection of parts that belongs to that" (CPJ 5:407/8). We have to resort to a teleological explanation—unlike an intuitive understanding.

Within the complex argument of section 77, the function of the contrastive concept of a synthetically universal or intuitive understanding is, therefore, as follows: it is designed to show that the teleological principle "does not pertain to the possibility of such things themselves (*even considered as phenomena*) in accordance with this sort of generation, but pertains only to the judging of them that is possible for our understanding" (CPJ 5:408; emphasis added). In the absence of any conception of this limiting faculty, we might think of this principle

---

44 For a different reading see Eckart Förster, *The 25 Years of Philosophy: A Systematical Reconstruction* (Cambridge, MA: Harvard University Press, 2012), 153.

(and its contrasting mechanistic maxim) as a *determining* principle—and the antinomy resulting from the tension between mechanistic and teleological explanation would be irresolvable.

The synthetically universal understanding's experiences, just like our own, take place in a world of phenomena—in fact, in our world. The positive analogy of the finite intuitive understanding to the discursive understanding is a strong one. However, the finite intuitive understanding is construed as an understanding that works with an *expanded concept of causality*—a concept of causality that is not unidirectional, but is able to incorporate a mutual dependence of cause and effect that can be used to mechanistically explain the dependence of the parts on the whole. Mechanistic thought, accordingly, cannot be in this context restricted to efficient causality as *we* understand it. It is instead characterized, purely negatively, as "a causal connection for which an understanding does not have to be exclusively assumed as a cause" (CPJ 5:406): such a connection could be put to a different use by finite beings different from us.

Those beings would, of course, have to be endowed with different forms of sensibility. For only then could the category of causality be nontemporally schematized—an alternative schematization that is of the utmost importance for the possibility of this alternative conception of causality. The temporal schematization of our category of causality consists in fixing the temporal order of cause and effect and it, accordingly, prohibits the mutual dependence of cause and effect required by the alternative conception of causality.[45] But this possibility of alternative forms of sensibility and, accordingly, an alternative way of schematizing categories was in play all along, as I showed in my discussion of the faculty of sense in the *CPR*.

Let me offer the following summary by way of conclusion. The organisms we find ourselves confronted with in nature give rise to an

---

45 See Rachel Zuckert, *Kant on Beauty and Biology: An Interpretation of the Critique of Judgment* (Cambridge: Cambridge University Press, 2007), 135–39.

antinomy for us, inasmuch as they seem to demand incompatible explanations—a mechanistic explanation insofar as they are parts of nature, and a teleological explanation insofar as they are thoroughly organized wholes. In order to reconcile mechanistic with teleological explanation, Kant introduces the contrastive concept of an *intellectual intuition* as a faculty generating unity of thought and being (or, alternatively, the concept of an infinite intuitive understanding). The cognitive activity of such a faculty would serve to ground a supersensible substratum in which teleology and mechanism are one. Due to the discursivity of our understanding we have to think of nature as an end of this supernatural ground. The contrastive concept of a finite, synthetically universal *intuitive understanding* on the other hand serves to restrict this assumption—and with it the necessity for a teleological description of the phenomenal world—to finite beings *of a certain kind*, that is, finite beings incapable of this kind of intuitive access to the world. In this way the contrast between mechanical and teleological description of nature can justifiably be ascribed to the reflective as opposed to the determining use of our power of judgment. Only together can these concepts of limiting faculties dissolve the antinomy of teleological judgment.

In post-Kantian German idealism, however, these limiting concepts are separated and put to use quite independently from each other. I will now turn to outlining some of the developments resulting from this separation.

4. German Idealism
4.1. Transcending Critical Philosophy

The transition from Kantian philosophy to the great systems of post-Kantian German idealism had a number of sources, which have been quite thoroughly investigated in the literature on the epoch. One of the most prominent reasons for this development was the criticism first raised by Karl Reinhold and soon shared by many of his contemporaries: that the Kantian philosophy was *incomplete* in a way that threatened the whole system.

This charge is sometimes simply reduced to the claim that Kant did not manage to realize one of the central aims he explicitly set for himself in the *Analytic* of the *CPR*—an aim he characterized as a privileged feature of all transcendental philosophy, that is, that it "find its concepts according to a principle" (CPR A67), where "a principle" is taken to mean "exactly *one* principle."[46] In the Kantian system, so the criticism goes, such a unifying principle is nowhere to be found.

There have been many different suggestions as to how to understand this criticism and, consequently, there are as many different ways of outlining the transition from Kant's critical philosophy to the systems of the German idealists.[47] One way to understand this criticism, which is particularly relevant to my present purposes, is as the claim that Kant's system provides no point of unification for the diverse faculties he has to postulate in the course of prosecuting his critical enterprise. Is this criticism justified?

There are two main divisions in the Kantian system that, at least at first glance, might seem particularly susceptible to this charge: one is the division between sensibility and understanding within the theoretical faculty of knowledge, and the other is the even more fundamental division between theoretical and practical faculties I briefly touched upon in discussing the *CPJ*'s limiting concept of a holy will.

I shall take up these potential sources of difficulty in turn. Kant's critical philosophy, as I have shown, does indeed rest on the doctrine of the two stems of knowledge. Sensibility and understanding are two

---

46 See the telling remarks in the *Critique of Practical Reason*, AA. 5:90/1, and the analysis in Eckart Förster, *The 25 Years of Philosophy: A Systematical Reconstruction* (Cambridge, MA: Harvard University Press, 2012), 166–67.

47 See for example Frederick C. Beiser, *German Idealism: The Struggle against Subjectivism 1781–1801* (Cambridge, MA: Harvard University Press, 2002), Terry Pinkard, *German Philosophy 1760–1860: The Legacy of Idealism* (Cambridge: Cambridge University Press, 2009), Paul Franks, *All or Nothing: Systematicity, Transcendental Arguments, and Skepticism in German Idealism* (Cambridge, MA: Harvard University Press, 2005). I follow Eckart Förster, *The 25 Years of Philosophy: A Systematical Reconstruction* (Cambridge, MA: Harvard University Press, 2012), as far as the overall picture is concerned.

distinct but necessary ingredients in our intentional or representational relationship to the world we conceive ourselves to be a part of. Now, one might argue that this distinction, far from being a problematic feature of the Kantian system, serves, among other things, the important philosophical purpose of explaining what Kant took to be the essential discursivity of our intentionality, while simultaneously providing a clear separation between the conceptual and the nonconceptual elements in our experience.

However, the idea that finite rational beings like ourselves have a discursive intentional relation to the world is not something we should take for granted in light of the development of the philosophy of German idealism. For the starting point for much of this development is precisely the denial of the essential discursivity of the understanding, which Kant was so convinced of. Given that one can indeed specify nondiscursive modes of representation and lend plausibility to the claim that we, as subjects of experience, can or even have to employ such nondiscursive modes of representation, Kant seems to have underestimated our epistemic capacities.

Furthermore, Kant did not show how the two stems of knowledge—sensibility and understanding—hang together, apart from noting that they may have a "common root that is, however, unknown to us" (CPR A15). Kant's inability to identify a common root of sensibility and understanding can thus be seen to indicate to a troublesome incompleteness in his system, since he cannot show whether and, if so, how these two sources of knowledge can have a common origin. Yet his own concept of a transcendental subject—at least in its theoretical guise—is the idea of a conjunction of representations that could not have any existence independently of the representations thus united.[48]

As such it is unfit to serve as a point of unification of the theoretical and the practical faculties either. Consequentially, it seemed to the

---

48 See Eckart Förster, *The 25 Years of Philosophy: A Systematical Reconstruction* (Cambridge, MA: Harvard University Press, 2012), 170.

German idealists that Kant had failed to reveal the common ground of theoretical and practical philosophy.[49] This charge may, at first glance, seem to be a wholly unfounded conclusion to draw from the aforementioned incompleteness. After all, in his *CPJ*, Kant does, at least to his own satisfaction, achieve a unity of theoretical and practical philosophy, in arguing that our power of judgment leads us to "look beyond the sensible and to seek the unifying point of all our faculties *a priori* in the supersensible: because no other way remains to make reason self-consistent [einstimmig mit sich selbst]" (CPJ 5:341).

This solution may, nevertheless, strike one as unsatisfying on a number of counts. First, it is only *reflective*, not constitutive, judgment that leads to this conclusion. Consequently, our anticipation of this point of unification cannot constitute *knowledge*, but must remain an assumption we are aware arises from the specific constitution of our mental faculties.[50] More important, this "point of unification" is not made comprehensible in any detail. We feel that we have to postulate it "to make reason self-consistent" (CPJ 5:341), but this is not an illuminating way of understanding what it actually involves. We have no way of exploring this point of unity and are, therefore, incapable of understanding it. We cannot reason our way to any identification of its structure as, for instance, we are able to do in practical philosophy with regard to our nonempirical self. (Though we are nonetheless on firmer ground in speaking of the unity of the practical and the theoretical than we are in talking about the "common root of sensibility and understanding." The latter conjecture was *mere* guesswork.) In the case at hand, creatures like us are compelled to quest after the supersensible as soon as they reflect on the possibility of certain phenomena (notably, organisms) *sub specie* their inexplicability by means of discursive understanding. Yet one may wonder just how safe these grounds are.

---

49 Eckart Förster, *The 25 Years of Philosophy: A Systematical Reconstruction* (Cambridge, MA: Harvard University Press, 2012), 163.

50 See Eckart Förster, *The 25 Years of Philosophy: A Systematical Reconstruction* (Cambridge, MA: Harvard University Press, 2012), 163.

All in all, one can understand the dissatisfaction of the post-Kantian philosophers with the completeness of the Kantian system. On the one hand it seemed to them incomplete insofar as it neglected nondiscursive faculties of genuine knowledge. On the other hand it seemed to provide no unified point of origin for the diverse faculties it postulated in the course of transcendental reasoning. It turned out that the resolutions to both these problems were intimately connected and that, by acknowledging the reality of such nondiscursive faculties, one would also provide the means to solve the problem of a common origin of the faculties.

In what follows, I will illustrate this way of introducing the problem and answering it by principally concentrating on Fichte's work. Schelling and Hegel, the other philosophers who worked out comprehensive idealist systems, will be discussed only by way of conclusion. This preferential treatment seems justified both by the fact that Fichte's philosophical system was the first of the three great idealist systems to be established and by the fact that, later on, his system served as a point of reference for the other German idealists, thereby laying the ground and setting the standards for future criticisms of Kant's philosophical system.

4.2. Fichte

Although Fichte was initially very sympathetic to Reinhold's demands,[51] he soon came to realize that no single proposition (or a family of propositions) expressing *a fact* could serve as a foundation of an entire critical system. Propositions expressing facts always presuppose, Fichte argued, a differentiation between subject and object, which has to be established by transcendental reasoning in the first place.[52] Consequently, it is this differentiation itself, conceived of as a *process*, to which any serviceable First Principle must give voice. Hence, Fichte's own approach sharply differed from Reinhold's way of solving the problem: since no

---

51 See Fichte, *Aenesidemus*, SW I, 20.
52 See Fichte, *Aenesidemus*, SW I, 9.

sentence describing a *fact* can do the trick, we must instead turn to an *act*. Accordingly, the original *Thathandlung* is an act of self-positing, not a fact.

Assuming that Fichte's turn to an act as opposed to a fact was indeed the right choice, might this spell trouble for Kant? Not necessarily. For the Kantian term "judgment" (*Urteil*) is itself ambiguous as to whether it designates a fact, for instance the uttered sentence, or an act (or activity), such as the utterance of a sentence. At the same time, Kant's focus on the activity of *judging* does presuppose that representations are given in advance. Both in the case of judgments and in the case of intuitions these would be *conscious* representations. But Fichte argues that no representations may be presupposed as given; they must rather be successively deduced from a first *act* that truly deserves to be called a first principle.[53]

FICHTE'S *THATHANDLUNG* AND INTELLECTUAL INTUITION

Fichte attempts to derive both theoretical and practical faculties from the very same original *Thathandlung*. His 1794 *Foundations of the Entire Science of Knowledge* outlines a complete circular movement of thought that starts with the act of self-positing and then moves back to this *Thathandlung* in a twofold manner, one covering the theoretical faculties, and the other the practical faculties of the subject that undertook this original act. He thereby lives up to his own methodological maxim that a principle is only completely exhausted if "the circle is really completed, and the investigator [Forscher] is left at exactly the point where he started."[54]

In what follows, however, I shall not try to sketch this complete circle.[55] I will instead concentrate on the theoretical branch of this twofold

---

[53] See Fichte, *Aenesidemus*, SW I, 9.
[54] Fichte, *Begriff*, SW I, 58/9.
[55] Eckart Förster took upon himself a careful analysis of the single steps that Fichte employed in his (not excessively clear) outline of this circle. See Eckart Förster, *The 25 Years of Philosophy: A Systematical Reconstruction* (Cambridge, MA: Harvard University Press, 2012), chs. 8 and 9.

circular movement—and, in particular, on some central aspects of Fichte's derivation of the forms of thought, which seem to me to highlight the differences between Kant's and Fichte's endeavors and thereby throw Fichte's particular treatment of faculties into greater relief.

Given that it is the task of theoretical philosophy to explain the possibility of objective representation, Fichte argues in a Kantian spirit, one must not presuppose any representations as given. But neither, Fichte argues, may the transcendental philosopher presuppose the existence of objects the representations would be representations of. Their existence must likewise be established through reflection on the conditions of the possibility of conscious experience, that is, by the method of transcendental philosophy.

This conception of an object, however, includes the conscious subject qua object of a self-representation. The reason for this is that, as soon as we become conscious of ourselves *as* ourselves, we turn ourselves into an object for ourselves. (We might term this the reflexive analysis of subjectivity.)[56] However, this transformation of the self into an object of thought seems to leave out the self qua subject. And this is unacceptable, since, as Fichte puts it,

> I can be conscious of any object only on the condition that I am also conscious of myself, that is, of the conscious subject. (*Attempt*, SW 1, 526/7)

In other words, consciousness of something as something—even consciousness of myself as myself—*presupposes* "real" or "pure" self-consciousness. The reflexive analysis of self-consciousness therefore takes for granted the antecedent separation of subject and object. But, again, this amounts to an illicit assumption given the methodological framework of transcendental philosophy. For this separation itself is supposed

---

56 Fichte's critique of the reflexive analysis of consciousness was first reconstructed in detail by Dieter Henrich, "Fichtes ursprüngliche Einsicht," in *Subjektivität und Metaphysik: Festschrift für Wolfgang Cramer*, edited by Dieter Henrich and Hans Wagner (Frankfurt: Klostermann, 1966), 193–97.

to be established through reflection on the conditions of the possibility of representation—and, ideally, through reflection on a *single principle* from which one begins and to which one returns in an argumentative circle. Since, as I have already pointed out, both Fichte and Kant require this principle to be an *act*, an act of real or pure self-consciousness would seem to be a promising candidate.

So, how can we uncover a real or pure consciousness that does not separate subject and object? In the *Grundlage* from 1794, Fichte presents us with a method for uncovering this *real* or *pure* consciousness in a systematic manner by abstracting from a given fact of empirical consciousness. In this way, he hopes to establish an original act of consciousness that he calls a *Thathandlung* one that is not itself just another fact of *empirical* consciousness but "rather lies at the basis of all [empirical] consciousness and alone makes it possible" (*Foundations*, SW 1, 91). This *Thathandlung* has to be an act of consciousness that does not separate between subject and object and, consequently, does not presuppose representations of any kind.[57]

Fichte's method of abstraction leads him to uncover the pure activity of positing oneself that underlies the basic proposition "I am."[58]

> The self's own positing of itself is thus its own pure activity.... It is at once the agent and the product of action; the active, and what the activity brings about; action and deed are one and the same, and hence the "I am" expresses an Act [Thathandlung]. (*Foundations* SW 1, 96)

So the original act of positing is a *Thathandlung* of pure consciousness. This consciousness is *pure* since we abstracted from all its empirical

---

57 A more detailed outline of Fichte's argument can be found in Johannes Haag, "Fichte on the Consciousness of Spinoza's God," in *Spinoza and German Idealism*, edited by Eckhard Förster and Yitzhak Melamed (Cambridge: Cambridge University Press, 2012), 100–120. The following reflections are a shorter version of the reconstruction of Fichte's argument in the *Foundations* presented there.

58 For a more detailed account of Fichte's abstraction see Johannes Haag, "Fichte on the Consciousness of Spinoza's God," in *Spinoza and German Idealism*, edited by Eckhard Förster and Yitzhak Melamed (Cambridge: Cambridge University Press, 2012), 104–7.

features. It is not a fact (*Thatsache*), since facts are already posited. And yet it is not a mere act (*Handlung*), since the act itself constitutes the existence of the self. Since the *Thathandlung* is unconditioned or absolute, Fichte calls the subject of this positing a *pure* or *absolute* self.

In this absolute *Thathandlung*, the separation of subject and object has not yet been carried out: it cannot have been, since all separation requires determinateness of the things to be separated. Thus the determinate consciousness of a self requires the positing of something that is *not* this self—a positing that is an act of *opposition* or *counter-positing*, that is, the second part of the complex tripartite original activity required for empirical consciousness. This act is likewise an *absolute* or *unconditioned* act, since the act of positing in no way entails the act of opposition. This act of counter-positing gives us a "mere contrary (*Gegentheil*) in general" (*Foundations* SW 1, 103), a *not-self*, and it is therefore, as I mentioned, a crucial aspect of the determination that is, in turn, necessary for empirical consciousness.

Counter-positing, though a condition of the possibility of determination, cannot itself be an act of determination. With positing and counter-positing we just have two acts of absolute positing related to each other in a way that is, as yet, *undetermined*.[59] What is needed, therefore, is a third step in the reconstructed generation of self-consciousness that somehow reconciles the two acts of positing and counter-positing and thereby determines both of them—thus providing us with empirical consciousness. The task for this third step is set through the first two acts: since the first two positings are *absolute* positings, they threaten the unity of the conscious subject in negating each other. The third step, accordingly, must consist in finding a way to reconcile positing and counter-positing that preserves the unity of the conscious subject. To do this, both acts have to be *limited* with respect to one another. In this way, the unity of consciousness is saved from the threat of disintegration. The resulting unity contains both a determinate self

---

59 See *Foundations*, SW 1, 109/10.

and a determinate not-self and can thus serve as a unity of consciousness in which empirical consciousness is possible as a consciousness of a determinate object—be it a limited not-self *or* a limited self.

This consciousness of the act of continuous self-positing, however, cannot be discursive or conceptual consciousness. For that would make it into reflexive consciousness of the self as an object again. It must instead be a nondiscursive form of knowledge or consciousness. Nondiscursive knowledge does not determine anything conceptually, but amounts to an undetermining awareness of something, that is, an awareness that cannot be an awareness of something *as* something.

Consequently, for this peculiar form of awareness we need a special kind of nondiscursive faculty. It has to be an awareness of something that would not be there as such without being the object of awareness. In this respect, although it is not creative in the way a divine intellect would be in producing the objects of its thought, it still bears a striking resemblance to Kant's limiting concept of an intellectual intuition as a productive unity of thought and being. For through this form of awareness we actualize a capacity that consists in this very act of actualizing. Fichte therefore concludes that this initial act of self-determination is given to us in the act of an *intellectual intuition*.[60]

But this act of awareness cannot be distinct from the *Thathandlung* it makes us aware of—for that would precisely lead us back into the regress we wanted to avoid. The intellectual intuition that makes us aware of the *Thathandlung* is, accordingly, itself just the *Thathandlung as carried out by subjects like us*. As Fichte writes in the later *Attempt*, "I am this intuition and nothing else. And this intuition itself is me" (SW 1, 529). We do not produce an object through this act, we rather actualize a subject-object that *is* consciousness in virtue of the specific kind of *Thathandlung* characteristic for subjects like us: a *Thathandlung*

---

60 Paul Franks helpfully puts this contrast in terms of Fichte's intellectual intuition, like Kant's, being *self-actualizing*, but, unlike Kant's intellectual intuition, not *creative*. See Paul Franks, *All or Nothing: Systematicity, Transcendental Arguments, and Skepticism in German Idealism* (Cambridge, MA: Harvard University Press, 2005), 311.

that *proceeds by intellectual intuition* and produces a self that is nothing but pure consciousness.

A PRAGMATIC HISTORY OF THE MIND: THE INTRODUCTION OF A FACULTY

Until the third step, which involves the synthesis of the acts of positing and counter-positing, we had not yet assembled or justified the elements necessary for an intentional relation to *any* determinate entities. Neither determinate objects nor determinate subjects were possible before this act of mutual limitation and, hence, determination. With the third step, we have achieved just that, though not (yet) in any explicit way. For it turns out that, from the perspective of the philosopher reflecting on it, the unelucidated form in which this synthesis is first introduced is insufficiently robust: on reflection it leads to a whole series of contradictions that must ultimately be resolved in order to make the original synthetic act self-consistent.

This elucidation in the *Foundations* proceeds by way of a dialectical process that runs through a whole series of successive acts of synthesis that turn out, in the end, to be but parts of the original act, whose proper elucidation they result from. Only after this process has been completed can Fichte say of his three foundational acts of positing: "What held good before in purely problematic fashion now has an apodictic certainty" (*Foundations* SW 1, 218).

The possibility of this kind of elucidation is obviously necessary if Fichte is to achieve his self-set aim of deducing the whole of the original conceptual inventory that characterizes the self-conscious subject. It would obviously exceed the limits of this chapter to even give an outline of the details of this extremely complex inquiry—an inquiry that, it turns out, theoretical philosophy could not finish on its own anyway, since a "full circle" must not only move through all of the theoretical faculties but also proceed through all the capacities that constitute the practical faculties as well. Only in this way can the common root of both theoretical and practical faculties truly be established and the worry about the unity of systematic philosophy finally laid to rest.[61]

Instead, I would like to address the turning point of the argument, that is, the point at which the synthetic-deductive reflection of the first chain of reflection—which consists exclusively of resolving the contradictions that arise from the third aspect of the original *Thathandlung*—is brought to an end by reaching noncontradictory rock bottom, whereupon the second chain of reflection begins. This second chain is now operating, as it were, on safe ground. It has a proper object to reconstruct, namely the consciousness of the thinking self (and not just the thinking self of the self-reflecting philosopher). In a way, the philosopher, from this point on, simply "observes" the motions of an I that has to go all the way back to a reflected consideration of the original *Tathandlung*. In the course of this process, the thinking self reenacts the actions that are presupposed in every conscious thought. Fichte calls this chain of reflection, accordingly, a "pragmatic history" (*Foundations* SW 1, 222) of the human mind or human consciousness.

The turning point that marks the beginning of this pragmatic history is the introduction of the faculty of imagination. Why is this faculty introduced at all? Fichte is adhering here to precisely the transcendental methodology he explicitly subscribes to. According to this methodology, a faculty (or ability) may only be introduced into our philosophical reconstruction if it is indeed a condition of the possibility of our being conscious of (i.e. representing) objects. And, after the first two steps—which consisted merely in acts of positing of indeterminate "somethings"—the third step had to unite the utter opposites that were posited and counter-posited.

Now, a condition of the possibility of such a unification of complete opposites is a faculty that is introduced solely for that purpose.[62] And it is this faculty that Fichte calls imagination. It has to be conceived as

---

61 See Eckart Förster, *The 25 Years of Philosophy: A Systematical Reconstruction* (Cambridge, MA: Harvard University Press, 2012), ch. 9, for the details of the practical side of the circle.
62 See *Outline*, SW 1, 386.

oscillating between opposite states of the subject—a state that Fichte figuratively describes as a "hovering" (*Schweben*; see *Foundations*, SW 1, 217). Imagination thus establishes the possibility of limitation and hence the determinability of an object[63]—it is not, as Fichte insists, the determination itself that precisely brings the "hovering" to an end. Such determination—as a *fixation*, a cessation of hovering—can only be a product of reason.[64]

The introduction of the imagination marks, in Fichte's exposition, the transition to the "pragmatic history of the mind." Fichte seems fully aware that, according to the methodological standards set by transcendental philosophy, a cognitive faculty may only be introduced in the context of a *quid iuris* question—that is, a question that asks for conditions of possibility as opposed to a *quid facti* question, which calls for a description of the presupposed psychological constitution of the human mind.

In Kant's philosophy, the imagination is introduced as the faculty that can both bridge the gulf between conceptual and nonconceptual representations *and* unite the manifold of sensibility, thus giving rise to conscious representations. Fichte, however, rightly insists that the very notion of representation (much less the distinction between conceptual and nonconceptual representations) may not be presupposed in this context *unless* it can be shown to be a condition of the possibility of conscious representation.

And, given Fichte's own work in the *Wissenschaftslehre*, it is no longer convincing that representation meets this criterion. For Fichte has, at least by his own standards, offered a plausible alternative explanation of the possibility of conscious representation without resorting to any ingredients that are not themselves justified by the demands of transcendental philosophy. This alternative invokes the faculty of imagination only as a unifier of opposites that in turn have been generated, *not* presupposed, by the thinking subject.

---

63 *Outline*, 1, 215.
64 See *Foundations*, SW 1, 216.

Of course, this alternative is viable only if Fichte's abstractive reconstruction of the original *Thathandlung* and the corresponding faculty of intellectual intuition can be convincingly established. But, again, Fichte provides arguments that can ultimately be understood as reflections on the possibility of experience—arguments that should therefore at least be taken into account by philosophers who subscribe to the methodology of transcendental philosophy.

5. Conclusion

Even sympathetic readers of Fichte tend to credit him with having found only the *subjective* dimension of the common root uniting being and consciousness: namely, insofar as the being of a self-conscious subject is founded in an original act of self-positing of this very subject. But given the Kantian background this problem has an *objective* dimension as well.[65]

The objective dimension can be brought into focus by grasping a common supersensible origin of nature and mind in order to account for the teleological structure we find in nature. For Kant, no such explanation can be hoped for, since this teleological description of the world is a consequence of the peculiar limitation of our epistemic faculties and, therefore, only a function of our reflective power of judgment. Schelling—and, following him, Goethe and Hegel—came to question this restriction.

It soon turned out, however, that the function of intellectual intuition in the subjective domain of self-actualizing consciousness could not simply be transformed into the positing (and hence the epistemic accessibility) of objective nature. For in thinking of nature as objective, we must precisely *abstract* from the activity of the thinking or positing subject. It is therefore impossible for us to intellectually intuit *objective*

---

65 A detailed account of the development sketched in the following remarks can be found in chs. 9–14 of Eckart Förster, *The 25 Years of Philosophy: A Systematical Reconstruction* (Cambridge, MA: Harvard University Press, 2012).

nature.⁶⁶ In a sense, we overreach in applying the concept of this limiting faculty directly to the whole of nature, for we thereby undermine the very objectivity of nature that we sought to make intelligible by invoking intellectual intuition in the first place. In its application to nature as a whole, the concept of intellectual intuition, as Kant clearly recognized, can only serve as contrastive concept that highlights our own epistemic limitations.

But Kant himself employs a different limiting faculty in his solution to the antinomy of teleological judgment, namely the faculty of (finite) intuitive understanding. Such a faculty operates with a synthetically universal intuition of a whole as such and thereby provides for the possibility that a finite rational being might enjoy an alternative mode of epistemic access to natural phenomena and, ultimately, even nature as a whole—though, for Kant, such a faculty could not be available to *discursive* rational beings like us.

It was left to none other than Johann Wolfgang Goethe to take up this idea of a faculty of intuitive understanding, unite it with Spinoza's idea of a *scientia intuitiva*, and thus try to uncover a method of observation that would allow even us finite beings to enjoy an intuitive understanding of nature and therein a supersensible reality of ideas. Hegel can then be seen as transforming these methodological insights in a manner that allowed him to apply this intuitive method in discovering the transitory world spirit—a "phenomenology of spirit"—and, thereafter, supersensible reality as a whole.

Fichte's early *Wissenschaftslehre* can thus serve as a prime example of the manner in which the German idealists were able to considerably extend the conception of our cognitive faculties, as finite rational beings, while simultaneously (at least in the early years) operating within the limits set by Kantian methodological constraints. And even in later stages of this development, including Hegel's 1807 *Phenomenology of*

---

66 See Eckart Förster, *The 25 Years of Philosophy: A Systematical Reconstruction* (Cambridge, MA: Harvard University Press, 2012), ch. 10.

Spirit, Kant's limiting concepts can be seen to structure much of the philosophical argumentation. Yet, in the case of intellectual intuition or intuitive understanding, what started as purely contrastive concepts of limiting faculties were recast, in the full course of this development, as actual faculties of finite rational beings like us.

Reflection
FACULTIES AND PHRENOLOGY
Rebekka Hufendiek and Markus Wild

In Quentin Tarantino's western *Django Unchained* (2012), the southern slave owner Calvin Candie, played by Leonardo DiCaprio, explains to his guests the unwillingness of slaves to rise up and take revenge by putting the skull of a recently deceased slave on the dinner table. "The science of phrenology," Candie candidly explains, "is crucial to understanding the separation of our two species." After partly sawing off the back of the skull, he points to what looks like a sizable cavity and clarifies that this part of the brain associated with "submissiveness" is significantly enlarged in black people. It seems that they are naturally submissive and therefore born to be ruled by white men. The science of phrenology explains and justifies slavery, or so the Europhile Candie points out with a grand illustrative gesture. Candie is neither a learned man nor very intelligent. He is a talkative, clever, emotionless, ruthless, and sadistic egoist who is fond of imitating European high culture. Phrenology appeals to this man not only because it justifies his way of life and his existence but also because it lends itself to visual corroboration and public display.

Why is phrenology so appealing? First, it rests on an easy line of reasoning: moral and mental faculties are to be found in specific organs of the brain, just as the perceptual faculties are connected with special organs (sight with the eyes, hearing with the ears, and so on). The more persistently such faculties prevail, the bigger the respective organ; the altered size of the organs leaves its imprint directly on the skull. You can literally read the mental makeup of individuals or groups from the bumps in their skull. Second, phrenology produces easy visual evidence. You are immediately able to see the intellectual abilities, emotional dispositions, and character traits in a person's skull. This powerful visible concreteness, extensively exploited by the advocates of phrenology in the nineteenth century, is what appeals to a man like Calvin Candie.[1]

What is the "science of phrenology"? Around 1800 the neuroanatomist Franz Joseph Gall (1758–1828) developed a research method he called *Schädellehre*, which came to be known (only later on) as "phrenology." The program was based on the idea that the mind consists of several independent mental faculties that can be located in different parts or "organs" of the brain.[2] Gall distinguished nineteen brain organs common to man and animals, and eight organs specific to the human brain. Among the first class, we find such powers as the instinct of reproduction; the instinct of self-defense; the carnivorous instinct; cleverness; pride; ambition; memory of things, facts, words, and people; the sense of place, colors, and sounds; and the architectural sense. The second class involves the sense for metaphysics; witticism; poetical talent; a moral sense; the faculty for imitation; the organ of religion; and, finally, steadfastness of purpose.

The main goal of phrenology is to establish a correlation between intellectual faculties and personal character traits on the

---

[1] Roger Cooter, *The Cultural Meaning of Popular Science: Phrenology and the Organization of Consent in Nineteenth-Century Britain* (Cambridge: Cambridge University Press, 1984).

[2] Franz Joseph Gall, "Des Herrn Dr. F. J. Gall Schreiben ueber seinen geendigten Prodromus ueber die Verrichtungen des Gehirns der Menschen und der Thiere, an Herrn Jos. Fr. von Retzer," *Wielands Neuer Teutscher Merkur* 12 (1798): 311–35.

one hand and cranial morphology on the other. Phrenology therefore amounts to reading character traits and mental abilities from the bumps in the skull. According to Gall, mental faculties and character traits reflect innate dispositions localized in specific areas or "organs" of the cerebral cortex. The development and prominence of these traits and faculties are different in each individual. Most important, they are expressed in the activity and hence the size of the respective cortical organ. Moreover, the size and potency of each cortical organ is reflected in the indentations and cavities of the overlying skull. A careful description of individual skull morphology could therefore systematically reveal its owner's intellectual and personal profile.

To develop and prove his theory, Gall (later joined by his assistant Johann Gaspar Spurzheim, 1776–1832) began to collect human and animal skulls and prepare colored wax molds of brains and plaster casts of heads. This collection and the public lectures that Gall gave in Vienna turned him into a local celebrity—hence "the man of skulls."[3] Between 1805 and 1807 Gall and Spurzheim traveled Europe for a lecture tour. Gall lectured on his organology while Spurzheim presented skulls or carried out dissections in front of huge audiences. At the same time Gall inspected local prisons and asylums in the cities he visited in order to further his research. Gall's unorthodox practices of collecting evidence and presenting his work in visually impressive ways to lay audiences were described by his critics as the mark of a charlatan.[4] Gall himself, however, claimed that his work could do away with metaphysical speculation about human nature and replace it with empirically founded claims about human psychology.

---

3 John Van Wyhe, *Phrenology and the Origins of Victorian Scientific Naturalism* (Burlington, VT: Ashgate, 2004), 24.

4 John Van Wyhe, *Phrenology and the Origins of Victorian Scientific Naturalism* (Burlington, VT: Ashgate, 2004), 72–95.

Gall's problematic set of ideas (that the activity of a cerebral organ varies with its size and the cranium reflects the organic structure of the underlying brain) was soon popularized in the pseudoscience of "craniology" or "phrenology." Its leading popularizers were Spurzheim, who gave a set of immensely influential lectures in Europe and the United States after Gall's departure, and George Combe (1788–1858), who published the enormously successful *System of Phrenology* (1824) and *The Constitution of Man* (1828).[5] At its high mark of popularization, the theoretical individualism of Gall's account was completely forgotten and phrenology had laid itself wide open for ideological interpretations in questions of gender, race, mental health, and criminology.

Phrenology was very influential in the nineteenth century despite the devastating criticism directed against it right from its very beginnings. Severe scientific blows were delivered in the works of William Hamilton (1827), Pierre Flourens (1845), and Paul Broca (1861). Phrenology was certainly not deemed by respected scientists to be a respectable science. Nonetheless, it achieved a status roughly comparable to that of psychoanalysis in the twentieth century, and, like psychoanalysis, continued to find a considerable audience even after serious science had refuted many of its central claims. Roger Cooter demonstrated in *The Popular Meaning of Science* (1986) that, in Great Britain, phrenology attracted, inspired, and reoriented social activists, reformers, and a large portion of the middle as well as working classes. Phrenology was in the air.

---

5 See Robert M. Young, *Mind, Brain and Adaptation in the Nineteenth Century: Cerebral Localization and Its Biological Context from Gall to Ferrier* (Oxford: Clarendon Press, 1970); Roger Cooter, *The Cultural Meaning of Popular Science: Phrenology and the Organization of Consent in Nineteenth-Century Britain* (Cambridge: Cambridge University Press, 1984); Edwin Clarke and L. S. Jacyna, *Nineteenth-Century Origins of Neuroscientific Concepts* (Berkeley: University of California Press, 1987); Michael Hagner, *Homo cerebralis: Der Wandel vom Seelenorgan zum Gehirn* (Frankfurt: Suhrkamp, 1992).

In the early decades of the nineteenth century, phrenology took America by storm. After Charles Caldwell's (1772–1853) *Elements of Phrenology* (1824), Johann Gaspar Spurzheim's lecture tour of 1832 and George Combe's tour of 1838–1840, phrenological societies sprang up all over the country. Despite the common tale that phrenology, after a very short period of respectability, turned into lowbrow entertainment and a pastime for scientific mavericks during the mid-1840s, phrenology was in fact highly influential also in the United States in the nineteenth century. As Norman Davies has argued, phrenology laid the foundations for psychology, criminology, health reform, neurology, and racial taxonomy, and provided a characterology and moral philosophy that was widely influential in mid-nineteenth-century literature, especially in the works of Edgar Allen Poe and Walt Whitman.[6] As an authoritative vocabulary of characterological descriptions, phrenology continued to be deployed for many decades by novelists, theologians, and artists (see also fig. 5a.1).[7]

Calvin Candie, the slave owner in Tarantino's last movie, is modeled on real-life historical figures, including the physician Charles Caldwell from Kentucky. Caldwell was one of the earliest experts in phrenology in the United States. In 1824 he published *Elements of Phrenology*, a book with a racist agenda. Between 1820 and 1851 the physician Samuel George Morton collected thousands of skulls. In *Crania Americana. A Comparative View of the Skulls of Various Aboriginal Nations of North and South America* (1939), Morton claimed that the races descended in natural mental worth in the following order: Caucasians, Asians, Native Americans, and, finally, Africans.[8] Not being a phrenologist himself, Morton was

---

6 See John Dunn Davies, *Phrenology: Fad and Science: A 19th-Century American Crusade* (New Haven: Yale University Press, 1955).

7 See Charles Colbert, *A Measure of Perfection: Phrenology and the Fine Arts in America* (Chapel Hill: University of North Carolina Press, 1997).

8 See Samuel G. Morton, *Crania Americana: A Comparative View of the Skulls of Various Aboriginal Nations of North and South America* (Philadelphia: J. Dobson, 1939).

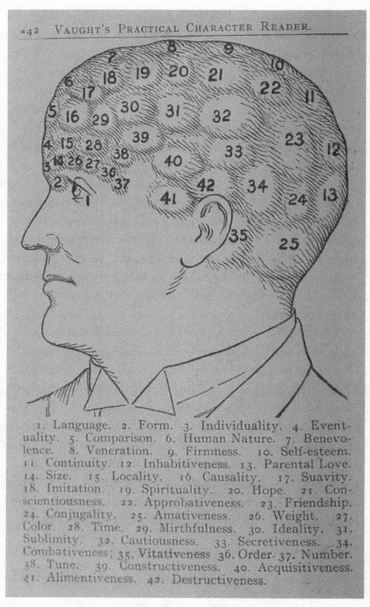

FIGURE Phrenological head from the manual *Vaught's Practical Character Reader* (Chicago: L. A. Vaught, 1902, 242).

nonetheless eager to emphasize the compatibility of craniometrical and anthropological work with phrenology. Scientific racism and phrenology were ready to go hand in hand with the slave owner's agenda.

However, it has to be pointed out that a majority of phrenologists were opposed to slavery, and in fact they used phrenology in order to promote their opposing stance. George Combe, himself an antislavery campaigner, explained that the very same faculties that make the slaves submit to their owners will guarantee that, once emancipated, they will not shed blood. Thus, their assumed submissiveness was used to counter fears that they would take revenge. Calvin Candie's argument cut both ways: because of the unwillingness to rise up and take revenge, the Africans could be set free without any danger to social stability. The premise, however, is still the same: the African is inferior to the Caucasian race in moral and mental abilities. However, according to Caldwell, the African is more suited for civilization than the Indian. Therefore, he might be partly liberated under the protection of the superior white race. It is clear, therefore, that scientific racism and antislavery campaigning were not mutually exclusive. On the contrary: the discourse of phrenology established a set of background beliefs about racial differences that came to figure in pro- and antislavery arguments alike.

CHAPTER SIX

Faculties and Modularity

Rebekka Hufendiek and Markus Wild

1. The Fate of the Faculties in the Nineteenth and Early Twentieth Centuries

Around 1800 the neuroanatomist Franz Joseph Gall (1758–1828) developed a research method that later on came to be known as phrenology. The program was based on the idea that the mind consists of several independent mental faculties that can be located in different parts or "organs" of the brain.[1] The kind of taxonomy of mental organs that includes, for example, an organ for poetic talent and a sense for architecture, looks strange to contemporary readers. What is more, the words "Gall" and "phrenology" immediately conjure up wild claims about how the mental abilities of a person are mirrored in the shape

---

[1] Franz Joseph Gall, "Des Herrn Dr. F. J. Gall Schreiben ueber seinen geendigten Prodromus ueber die Verrichtungen des Gehirns der Menschen und der Thiere, an Herrn Jos. Fr. von Retzer," *Wielands Neuer Teutscher Merkur* 12 (1798): 311–35.

and size of the brain (such that the skull of Kant would show a remarkably huge metaphysical organ).

Yet, Gall's work includes fundamental claims that, as we will aim to show in what follows, became part and parcel of the reasoning about mental faculties from the beginning of the nineteenth century onward. This becomes evident when comparing Gall's theory with the work of the anatomist Samuel Thomas Soemmerring (1755–1830), whose treatise *On the Organ of the Soul* was published at about the same time that Gall was developing his method. Soemmerring suggested that the organ of the soul was located in the fluid of the cerebral ventricles and thus attempted to offer a solution to the age-old question about the location of the soul inside the brain. The book stirred up some considerable controversy. (Kant himself wrote an afterword for the book in which he claimed that the soul cannot be located at all.)

Soemmerring's ideas about the soul's locus in the brain can be understood as the end point of a tradition and Gall's work on the multiple mental organs in the brain as the beginning of a new research program.[2] While Soemmerring joins the long tradition of authors who tried to locate the soul or the place of the interaction between body and soul in one particular part of the brain, Gall remained entirely agnostic about the mind-body problem. It seems that he simply took it for granted that the distinct mental faculties could be localized in the brain and that they altogether made up an intelligent system, even if constituting largely independent parts. Not only did Gall introduce the idea that the relation between mental functions and their location in specific brain areas should be subjected to an empirical investigation, he also rejected the speculative traditional faculties of imagination, reason, memory, and so on, and instead proposed more specific powers. This establishes Gall as the grandfather of the functional decomposition of the mind, as it grew influential in cognitive science from the middle of the twentieth century onward.

---

[2] For a more nuanced view Michael Hagner, *Homo cerebralis: Der Wandel vom Seelenorgan zum Gehirn* (Frankfurt: Suhrkamp, 1992).

While discussion about the localization of mental faculties in the brain remained alive in neuroanatomy all through the nineteenth century and the early twentieth, traditional accounts of the mental faculties lost their explanatory status and role in philosophy at the same time. The reasons for this decline are complex. Before we turn to the resurrection of faculty psychology in the second half of the twentieth century, we will sketch some of those reasons.

As was shown in chapter 4, there is an eliminativist tendency in Hume's account of the faculties. Hume's associationism leads him to propose a reduction of mental faculties to patterns of regularly succeeding perceptions and even compels him to deny that there should be any cognitive powers at all. The principle of the association of ideas, which was first introduced by John Locke,[3] states that mental processes are not to be explained as the products of the activity of mental powers, but instead in terms of the relation in which mental states stand with their mental successor states. David Hume's systematic and extensive use of this principle inspired the British associationist school (including David Hartley, James Brown, Alexander Bain, John Stuart Mill, and Herbert Spencer) to dismiss the faculties in favor of investigating the nature and laws of association.[4]

The principles of association explain both the formation and retrieval of different kinds of complex ideas, and the various kinds of relations obtaining between those ideas. Consequently, there does not appear to be any need for ascribing the formation and retrieval of ideas to faculties. A memory, for example, is not something formed by the relevant faculty and stored and retrieved by it, but is an idea or chain of ideas related to past events, which is produced in response to a certain external stimulus or an occurrent internal mental state. Of course, one might still want to call the set of ideas so produced the "faculty of

---

[3] The chapter "The association of ideas" was inserted in the fourth edition of the *Essay concerning Human Understanding* in 1700.

[4] Théodule Ribot, *La psychologie anglaise contemporaine* (Paris: Librairie philosophique de Ladrange, 1870); Howard C. Warren, *A History of the Association Psychology* (New York: Scribner, 1921).

memory"—however, this concept would no longer refer to an underlying power. Thus, associationism did away with inborn faculties (e.g., memory) and, instead, investigated how the functions allegedly performed by the traditional faculties could emerge from a combination of elementary psychic elements like ideas or sense impressions. Moreover, once the faculties were dismissed, there was no need for supposing any longer a soul or a mind acting as a vehicle for the faculties, while remaining itself inaccessible to empirical investigation. Associationism was conceived as offering something like a paradigm for scientific psychology, because it allowed getting rid of the faculties of the mind—just as Newtonian physics was regarded as the paradigm for a science built on a few mechanical laws, because it eliminated the need for ascribing active powers and qualities to matter. Moreover, associationism was able to provide a framework for the emerging experimental study of the mind. In the second half of the nineteenth century the highly influential psychologist Théodule Ribot provided the foundations for a scientific psychology by taking associationism as a starting point.[5] He amended this empiricist foundation, first, by focusing on heredity as a mechanism for producing innate dispositions that favor certain kinds of association,[6] and second, by proposing the scientific study of mental diseases.[7] Ribot thus founded the French school of psychology, which was subsequently developed by his followers Jean-Marie Charcot and Pierre Janet.

By the end of the nineteenth century associationist psychology was thus very much aligned with experimental psychology and the introspective method. Significantly, however, proponents of the introspective method were not much in favor of mental faculties either. Wilhelm

---

[5] Théodule Ribot, *La psychologie anglaise contemporaine* (Paris: Librairie philosophique de Ladrange, 1870).

[6] Théodule Ribot, *L'hérédité: Étude psychologique sur ses phénomènes, ses lois, ses causes, ses conséquences* (Paris: Librairie philosophique de Ladrange, 1873).

[7] Théodule Ribot, *Les maladies de la mémoire* (Paris: Félix Alcan, 1881); Théodule Ribot, *Les maladies de la volonté* (Paris: Félix Alcan, 1882); Théodule Ribot, *Les maladies de la personnalité* (Paris: Félix Alcan, 1885).

Wundt, one of the founders of introspection, identifies consciousness with immediate inner experience. Consequently, he holds that the sole aim of experimental psychology is the exact description of consciousness.[8] Experience, and nothing above or beyond it, constitutes the subject matter of psychology. Accordingly, the objects of inquiry are introspectible mental *contents*, and not a mental substance or mental faculties. Introspection detects and identifies nothing but conscious thoughts and experiences, where previously mental faculties were supposed to be the powers that produce these conscious mental states in a structured way.[9] The mental faculties are not viewed as powers that produced the mental phenomena, but as nominal sets uniting a group of inner experiences. Thus, when talking about imagination or memory, authors such as Hume, James, and Wundt do not refer to mental powers as producers of different kinds of mental states, but rather to terms for unifying classes of inner experiences.

At the beginning of the twentieth century critical voices against the introspective method grew increasingly louder and, with behaviorism, a method turned mainstream that was quite opposed to introspective psychology. According to behaviorism in the strict sense, psychology is not a science of the mind, but rather a science of behavior.[10] In principle, behavior can be described and explained without any reference to internal mental powers and processes. Concepts for the mental should be replaced by descriptions of the input and behavior patterns that are observable. While introspective psychology relied on the introspection of one's own perceptions and experiences, behaviorism decried this procedure as a pseudoscientific method and sought to define

---

8 Wilhelm Wundt, "Die Aufgaben der experimentellen Psychologie," in *Unsere Zeit III* (Leipzig: F. A. Brockhaus, 1882).

9 Despite the fact that William James adopts a number of different methodological approaches in *The Principles of Psychology*, James tells the reader that he will follow the method of introspection, which he characterizes as "the looking into our own minds and reporting what we there discover" (see William James, *The Principles of Psychology* [New York: Dover, 1980], 185). Given this methodological starting point and the Jamesian conception of the mind as a stream of consciousness, it comes as no surprise that the mental faculties have no role to play in the *Principles*.

10 See John Watson, "Psychology as the Behaviorist Views It," *Psychological Review* 20 (1913): 158–77.

scientific approaches to the mind restrictively in terms of phenomena, such as sensory input and behavioral output, that are observable from a third person point of view. Although behaviorism can be seen as radically opposed to introspective psychology, it avoids any reference to traditional faculty psychology just as much as introspective psychology does, though for slightly different reasons.

As we have shown, the status of the concept of mental faculties has been steadily in decline throughout the development of modern psychology. The same is true in relation to the major currents of philosophy in the twentieth century, albeit for rather different driving forces. Many influential philosophical movements of the twentieth century can be characterized roughly by a reorientation toward the practical on the one hand and by what is called the "linguistic turn," on the other. Pragmatism and existentialism can claim to be the philosophical movements that mediate theory and praxis in the aftermath of Hegelian thought.[11] The principle of pragmatism identifies the "practical consequences" of a theory, concept, or hypothesis by describing how it functions as an instrument in thought, analysis, and practical deliberation. It is the practical consequences that determine the nature of thoughts and concepts, and not their origins in mental faculties. The existentialism of Heidegger and Sartre rejects the assumption of faculties, because it prioritizes existence over essence. What is essential to human beings is not fixed by their essence but by what they do and make of themselves, in contrast to other entities whose properties and dispositions are determined by the kind of thing they are. Human consciousness amounts to a practical interaction with the world and is not the result of the actions of some inner powers.

Analytic philosophy, in turn, can be characterized by its focus on language. Michael Dummett famously defined analytic philosophy in claiming "that a philosophical account of thought can be attained

---

[11] Karl-Otto Apel, *Charles S. Peirce: From Pragmatism to Pragmaticism* (Amherst: University of Massachusetts Press, 1981).

through a philosophical account of language, and, secondly, that a comprehensive account can only be so attained."[12] Once the philosophy of language is accorded primacy in philosophical investigations, the mental faculties inevitably lose their status as the *explanans* of the mental, if for quite a different reason: it is now due to the methodological decision to account for all kinds of mental states predominantly by reference to linguistic systems or linguistic practices. Putting language first in terms of taking into account linguistic practices calls for a blend of the practical orientation and the "linguistic turn."[13] This combination, in alliance with the behaviorist turn in psychology, led to a widespread rejection of the view that thought and other mental processes were based in mental faculties. A very influential example of this approach and, consequently, of a complete rejection of mental faculties is provided in Gilbert Ryle's book *The Concept of Mind*. Ryle devoted his philosophical work to contesting the unhappy tendency of philosophers to hypostatize the supposed referents of their own concepts. He opposes the idea that minds and mental faculties must exist if there is to be a causal explanation of intelligent behavior, and the book is therefore an examination of various mental concepts, such as knowing, learning, imagining, pretending, hoping, wanting, doing voluntarily, doing deliberately, perceiving, remembering, and so on. As an alternative, Ryle tries to establish connections between mental predicates and behavior by proposing that statements involving mental terms can be translated into subjunctive conditionals about what the individual will do in various circumstances. Thus, Ryle offers a dispositional analysis of statements involving mental terms into dispositional-behavioral

---

12 Michael Dummett, *Origins of Analytical Philosophy* (London: Duckworth, 1993), 4.

13 This combination of theoretical commitments emerges clearly in Wilfrid Sellars's work. Sellars thinks that the revolutionary insight of classical pragmatism can be captured if its central doctrine is not confused with a rather abstruse instrumentalist theory about linguistic meaning and instead it "is reformulated as the thesis that the language we use has a much more intimate connection with conduct than we have yet suggested, and that this connection is intrinsic to its structure as language, rather than a 'use' to which it 'happens' to be put." Wilfrid Sellars, *Science, Perception and Reality* (London: Routledge and Kegan Paul, 1963), 340.

statements and tries to show that, in using dispositional predicates, we do not refer to certain internal mental powers. A radical scientific version of behaviorism was also developed by Burrhus Frederic Skinner.

Cognitive science challenged both the behaviorist outlook and the primacy of linguistic analysis and language use, on which philosophers like Ryle and tough-minded behaviorists like Skinner had relied, and instead offered a new framework for understanding the mind—a framework, nonetheless, that made space once again for the concept of mental faculties.[14] In the 1950s researchers in psychology, computer science, and linguistics began to develop theories of the mind that were based on the idea that the mind is a symbol-processing machine akin to the computer. Pioneers of computer science, such as John McCarthy, Marvin Minsky, Allen Newell, and Herbert Simon, developed programs, for example the General Problem Solver,[15] that were meant to imitate human reasoning. At about the same time, Noam Chomsky rejected the behaviorist idea that language could be an acquired habit and proposed that one could only explain language comprehension in terms of innate mental grammars.[16] The main argument he marshaled in favor of this claim and in opposition to behaviorism was concerned with the concept of the "poverty of the stimulus": Chomsky argued that the spoken linguistic data that children are exposed to while learning a language hardly suffice for inferring grammatical rules. He suggested that children would be unable to distinguish grammatically correct from incorrect statements if they did not dispose of some kind of innate grammatical knowledge. This illustrates

---

14 While today "cognitive science" is simply the label used for the interdisciplinary study of (the animal, human, and artificial) mind and intelligence, involving disciplines such as philosophy, psychology, artificial intelligence, neuroscience, linguistics, and anthropology, the aims and methods of this science were narrower in the beginning, mainly comprising psychological, linguistic, and computer science research.

15 See Allen Newell, John Shaw, and Herbert Simon, "Report on a General Problem-Solving Program," *Proceedings of the International Conference on Information Processing* (Paris: UNESCO, 1959), 256–64.

16 See Noam Chomsky's famous review of Skinner's *Verbal Behavior* in his "Review of *Verbal Behavior*," *Language* 35 (1959): 26–58, and the subsequent development of his ideas about innate linguistic knowledge in Chomsky, *Aspects of the Theory of Syntax* (Cambridge, MA: MIT Press, 1964).

how the return to rationalist assumptions about the mental had partly an explanatory purpose: Chomsky claimed that behaviorist theories have no satisfying explanation to offer for how we acquire language and that we need to presuppose some inner structure of the mind in order to construct a better theory. However, this particular attack on Skinner's behaviorist theory of language use was not the only reason for the decline of behaviorism during the second half of the twentieth century. Quite generally, scientists working on human psychology and animal ethology grew increasingly interested in cognitive states that do not entertain a direct relation with overt behavioral patterns and in cognitive activities that do not depend on prior learning (as exemplified in many animals).[17] Vision science is a case in point, as there are many phenomena in visual perception (such as visual illusions, constancy mechanisms, and depth perception) that can be explained purely by reference to the mechanisms that underlie visual perception.[18]

The return to the inner in early cognitive science was also motivated by developments in computer science and artificial intelligence. Pioneers in this field, such as Newell and Simon, maintained that the latest computer models were symbol-processing machines and, as such, could be understood as concrete proof for the nature of the mind: the mind is a symbol-processing software and the brain its machine implementation. The respective developments in artificial intelligence can therefore be seen as the background for new naturalistic approaches to the mind. Whereas nineteenth- and early twentieth-century psychology had been trying to come up with scientific methods for

---

17 See the classical article by Breland and Breland, containing the following opening passage: "There seems to be a continuing realization by psychologists that perhaps the white rat cannot reveal everything there is to know about behavior.... Perhaps this reluctance is due in part to some dark precognition of what they might find in such investigations, for the ethologists Lorenz...and Tinbergen...have warned that if psychologists are to understand and predict the behavior of organisms, it is essential that they become thoroughly familiar with the instinctive behavior patterns of each new species they essay to study." Keller Breland and Marian Breland, "The Misbehavior of Organisms," *American Psychologist* 16 (1961): 681.

18 See Richard Gregory, *Eye and Brain: The Psychology of Seeing* (London: Weidenfeld and Nicolson, 1967).

observing the mind (from an inner or outer point of view), artificial intelligence offered an entirely novel model for thinking about the mind: by trying to rebuild it, or rather, by the method of reverse engineering as a way of analyzing the mind. Consequently, proponents of early cognitive science put forward metaphysical claims about the nature of the mind that were based on their strong engineering convictions. As a result, the inner did not seem to be mysterious anymore.

The computer theory of the mind, that is, the idea that thinking is a kind of syntactically structured symbol-processing, comes hand in hand with theorizing about the "cognitive architecture" of the mind, understood in the same way as the architecture of computers. More specifically, talk about cognitive architecture refers to the functional decomposition of the mind, the architecture's relation to the realizing brain (conceived as a machine) and the general constraints that this machine/brain puts on what the mind is able to accomplish.[19] In many early artificial intelligence models, the famous von Neumann architecture functions as the prototype for the basic architecture of the mind, consisting of input and output units, a memory unit, and a central processing unit.[20]

Discussions revolving around "cognitive architecture" can be understood as a modern and naturalized version of talk about mental faculties. Nevertheless, apart from the analogy with the computer, this way of thinking about the architecture or structure of the mind also involves a distinctly different approach, given its focus on subconscious processing levels. These levels can be partly reconstructed theoretically by employing research methods based on an objective third person perspective, as in neuroanatomical research on brain damage and resultant cognitive impairments of, for example, speech perception and face recognition; in psychological studies focusing on unconscious, subpersonal processing, for example different kinds of memory; and so on.

---

19 Zenon Pylyshyn, *Computation and Cognition* (Cambridge, MA: MIT Press, 1984).
20 See John Von Neumann, *First Draft of a Report* (Philadelphia: Moore School of Electrical Engineering, University of Pennsylvania, 1945).

Traditional faculty psychology drew intuitive distinctions between different mental states like imagining and memorizing and hypothesized that they would originate in different faculties of the soul. Cognitive science claims instead that, for example, speech perception must be a distinct mental ability residing in a particular part of the brain, given that it can be lost due to brain damage while all other abilities remain intact. In the context of twentieth-century cognitive science, this shift from the personal to the subpersonal level is probably the most significant change in theorizing about the faculties.

We have provided a short sketch of the reasons for the decline of faculty psychology in the late nineteenth and early twentieth centuries. In the aftermath of Hume's empiricist program, psychology—in the shape of associationism, introspective psychology, and behaviorism—dismissed mental faculties. The same was true for the major currents in philosophy in the twentieth century, due to their practical or linguistic orientation or both. Yet, within the framework of the cognitivist paradigm in the cognitive science of the mind, the mind's supposedly innate powers, inner structure, and architecture moved, once again, to the center of interest. Gall's core idea to correlate mental faculties, that are also more fine-grained than traditionally assumed, with identifiable segments of a given cognitive architecture (which is itself localized in some specific brain areas, as empirically established), re-emerges emphatically within the framework of cognitive science. The birth of cognitive science as well as of contemporary philosophy of mind marked the renaissance of a robust philosophical interest in the mind and its powers. In contrast to the philosophical investigations of the mind during past centuries, theorizing about the faculties is now taking place within the framework of cognitive science. This new orientation offers the opportunity for establishing a naturalistic understanding of the mind and the mental faculties.[21]

---

21 There are also other areas of research in contemporary philosophy that rely on mental faculties. Take, for example, contemporary virtue epistemology. Virtue epistemologists consider knowledge as something we obtain by exercising our intellectual virtues. In the broadest sense, a virtue is an

2. Modularity, Cognitive Science, and Naturalism

We now want to argue that the notions of module and modularity, as they are currently used in cognitive science, are naturalized successor concepts that replace the traditional mental faculties. This claim should be understood in the following way: In the broad Aristotelian tradition that runs from ancient to early modern philosophy, the faculties of the soul can be divided into the lower ones, namely the sensory and the appetitive faculties, and the higher ones, namely the intellectual and volitional faculties.[22] The lower faculties comprise not only the appetites and the passions as well as the visual, auditory, tactile, and other kinds of perception but also memory, imagination, and estimation.[23] The higher faculties comprise the capacities for forming concepts, beliefs, and inferences; for making choices and for taking decisions. The subdivision is real because each faculty is supposed to be dedicated to specific objects; in modern parlance, the faculties are "domain-specific."[24] For instance, the visual faculty deals with colors, the auditory faculty with sounds, the intellectual faculty with concepts, and so on. Roughly, humans and the higher animals share the lower faculties, yet the human mind is distinctive in that it also features the higher faculties. There has been some debate in early modern philosophy about whether the

---

excellence of some kind (normally an ability or a character trait). In epistemology, the relevant kind of excellence will be intellectual. Many virtue epistemologists characterize the intellectual virtues as mental faculties or powers for producing beliefs that are true (see John Greco and John Turri, *Virtue Epistemology: Contemporary Readings* [Cambridge, MA: MIT Press, 2012]). These faculties include good eyesight, well-functioning memory, introspection, and logical reasoning. More specifically, these faculties are virtues precisely because they are considered to be stable and reliable dispositions for producing true beliefs. What is more, their reliable character renders them virtues relative to the actual world—even if they are not considered to be reliable relative to any other possible world. Nevertheless, the mental faculties are only rarely discussed directly in virtue epistemology and instead tend to be merely presupposed. This contrasts markedly with discussions revolving around the modularity of the mind. We will therefore focus instead on modularity as a prime example of how the mental faculties have fared in contemporary philosophy.

22 To be sure, in the Aristotelian framework, there are vegetative faculties; however, they do not amount to cognitive faculties.

23 We will not deal with the question of the modularity of the passions and the emotions here; but see Luc Faucher, ed., *The Modularity of Emotions* (Calgary: University of Calgary Press, 2008).

24 On this criterion of division see chapter 3, section 3.

higher faculties characteristic of the human mind are part of the natural (material) world. Descartes and Malebranche were opposed to this idea, while Hobbes and Hume were its defenders. As we will show, this question is still pertinent within the naturalistic framework of the contemporary philosophy of mind. While Jerry Fodor argues that higher cognitive abilities cannot be understood as modular and that cognitive science has so far no means at its disposal for properly conceptualizing the central cognitive system, Peter Carruthers and other authors claim that, in order to understand the mind in a naturalistic framework, one needs to assume that it is modular all the way up.[25]

Within contemporary cognitive science, the traditional division between higher and lower faculties is reproduced in the distinction between low-level and high-level cognitive systems. As we have already shown, the important difference between traditional theorizing about higher and lower faculties and modern proposals about higher and lower cognition is that modern cognitive science focuses most of all on third person methods of investigating subpersonal processes in order to try and verify claims about the underlying realizing mechanisms. One important reason for supposing that there exists a specific difference, for example, between perception and higher cognition, stems from research on perceptual illusions, which typically persist even when the viewer knows the real character of the stimulus. For example, if a subject knows that the two lines of the Mueller-Lyer illusion are the same length, the subject still persists in perceiving them to be of unequal length. The perceptual representation of the stimulus by the visual system is, as is aptly said, "cognitively impenetrable" by the subject's knowledge stored in the central cognitive systems.[26] The visual system apparently operates in isolation and produces the percept of the

---

25 See Jerry Fodor, *The Modularity of Mind: An Essay on Faculty Psychology* (Cambridge, MA: MIT Press, 1983); Peter Carruthers, *The Architecture of Mind* (Oxford: Oxford University Press, 2006).

26 See Zenon Pylyshyn, "Is Vision Continuous with Cognition? The Case for Cognitive Impenetrability of Visual Perception," *Brain and Behavioral Sciences* 22 (1999): 341–423.

Mueller-Lyer illusion untouched by the potential influence of reason or some other higher-level system(s).

Attempts at capturing phenomena, such as the independence of visual processing, and the respective theorizing about independent faculties or relatively independent functional units have led to the concept of the modularity of the mind. Within this paradigm, the functional units of the mind that realize certain abilities, like speech perception or face recognition, have now been labeled "modules." In effect, those modules can be understood as the naturalized successor concepts of the concept of the faculties. They are what constitutes today's supposed architecture of the mind, even if they reside at a subconscious and more fine-grained level than the traditionally supposed faculties of the soul.

Fodor was the first to establish the notion of the modularity of the mind in philosophy, with the explicit goal of reviving faculty psychology. Very roughly, a module can be characterized as a dissociable and relatively autonomous cognitive mechanism with a certain function or purpose. It is central to Fodor, but not to other defenders of modularity, that the lower cognitive systems do not receive input from the higher cognitive systems. They are, in this sense, cognitively impenetrable. Yet the higher cognitive systems are not modular, for they do receive informational input from the low-level systems and other higher systems. His proposal can therefore be labeled the "moderate modularity" hypothesis.[27]

According to other philosophers and cognitive scientists,[28] however, the (human) mind consists more or less exclusively of numerous

---

[27] In principle, there is space for another option here. Cundall argues that cognition is best viewed as a continuum of cognitive processing stretching from modules into central systems as opposed to a discrete architectural division between peripheral and central systems; see Michael K. Cundall, "Rethinking the Divide: Modules and Central Systems," *Philosophia* 34 (2006): 379–93.

[28] Leda Cosmides and John Tooby, "Cognitive Adaptations for Social Exchange," in *The Adapted Mind*, edited by Jerome Barkow, Leda Cosmides, and John Tooby (Oxford: Oxford University Press, 1992), 163–228. Henry Plotkin, *Evolution in Mind* (London: Alan Lane, 1997); Stephen Pinker, *How the Mind Works* (New York: Norton, 1997); Peter Carruthers, *The Architecture of the Mind* (Oxford: Oxford University Press, 2006), and "The Case for Massively Modular Models of the Mind," in *Contemporary Debates in Cognitive Science*, edited by Robert Stainton (Oxford: Blackwell, 2006), 3–21.

cognitive systems, both low-level and high-level. Each system communicates with a limited number of other systems while having little influence on the processes going on inside other systems. Put differently, the (human) mind is entirely made up of modules. This means that both peripheral low-level cognitive systems and central high-level cognitive systems are modular. This is called the "massive modularity" hypothesis or, more critically, the "promiscuous modularity" hypothesis.[29] As will become clear, the concept of massive modularity both weakens the notion of what a module is and blurs the traditional distinction between lower and higher faculties. The difference between defenders of moderate modularity—most notably Fodor himself—and defenders of massive modularity is that the latter apply adaptationist reasoning to the structure of the mind, while the former shrink back from Darwinian interpretations of modularity.

Despite these differences, both theories of modularity agree that the faculties have to be explained naturalistically and both agree that all or at least some of the traditional and the folk-psychological notions of the faculties refer to subpersonal (and perhaps naturally purposeful) mechanisms that account for the cognitive capacities of higher animals and humans. In addition, there may be modules that do not feature in folk-psychological or traditional theories of the mental faculties, such as modules for speech perception, mind reading, and the notorious "cheater-detection-module." In general, current theorizing in cognitive science about the architecture of the mind differs most significantly from traditional approaches in its tendency to multiply mental faculties at the subpersonal level. This tendency can ultimately be traced back to Gall and the dawn of phrenology in the late eighteenth century.

---

29 Richard Samuels, "Evolutionary Psychology and the Massive Modularity Hypothesis," *British Journal for the Philosophy of Science* 49 (1998): 575–602; Daniel Sperber, "In Defense of Massive Modularity," in *Language, Brain, and Cognitive Development: Essays in Honor of Jacques Mehler*, edited by Emmanuel Dupoux (Cambridge, MA: MIT Press, 2002), 47–57; Peter Carruthers, *The Architecture of Mind* (Oxford: Oxford University Press, 2006); David Buller and Valerie Hardcastle, "Evolutionary Psychology, Meet Developmental Neurobiology: Against Promiscuous Modularity," *Brain and Mind* 1 (2000): 302–25.

The connection with the traditional concept of the mental faculties seems obvious: The moderate modularity thesis claims that the traditional lower faculties are modular (in the strong sense) whereas the traditional higher faculties are not. By contrast, the massive modularity thesis claims that both the lower and the higher faculties are modular (in the weak sense). All contemporary theories of modularity deny that the traditional lower and higher faculties are faculties of the *soul*, since souls of any kind are not a proper part of a naturalistic picture. Modularity theories conceive of faculties as psychological capacities of organisms (higher animals and humans) that are realized in the brain. Moreover, both the moderate and the massive modularity theses claim that the faculties referred to by folk-psychological notions are constituted by subpersonal capacities.

Hence, "modules" is a *successor* concept to the traditional "mental faculties," though modules tend to be defined in a more fine-grained manner and are said to be located at the subpersonal level. But what about the idea that "modules" is a *naturalized* successor concept? First, one should acknowledge that the term "naturalism" does not have a precise meaning. Generally speaking, naturalists hold that reality entirely consists of nature as specifically defined by the natural sciences and hold that scientific methods must be used for investigating any part of reality, including the human mind. Thus, naturalism entails both an ontological and a methodological claim. The ontological aspect revolves around the idea that reality has no place for "supernatural" entities; the methodological aspect involves attributing a kind of general authority to the scientific method over other methods in investigating reality. Both aspects are captured in the slogan that philosophy has to be continuous with the natural sciences. Thus, for mental faculties to form a part of nature, as understood by the natural sciences, they have to be able to make a causal difference in the spatiotemporal world. And for the mental faculties to be natural faculties, they have to be realized by concrete mechanisms in living organisms.

Traditionally, the mental faculties denominate specific capacities with a cognitive purpose. However, from a naturalistic perspective, discussions about cognitive capacities and cognitive purpose remain superficial as long as no account is provided of the particular mechanisms that supposedly realize the cognitive capacities in question. As we showed in the preceding section, the particular framework employed in cognitive science facilitates investigations of the mind in terms of cognitive architecture rather than mental faculties and thus encourages research into different kinds of information processing units that are realized in the brain and can solve certain tasks.

The representational states produced by modular cognitive systems are said to be sensitive only to certain kinds of information and to operate in relative isolation from other cognitive systems. As discussed previously, a module can be very roughly characterized as a dissociable and relatively autonomous cognitive mechanism with a certain function or purpose. This idea of a module is well embedded in traditional cognitive science, given that it is firmly committed to the modular approach in the study of cognition. The mind is taken to be a computational device composed of functionally specifiable and detachable mechanisms, and functional decomposition is therefore a central aim of classical cognitive science.[30]

Yet, despite the fact that the notion of modularity is well embedded in cognitive science, there seems to be no generally accepted understanding of what a module is and what the thesis of the modularity of mind is supposed to amount to.[31] We have already seen that there are two versions of the modularity thesis: *moderate modularity* and *massive modularity*. However, there are also two interpretations of what a module is supposed to be: a weaker and a stronger one, both of which

---

30 Richard Samuels, "Is the Human Mind Massively Modular?," in *Contemporary Debates in Cognitive Science*, edited by Robert Stainton, 37–56 (Oxford: Blackwell, 2006).

31 For instance, Pylyshyn notes that his thesis about the cognitive impenetrability of visual perception "is closely related to what Fodor (*The Modularity of Mind*) has called the 'modularity of mind.'" And he adds: "Because there are several independent notions conflated in the general use of the term 'module', we shall not use this term to designate cognitively impenetrable systems in this article." Zenon Pylyshyn, "Is Vision Continuous with Cognition? The Case for Cognitive Impenetrability of Visual Perception," *Brain and Behavioral Sciences* 22 (1999): 364.

we will discuss later in more detail. Finally, there are also two different ways of integrating modules into a naturalistic framework: either one emphasizes that modules are cognitive *mechanisms* and, hence, relies methodologically on *cognitive psychology* in realizing the naturalization project; or one highlights that modules must have a *purpose* and therefore relies on *evolutionary biology* in accomplishing the naturalization project.

So, what exactly is meant by the term "module"? And what precisely is the thesis of the modularity of the mind? We will first sketch an answer to both of these questions in the terms of moderate modularity, as it is defended by Fodor. Subsequently, we will turn to the standard reply given by defenders of massive modularity, and specifically by Carruthers.

3. Fodor: The Modularity of (Some Parts of the) Mind

The most prominent attempt of reestablishing faculty psychology within philosophy of mind and cognitive science in recent decades has been the approach developed by Jerry Fodor in his *The Modularity of Mind*. The modularity thesis defended by Fodor represents a particular version of faculty psychology, which emphasizes how the "lower" faculties have only restricted access to information, and distinguishes them from more global "higher" cognitive processes.

According to Fodor, the variety and special functions of our cognitive activities can be explained with reference to the existence of such subpersonal task-specific modules. Fodor subscribes to the idea that, in order to explain specific cognitive activities, one needs to appeal to the "functional architecture" of the mind as the engine of all these activities. The building blocks for such an architecture comprise faculties that can be individuated with respect to their causes and effects (rather than with regard to their location in the brain or the kind of innate ideas they might entail). Classical functionalist approaches to the mind characterize cognitive states in terms of their causal roles. (To give a somewhat simplified example, pain is the neuronal state that is caused by nociceptive input and normally causes pain behavior. If there is pain perception without actual nociceptive input, perhaps accompanied by abnormal behavior, pain is identified with the activation of those neuronal states that usually mediate between pain inputs and outputs.) Fodor straightforwardly applies this view to faculties, claiming that "the language faculty is whatever is the normal cause of one's ability to speak."[32] However, such a claim should not lead to the conclusion that a faculty exists for every single capacity we possess. This would evidently lead to the absurd consequence that we would need to stipulate a faculty that underlies, say, the ability for adding one and one, and another ability for adding one and two, and so on. For not every behavior we display differs in its function and etiology fundamentally from every other one. A functionalist faculty psychology is therefore tasked to find the causal uniformities in behavior underneath the heterogeneity of surface appearances. The guiding idea in the background, therefore, is to apply a kind of reverse engineering to the mind: while we could of course conceive of a single machine or mechanism that accomplishes all kinds of additions and even arithmetic operations, it would appear unlikely that this very mechanism should also be able to detect colors or trigger flight behavior.

While the insistence on the mind's functional architecture distinguishes Fodor's view from other approaches, such as Chomsky's theory, not all functionalists are committed to the kind of faculties that Fodor stipulates. However, what would appear to be uncontroversial among functionalists is the claim that the mind shows a kind of functional decomposition, that is, that the mind contains systems that can be distinguished by the functions they serve. Nevertheless, even though Fodor's modularity hypothesis is a claim about what kinds of systems there are and what they are like and, therefore, can only be developed

---

[32] Jerry Fodor, *The Modularity of Mind: An Essay on Faculty Psychology* (Cambridge, MA: MIT Press, 1983), 26.

within a functionalist framework, his proposal is not the only interpretation that such a functionalist framework permits.

Fodor suggests capturing the architecture of the mind by way of a threefold functional taxonomy that distinguishes transducers, input systems, and central systems, as follows.

Transducers are subsidiary systems that have the function of providing the system with information about changes in the environment. While a Turing machine, for example, is a closed system operating merely on the restricted amount of information that it receives from its tape, living organisms steadily exchange information with their environments and, thus, their computations are continuously affected by what happens around them. Such transducers are always organs that convert energy impinging on the organism's surface, such as the retina or cochlea, into nerve signals. The outputs of transducer systems specify the distribution of proximal stimuli at the organism's surfaces, without producing inferences about the distal objects causing the stimulation.

Input systems are designed to deliver information to the central systems; more specifically, they mediate between transducer outputs and central cognitive mechanisms by producing mental representations out of the data delivered by the transducers and presenting them to the central cognitive mechanisms for further processing. According to this view, there are substantially more mechanisms that can be identified as different input systems than just the five senses. What Fodor has in mind are highly specialized computational mechanisms that generate hypotheses about the distal sources of proximal stimulation, such as mechanisms for color perception or the analysis of shapes or three-dimensional relations, for example in the case of vision.[33]

Such input systems constitute a natural kind, that is, a class of entities that share many scientifically interesting properties and cannot be

---

33 See Jerry Fodor, *The Modularity of Mind: An Essay on Faculty Psychology* (Cambridge, MA: MIT Press, 1983), 47.

reduced to other more fundamental faculties of the mind. What input systems have in common as a natural kind can be summarized in a simple phrase: input systems are modules. That is, input systems are exactly the kind of objects Gall was right about. Fodor lists specifically nine characteristic criteria that identify a module:

1. *Domain specific*: Fodor harks back to Gall's idea that there are distinct psychological mechanisms corresponding to distinct stimulus domains and argues that Gall was correct with regard to the input processing systems.[34] The question of how many modules there are (partly) depends on the question of how many mechanisms there are that respond only to stimuli from a certain domain. The latter is essentially an empirical question.

2. *Informationally encapsulated*: Modules are restricted with regard to the information they take into account before producing an output. A perceptual hypothesis, for example, about the distance or size of an object, may be based on considerably less data than the organism as a whole has access to. The operations of input systems are effectively unaffected by top-down feedback, such as theoretical knowledge with regard to how the distance and size of objects tend to vary under certain perceptual conditions. Informational encapsulation therefore explains the persistence of perceptual illusions. Domain specificity and informational encapsulation are different features of a module and could, at least in principle, come apart. There could be a system that only reacts to a certain kind of stimulus but is sensitive to all kinds of top-down feedback; yet there could also be a system that reacts to all kinds of stimulus without, however, integrating

---

34 Jerry Fodor, *The Modularity of Mind: An Essay on Faculty Psychology* (Cambridge, MA: MIT Press, 1983), 48.

any top-down feedback concerning the system's background knowledge. According to Fodor, the intriguing aspect about modules is that they always display both features concurrently.
3. *Mandatory*: Modules operate in an automatic mode. An individual cannot help but read the letters she sees, hear an utterance as an utterance, and be afraid when hearing a sudden loud noise. Input systems are constrained to operate unfailingly whenever an opportunity presents itself. This is what distinguishes them from central representational capacities that are under "executive control": we apply them in a manner that is conducive to the satisfaction of our goals, while perception operates purely automatically without regard to any immediate concerns.
4. *Fast*: Processing in input systems is fast when compared with the relatively slow processing occurring in paradigmatic central systems like problem-solving. Fodor classifies cognitive processes as fast if they take place in half a second, at most. Remarkably, according to this claim, being fast is a direct result of being mandatory and informationally encapsulated. However, automatic processes are therefore in a certain sense unintelligent—they can operate fast because they merely consider a stereotyped subset of the whole range of computational options available to the organism.
5. *Inaccessible*: Input analysis typically involves mediated mappings from transducer outputs onto percepts. A system is inaccessible if these intermediate levels of processing are not available to consciousness and explicit reports. Central systems like memory can freely access the content of modules only at their output level. Inaccessibility and encapsulation are therefore two sides of the same coin: while inaccessibility involves restrictions on the flow of information emanating

from a module, encapsulation involves restrictions on the information flow entering the mechanism.[35]

6. *Shallow*: Since modules are informationally encapsulated, we should not expect their outputs to be theoretically demanding concepts. The output of the visual detectors, for example, is shallow, that is, there should be visual output representations that do not categorize visual stimuli in terms of biological or chemical kinds, but form a level of representation by some criterion independent of theoretical knowledge. Fodor suggests that the outputs of modular systems are understood as basic-level concepts,[36] which can be acquired during direct observation rather than by inferential reasoning.

7. *Localized:* Fodor assumes that there is a characteristic neural architecture associated with the input systems. Hardwired connections are supposed to facilitate the information flow between different neural structures inside a module, but they thereby also restrict it to the module. Neural architecture is therefore the natural concomitant of informational encapsulation.

8. *Subject to characteristic and specific breakdown patterns*: A suitable criterion for a system's functionally dissociable character is whether it can be impaired, for example as the result of a brain lesion, without other cognitive systems being significantly impacted. Disorders such as prosopognosia (impaired face recognition), achromatopsia (total color blindness), and agrammatism (loss of syntax) occur in individuals independently of any other impairments.[37] Specific breakdown patterns are also good evidence for neural

---

35 Philip Robbins, "Modularity of Mind," in *Stanford Encyclopedia of Philosophy*, http://plato.stanford.edu/entries/modularity-mind/, 2009, accessed September 3, 2014.
36 Eleanor Rosch et al., "Basic Objects in Natural Categories," *Cognitive Psychology* 8 (1976): 382–439.
37 But see an objection in Jesse Prinz, "Is the Mind Really Modular?," in *Contemporary Debates in Cognitive Science*, edited by Robert Stainton (Oxford: Blackwell, 2006), 22–36.

localizability, since the breakdown of one particular module in the course of a lesion in a certain part of the brain renders it highly likely not only that the system in question is localized in the area in question but also that this area is dedicated *exclusively* to the realization of the very system.[38]

9. *Ontogenetically determined:* The ontogeny of input systems exhibits a characteristic pace and sequencing. Modules are innate faculties that are either present already shortly after birth, as seems the case with visual categorization, or develop "according to specific, endogenously determined patterns under the impact of environmental releasers."[39] The commitment to strong claims about innateness forms part and parcel of faculty psychology from Gall through to Chomsky.

Central systems are the nonmodular systems that constitute the "higher cognitive mind" and enable creative and holistic reasoning processes. Fodor assumes that in addition to the input systems there must also be nonmodular systems that evaluate and exploit the information provided by the input systems. These higher cognitive systems are neither domain-specific nor encapsulated, but instead cut across domains and also have global access to information. Even though candidates might include, for example, choice formation and decision-making systems or what was traditionally called "the will," Fodor focuses entirely on the formation of belief based on prior perceptual processing, which he describes as an evaluation of how things look in the light of background information.

Fodor takes it that, given the cognitive abilities we have, it is necessary to assume the existence of nonmodular systems where the representations

---

[38] Philip Robbins, "Modularity of Mind," in *Stanford Encyclopedia of Philosophy*, http://plato.stanford.edu/entries/modularity-mind/, 2009, accessed September 3, 2014.

[39] Jerry Fodor, *The Modularity of Mind: An Essay on Faculty Psychology* (Cambridge, MA: MIT Press, 1983), 100.

provided by input systems can interface. The mechanisms of belief formation cannot be modular because it is precisely the point of such mechanisms to ensure that what the organism believes is determined by a process that tests and corrects the information that is provided by the input systems in the light of all the background knowledge the individual has.

The problem is that in the case of a domain-general system where every representation is sensitive to any other one we have no conception of how to build such a structure. Given how clueless reverse engineering appears to be when faced with a global mechanism of belief formation, Fodor postulates what he—with a twinkle in the eye—calls "Fodor's First Law of the Nonexistence of Cognitive Science," namely that "the more global...a process is, the less anybody understands it."[40]

The main function of a threefold architecture, such as the one just sketched, is to isolate perceptual analysis from certain effects of background belief. Input analysis is therefore thought to take place in modules, which, according to Fodor, represent a functionally definable subset of the mind and share a certain functional role. They receive input from a transducer and integrate this information in order to produce a distal representation of the external stimulus.

Fodor's explicit goal is to come up with "an overall taxonomy of cognitive systems,"[41] as a way of developing a contemporary version of faculty psychology. Yet, when comparing Fodor's results with traditional positions, it is striking that his account is mainly dedicated to drawing a distinction between perceptual input analysis and belief formation. Fodor therefore leaves many questions entirely unanswered as regards the individuation of faculties. First of all, it is unclear how many modules there are supposed to be. It seems patent that modules do not mirror common-sense psychology and that we should expect modules

---

40 Jerry Fodor, *The Modularity of Mind: An Essay on Faculty Psychology* (Cambridge, MA: MIT Press, 1983), 107.
41 Jerry Fodor, *The Modularity of Mind: An Essay on Faculty Psychology* (Cambridge, MA: MIT Press, 1983), 111.

to represent mechanisms that are more fine-grained than, for example, the five senses, memory, imagination, and so on. Examples of modules that Fodor mentions include mechanisms dedicated to color perception, mechanisms that assign grammatical descriptions to token utterances, and mechanisms for face-recognition.[42]

Even leaving to one side the question of how to individuate input systems, there might also be modules with entirely different functions beyond input analysis, such as those involved in triggering motoric behavior. While these systems would probably share some characteristics with input systems, in terms of being fast, automatic, and domain-specific, other standard criteria simply do not seem to apply to motor behavior systems, such as the property of "producing shallow output." Fodor does not say anything about how to individuate these systems and where to situate them in his threefold architecture.

Furthermore, Fodor remains silent about the conative aspect of the higher cognitive mechanisms, such as different kinds of choice formation and decision-making, or what has traditionally been labeled "the will." Finally, there are the traditional lower faculties in addition to the senses or input-systems, such as memory and imagination, where it is hard to see how they could fit into Fodor's architecture of the mind at all. According to traditional faculty psychology, they belong to the lower faculties, yet they hardly fit Fodor's central criteria for being a module, namely, that they should be domain-specific and informationally encapsulated. But even if Fodor would alternatively include them with the higher cognitive faculties, he would need to say something more about how their particular function distinguishes them from belief formation. However, Fodor does not touch on any of these questions.

It should be evident from the foregoing remarks that Fodor does not really aim at a complete taxonomy of the mental faculties. He is rather *using* faculty psychology as a background theory for establishing two

---

42 Jerry Fodor, *The Modularity of Mind: An Essay on Faculty Psychology* (Cambridge, MA: MIT Press, 1983), 47.

claims: first, that perception is cognitively impenetrable and, second, that cognitive science cannot account for higher cognitive processing.

As regards the first claim, Fodor employs faculty psychology in drawing a clear-cut distinction between perception and cognition, thereby accounting for the cognitive impenetrability of perception. Such a clearcut distinction is meant as an argument against cognitivist tendencies in psychology that focus on top-down processing running from cognition to perception.[43] Fodor objects that such an account blurs the difference between perception and cognition and suggests instead that perception should be perceived as a *tertium quid*: Perception is "smart," like cognition, in that it is usually inferential, yet it is also "dumb," like reflexes, in that it is informationally encapsulated.[44] The functional conflict between inference and encapsulation that might arise from this situation is resolved by assuming that such mechanisms have only sharply delimited access to background theories. The essential criterion of modularity, that is, informational encapsulation, therefore remains the key feature in explaining why and how perception must be strictly distinguished from global cognitive processing. What Fodor is objecting to here is the idea that perception could be theory-relative, namely, in itself biased by the background knowledge the individual has. Relativism, Fodor argues, overlooks the predetermined structure of human nature and underestimates our capacity for securing objective information about the world via our perceptual systems.[45]

The second claim to the effect that cognitive science cannot account for higher cognitive processing is articulated in the already mentioned

---

43 Richard Gregory, *The Intelligent Eye* (New York: McGraw Hill, 1970); Jerome Bruner, "On Perceptual Readiness," in *Beyond the Information Given*, edited by Jeremy M. Anglin (New York: Norton, 1973), 7–42.

44 Jerry Fodor, "Precis of the Modularity of Mind," *Behavioral and Brain Sciences* 8 (1985): 1–42.

45 Jerry Fodor, "Precis of the Modularity of Mind," *Behavioral and Brain Sciences* 8 (1985): 1–42. For a discussion of whether perception is hard-wired and theory-neutral or plastic and theory-dependent and for the epistemological dimension of this question, see the debate between Jerry Fodor, "A Reply to Churchland's 'Perceptual Plasticity and Theoretical Neutrality,'" *Philosophy of Science* 55 (1988): 188–98, and Paul Churchland, "Perceptual Plasticity and Theoretical Neutrality: A Reply to Jerry Fodor," *Philosophy of Science* 55 (1988): 167–87.

"Law of the Nonexistence of Cognitive Science." Cognitive science, according to Fodor, has made good progress in explaining how modules work. But central systems cannot be explained as modular systems, since:

(1) Central systems are in charge of belief formation.
(2) Belief formation is a global process that can access all the information the system has.
(3) Global processes cannot be modular since modules are domain-specific and encapsulated.
(4) Therefore, the central system cannot be modular.

The problem with cognitive science seems to be that it has often treated central systems as if they were modular: "Intellectual capacities were divided into what seem, in retrospect, to be quite arbitrary sub-departments (proving theorems of elementary logic; pushing blocks around; ordering hamburgers).... What emerged was a picture of the mind that looked rather embarrassingly like a Sears catalogue."[46] It is therefore not surprising that Fodor does not distinguish different kinds of higher cognitive faculties and their functions. According to Fodor, such finer taxonomies can be tentatively devised in relation to modules by taking the relevant psychological research into account. However, psychological reasoning about the higher cognitive abilities of the mind appears to be so unconvincing in Fodor's view that he prefers to remain altogether silent about the details of higher cognitive processing, adhering instead to the maxim: better no explanation than a bad explanation.

Nevertheless, such a radical line of argument raises more questions than it seems to answer. In dividing cognitive faculties into input systems that can be explained by reverse engineering and higher cognitive faculties that cannot be explained at all, Fodor's account shows a

---

46 Jerry Fodor, *The Modularity of Mind: An Essay on Faculty Psychology* (Cambridge, MA: MIT Press, 1983), 126.

remarkable resemblance to Descartes's distinction between the body-dependent and the pure cognitive faculties.[47] It seems rather surprising, however, that such a resemblance should emerge from within a naturalistic framework. For Descartes, the pure intellect is part of a distinct substance, namely, the *res cogitans*, which is immortal and not extended in space. Fodor certainly does not aim at reestablishing Cartesianism in this comprehensive sense. Yet, again, the only thing this point illustrates is that Fodor does not aim at establishing any comprehensive approach to the faculties of the mind at all. Instead his concept of modules is meant to establish that input systems cannot be penetrated by global cognition and that we are far away from having a plausible theory of how global cognition works.

4. Massive Modularity

It is remarkable that Fodor's outlook should not only question the explanatory power of cognitive science but also the naturalist framework of explanation. This approach is based on his idea that there can be no cognitive science explanation of the central system and that computational-cum-representational explanations are the only viable scientific explanations of mental processes. If it is true that we do not have the slightest idea of how to describe the structure of higher cognitive processes in terms of reverse engineering, then at least those parts of the mind that are central or global simply cannot be captured in terms of computation. And, according to Fodor, cognitive scientists who have tried to explain the whole mind in computational terms have so far just chased it back further into the machine without really understanding its way of functioning.[48]

---

47 A fairly similar argument for the irreducibility of global systems to underlying mechanisms can be found in Descartes's *Discourse on the Method*, part V; see Descartes, *The Philosophical Writings*, vol. 1, 43–46.

48 Jerry Fodor, *The Modularity of Mind: An Essay on Faculty Psychology* (Cambridge, MA: MIT Press, 1983), 127.

Evolutionary psychologists have questioned this view by arguing that higher cognitive abilities might likewise be described as modular, once we just loosen the criteria for what it takes to be a module. The basic argument for massive modularity applies the general structure of evolutionary and adaptive processes to the architecture of the mind.[49] Accordingly, we should not expect the mind to have one central system running on one general-purpose rule of reasoning or one kind of representational format, since "different adaptive problems frequently have different optimal solutions, and can therefore be solved more efficiently by the application of different problem-solving procedures."[50] According to this view, natural selection is likely to have produced many different specialized mental rules for reasoning about various evolutionarily important domains.

The main argument for massive modularity can therefore be called the "argument from design";[51] it can be represented in the following form:

(1) Biological systems are designed systems that are constructed incrementally.
(2) Such systems, if complex, need to have massively modular organization.
(3) The human mind is a biological system, and it is complex.
(4) So, the human mind will be massively modularly organized.

---

49 Leda Cosmides and John Tooby, "Cognitive Adaptations for Social Exchange," in *The Adapted Mind*, edited by Jerome Barkow, Leda Cosmides, and John Tooby (Oxford: Oxford University Press, 1992), 163–228; Daniel Sperber, "The Modularity of Thought and the Epidemiology of Representations," in *Mapping the Mind*, edited by Lawrence A. Hirschfeld and Susan A. Gelman (Cambridge: Cambridge University Press, 1994), 39–67; Daniel Sperber, *Explaining Culture: A Naturalistic Approach* (Oxford: Blackwell, 1996); Daniel Sperber, "In Defense of Massive Modularity," in *Language, Brain, and Cognitive Development: Essays in Honor of Jacques Mehler*, edited by Emmanuel Dupoux (Cambridge, MA: MIT Press, 2002), 47–57; Stephen Pinker, *How the Mind Works* (New York: Norton, 1997); Harold Clark Barrett, "Enzymatic Computation and Cognitive Modularity," *Mind and Language* 20 (2005): 259–87.
50 Leda Cosmides and John Tooby, "Cognitive Adaptations for Social Exchange," in *The Adapted Mind*, edited by Jerome Barkow, Leda Cosmides, and John Tooby (Oxford: Oxford University Press, 1992), 179.
51 Peter Carruthers, *The Architecture of Mind* (Oxford: Oxford University Press, 2006).

According to Carruthers, the design argument relies on a further argument, which states that our minds have evolved gradually from the minds of other animals. It can therefore be labeled "the continuity argument":

(1) The minds of nonhuman animals are massively modular in their organization.
(2) Evolution is characteristically conservative, preserving and modifying existing structures rather than starting afresh.
(3) We can expect that the human mind should be organized along massively modular lines.

By stressing the continuity between animal and human minds, the continuity argument supports the central claim advanced in the third step of the argument from design, namely, that the human mind is a biological system. A central system, such as the one described by Fodor, is not completely impossible but is highly unlikely, according to this kind of reasoning, since it would require higher cognition to have been the product of one single macro-mutation, instead of being the result of a complex incremental process that developed many task-specific mechanisms over time.

While it seems plausible that the principles of evolutionary development speak in favor of a mental architecture that is modular all the way up, one might wonder about how Fodor's notion of modularity relates to Carruthers's claims. As we have shown, Fodor's notion is meant to fit only input systems. By contrast, advocates of massive modularity tend to use the notion of modularity in a much weaker sense. The weakest sense that underlies many accounts in evolutionary psychology takes a module simply to be a dissociable functional component of the mind. Such a conception of modularity is not even restricted to the mind, but is rather meant to describe the whole living organism as an organization composed of functional components that themselves consist of assemblies of subcomponents, reaching from individual organs at the top level down to cellular assemblies and processes involving genes at the bottom level.

Once we define a module in this way, however, it is questionable whether the proposal that the mind is modular all the way up remains a controversial claim. Fodor might perfectly well agree that higher cognitive abilities divide into subparts, such as the intellect and the will, and that these parts have different functions—provided only that these functional components of the mind are not perceived as domain-specific, encapsulated, and so on, but instead as global general-purpose mechanisms. It turns out therefore that the conclusions of Carruthers's arguments, instead of directly contradicting Fodor's approach, in fact use a weaker notion of modularity.

Nevertheless, the concept of massive modularity is far from being uncontroversial and is explicitly rejected not only by Fodor but by several others. As we will show, however, massive modularity is not controversial merely because of how modules are defined, but because of the way the concept is used in dividing up the whole mind—including higher cognition—into a range of different systems and subsystems that are perceived as adaptations to particular problems that our ancestors faced. What renders massive modularity controversial is therefore not the claim that the mind is modular all the way up—since this claim simply relies on a weaker definition of what a module is—but rather the way in which specifically biological explanations are applied to reasoning about the architecture of the mind.

A closer look at the account developed by Carruthers will underline this point. He argues that our minds have the general structure of a perception-belief-desire-planning-motor-control architecture, which is of "ancient ancestry" and can already be found in insects and spiders.[52] To be a believer-desirer in this sense means that one possesses distinct content-bearing belief and desire states that are discrete, structured, and causally efficacious in virtue of their structural properties.

---

52 Peter Carruthers, *The Architecture of Mind* (Oxford: Oxford University Press, 2006), 65.

Just as does Fodor, Carruthers therefore subscribes both to a realist position with regard to belief-desire psychology and to an overall model of the mind that Susan Hurley has labeled "the sandwich model of mind."[53] This model takes perception and action to be distinct units (or peripheral systems) and cognition to be interposed between these two units. In claiming that the belief-desire psychology is realized in the brain, both Fodor and Carruthers offer strong, rich ontological claims about the localization and structure of the mental faculties. According to Carruthers, such a rich mental structure must in fact be ascribed to all organisms that show at least some behavior that does not reduce to fixed action patterns.[54] The fact that even bees and spiders show apparently various forms of spatial reasoning and planning forces us to conceive of them as believers-desirers in the minimal sense, including any ontological consequences that such a claim might entail. Belief-desire architectures, in the sense of distinct, causally efficacious structures, must be ascribed to all organisms with a central nervous system in order to account for the flexibility of their behavior. Mental capacities cannot be explained merely as a result of associations, nor flexible behaviors as fixed action patterns or conditioned responses. As an approach to the overall structure of the mind, behaviorism and associationism are therefore already excluded at the level of insects. Elucidating the capacity for learning and the various abilities of organisms with a central nervous system presupposes that we assume a rich inner structure.

This then raises the question of what the structure of the mind looks like at a more fine-grained level. Perception, belief formation, desire, planning, and motor control systems can be understood to represent systems that are reminiscent of the traditional faculties. Yet, even if all of these decompose into many different functional subsystems, such modules should not be thought of as physically distinct objects, but

---

53 Susan Hurley, "Action and Perception: Alternative Views," *Synthese* 129 (2001): 3–40.
54 Peter Carruthers, *The Architecture of Mind* (Oxford: Oxford University Press, 2006), 73.

instead need to be understood as cognitive systems. However, this does not entail that modules should not be localizable in the brain; on the contrary, even though they must not be assumed to be localizable in one single place, they can be spread over various parts of the brain. While modules are supposed to be realized somewhere in the brain, however, they are still individuated by their function and not by their location in the brain. Up to this point, then, Carruthers's and Fodor's accounts are in agreement.

A major difference between Fodor and Carruthers opens up, however, when it comes to Carruthers's particular focus on the general rules of evolutionary design processes and the way he extrapolates from this background model several design constraints with regard to the mind's architecture. When thinking about the design of modules, we have to keep in mind, according to Carruthers, that modules are biologically derived systems that developed by co-opting and connecting resources in novel ways. Evolution needs to be able to add new functions without disrupting existing ones; and it needs to be able to tinker with the operations of a given functional subsystem. We should expect different modules therefore to have complex input and output connections with one another and in fact to be sharing parts on a massive scale. Since brain processing is relatively slow, we should furthermore expect massive parallelism and duplication of structure whenever signaling distances increase beyond a certain point or different sorts of information need to be processed within the same restricted time frame.

According to Carruthers, Fodor's criteria of a module's having proprietary transducers and shallow outputs will have to be dropped simply because they only apply to input systems. Carruthers furthermore denies that fast processing is an interesting criterion for distinguishing between different mental systems. Although he is in favor of strong nativism, Carruthers also drops innateness as a necessary criterion, simply because there *might* be *some* modules that are to a large degree acquired through learning. Nevertheless, interestingly, Carruthers adopts

Fodor's criteria that modules are domain-specific, that is, that they process only a certain kind of input and are mandatory in the sense that they automatically process any input that matches their target domain.[55] In claiming that all modules work in this way, including those that are part of higher cognitive decision-making, Carruthers effectively denies that there is any interesting concept of free will or any significant difference between mandatory cognitive processing on lower and higher levels.

In his most important departure from Fodor, however, Carruthers also rejects the claim that encapsulation is necessary, where of course Fodor very clearly insists that encapsulation is the most important property identifying a module.[56] Nevertheless, on closer inspection, it turns out that Carruthers merely modifies Fodor's claim rather than denying it outright: both Carruthers and Fodor agree that we should think of the design of the mind in terms of "computational frugality." It is impossible that every system of the mind should have access to every kind of information processed elsewhere. Nevertheless, Carruthers doubts that frugality requires encapsulation in such a strong sense. The idea of encapsulation derives from a tradition of thinking about the mind according to which information search was deemed to be exhaustive and algorithms were supposed to be designed so as to be optimally reliable. With reference to the work of Gerd Gigerenzer and his colleagues,[57] Carruthers suggests that we instead think about mental processes in terms of simple heuristics. Processing rules of the mind have been designed to be good enough, but not perfect. To

---

55 Domain specificity becomes the most important criterion of a module in evolutionary psychology in general. See also Daniel Sperber, "The Modularity of Thought and the Epidemiology of Representations," in *Mapping the Mind*, edited by Lawrence A. Hirschfeld and Susan A. Gelman (Cambridge: Cambridge University Press, 1994), 39–67; Daniel Sperber, "Modularity and Relevance: How Can a Massively Modular Mind Be Flexible and Context Sensitive?," in *The Innate Mind: Structure and Content*, edited by Peter Carruthers, Stephen Laurence, and Stephen Stich (Oxford: Oxford University Press, 2005), 53–69.

56 Jerry Fodor, *The Mind Doesn't Work That Way* (Cambridge, MA: MIT Press, 2000).

57 Gerd Gigerenzer, Peter Todd, and the ABC Research Group, *Simple Heuristics That Make Us Smart* (Oxford: Oxford University Press, 1999).

repeat, while Carruthers's notion of a module is much weaker than Fodor's, the background idea is to suppose certain design constraints and then produce a model of the mind that satisfies them. In this context, heuristics is seen as the outcome of selective processes that mediate a compromise between speed and reliability. Certain memory systems or social skills are encapsulated not in the narrow sense of having no access at all to information external to the system. Rather, these modules need to be described as being encapsulated in a wide sense: they have access to a very limited amount of information outside the system, which they access via structure-sensitive searching rules. According to Carruthers, the Fodorian argument for computational tractability does not warrant a claim about encapsulation as traditionally understood. It merely warrants what could be labeled "the wide-scope version" of the encapsulation claim: The mind should be constructed entirely out of systems that are frugal in the way they access information of other systems.

These design constraints also shed light on how modules should be individuated in principle. Each reliably recurring function that the human mind is requested to perform is apparently realized by an underlying system. Whenever an executed function is complex, the system in question is structured as an array of subsystems.

It is impossible to specify precisely how many modules there are, since the traditional faculties all divide into multiple subsystems. Not only can the memory system already be partitioned into working memory and long-term memory, but long-term memory can again be divided into explicit and implicit memory; explicit memory in turn separates out into episodic and semantic memory, where semantic memory, once again, splits off into multiple subsystems, and so on. Such a multiplication of faculties at the subpersonal level mirrors a general tendency in modern cognitive science when compared with traditional faculty psychology. Nowadays, traditional faculties, such as memory, tend to be divided into more fine-grained subsystems, which are not meant to be accessible from the first person perspective but are instead presupposed with respect to certain theoretical background assumptions and empirical research methods.

While the tendency to trade in traditional faculties for more fine-grained, subpersonal mechanisms is already present in Fodor, and in cognitive science quite generally, massive modularists depart from faculty psychology in yet another sense. Where Fodor left the distinction between higher and lower cognitive faculties untouched and mirrored Cartesianism in the belief that higher faculties cannot be explained as computational mechanisms, massive modularists tend to blur the distinction between lower and higher faculties by suggesting instead a system of gradual differences extending between human and animal minds. Rather than simply assuming one central system, Carruthers lists more than twenty uniquely human capacities that point to various specifically human adaptations of the mental structure. The capacities that Carruthers cites range from a capacity for folk-physics, which facilitates deeper causal reasoning, to the ability to produce gossip, learn social norms, or acquire complex skills.

The most important additions to the human brain that account for the anthropological difference are a mind-reading system, capable of attributing mental states to others and oneself; a language learning system, designed to build modular production and comprehension systems suited to the surrounding language; and a normative reasoning and motivation system, which assists in forming judgments about what is permitted or prohibited and also generates relevant motivations. Furthermore, humans have an innate disposition for creatively generating and rehearsing action plans by utilizing a variety of heuristics and constraints.

The general reasoning capacity, which, according to Fodor, is implemented by the central system, is mainly realized in operational cycles of existing modular systems and their mutual interactions. The language module plays a particularly prominent role in this process, since Carruthers takes it to be responsible for the seemingly unlimited content flexibility of the human mind. The particular human kind of reasoning is also slower, and conscious reasoning is accomplished in virtue of the "global broadcasting" of sensory representations of utterances in inner speech as well as other action rehearsals.

Carruthers manages to develop a comprehensive account of the mental architecture mainly by applying adaptationist reasoning. His particular strategy involves multiplying traditional faculties in terms of task-specific mechanisms that can only be individuated at a subpersonal level and redirecting the theoretical focus away from the strict traditional distinction between lower and higher faculties and toward an architecture of the mind that is incremental all the way up. Yet Carruthers's account and the approach to modularity of evolutionary psychology in general has not remained unchallenged.

5. The Critique of Modularity and Contemporary Reasoning about Faculties

Many people argue that, while Fodor's notion of a module is too strong, proponents of massive modularity, like Carruthers, Daniel Sperber, and Leda Cosmides and John Tooby, are weakening this notion to such a degree that to claim that the mind is modular becomes uninformative. Evolutionary psychology has been, furthermore, heavily criticized quite generally for its highly speculative approach to the mind.[58] It is therefore not surprising that the notion of the mental module has rarely been deployed in recent years. Yet the question of how to carve up the mind remains an open project. While there has been little explicit debate about mental faculties lately, many discussions have instead revolved around the details of functional decomposition, realizing mechanisms, and the supervenience base of mental abilities. These debates take place in newly transformed scientific contexts that now rely on neuroscientific studies and biological research on niche construction,

---

58 Jesse Prinz, "Is the Mind Really Modular?," in *Contemporary Debates in Cognitive Science*, edited by Robert Stainton (Oxford: Blackwell, 2006), 22–36; Fiona Cowie, "Us, Them and It: Modules, Genes, Environments and Evolution," *Mind and Language* 23 (2008): 284–92.

or form part of the work on embodied and situated cognition. These debates also open up new ways of thinking about mental faculties, and we will therefore provide a brief sketch of these developments in the following.

Jesse Prinz articulates a critique both of the Fodorian approach and evolutionary psychology, arguing that the neuroscientific perspective suggests instead that the mammalian brain uses the same areas for different functions and that neither domain-specific rules nor many interesting cases of encapsulation are to be found. Significantly, sensory cells have turned out to be often bimodal and to be used, for example, for both vision and touch. There is also plenty of evidence on cross-modal perception, on top-down effects in perception, and so on.[59] An ability such as mind reading, which, according to Carruthers and others, represents a prime example of a module, has been shown in neuroimaging studies to recruit language centers in the left frontal cortex, visuo-spatial areas in the right temporal-parietal regions, the amygdala, and the precuneus (involved in mental image assessments)—that is, mind reading seems to exploit a large network of structures that contribute also to many other capacities. The upshot of Prinz's argument is that neither Fodor nor Carruthers carve out interesting divisions within the mind. Instead, the mind ought to be described as a network of interconnected systems and subsystems and not as a collection of encapsulated domain-specific modules.

Many people have criticized the method of evolutionary psychology quite generally and its assumptions about the modularity of the mind more particularly. It has been argued that brain evolution would not design a brain that consisted of numerous prefabricated adaptations, as evolutionary psychology would have it, but instead has produced a brain that is capable of adapting to its local environment.[60] In this vein,

---

59 Jesse Prinz, "Is the Mind Really Modular?," in *Contemporary Debates in Cognitive Science*, edited by Robert Stainton (Oxford: Blackwell, 2006), 22–36.
60 David Buller, *Adapting Minds* (Cambridge, MA: MIT Press, 2005).

various authors have pointed out that evolutionary psychology both overestimates the role of hardwired, genetically determined mechanisms and underestimates the roles of the ecological niche, of learning abilities, and of cooperative foraging being practiced across generations within an ecological niche.[61]

While Prinz criticizes mainly the narrow notion of a module, suggesting a more liberal way of conceiving of the functional decomposition of the mind, recent years have also seen a more radical critique of the modularity approach. This criticism is part of a general turn in cognitive science away from the computationalist paradigm and toward a focus on the interrelation between brain, body, and world. Embodied and situated cognition approaches have criticized the attitude of identifying the mind with the brain only and of conceptualizing the brain as a symbol-processing computer. The mind, according to these approaches, should rather be understood as a collection of abilities that evolved in agents being endowed with particular bodies within specific surroundings: accordingly, language development is claimed to be dependent on the use of material symbols and the bodily ability of gesturing, perception is said to be closely intertwined with action, and memory to be bounded by certain contexts, including distinct social relations.[62] This way of thinking about the mind calls the traditional view about the architecture of the mind and also the concept of a module into question. Dynamical systems theory rejects the decomposition of the mind into separate modules and instead highlights the close couplings between various parts of the brain and between the

---

61 Fiona Cowie, "Us, Them and It: Modules, Genes, Environments and Evolution," *Mind and Language* 23 (2008): 284–92; Sterelny, "Language, Modularity and Evolution," in *Teleosemantics*, edited by Graham MacDonald and David Papineau (Oxford: Oxford University Press, 2003), 23-41; Kim Sterelny, *The Evolved Apprentice: How Evolution Made Us Human* (Cambridge, MA: MIT Press, 2012).

62 Andy Clark, *Supersizing the Mind: Embodiment, Action, and Cognitive Extension* (Oxford: Oxford University Press, 2008); Susan Hurley, *Consciousness in Action* (Cambridge, MA: Harvard University Press, 2002); John Sutton and Kelly Williamson, "Embodied Remembering," in *Handbook of Embodied Cognition*, edited by Lawrence Shapiro (New York: Routledge, 2014), 315–26.

body and the environment.[63] It has been suggested in this vein that on the subpersonal level the distinction between cognition and emotion would not make any sense since emotional and cognitive activities both result from the activity of a variety of brain areas, none of which is exclusively dedicated to emotional or cognitive processing.[64] The general idea that traditional faculties, such as emotion and cognition, have no counterpart at the subpersonal level has been developed in the work of Hurley, who has established this claim with respect to perception and action.[65] Hurley, furthermore, criticizes what she calls the "sandwich model" of the mind, that is, the view that perception and action are distinct elements each lodged at the periphery of the mind while cognition forms the "hearty filling." Hurley argues that perception and action need to interact closely in order for percepts and intentions to have any content at all. She furthermore suggests that we might think of the modularity of the mind not so much in terms of, for example, perception, cognition, and action being separate parts of the mind that process information in a one-way linear order. (She labels this concept "vertical modularity" and attributes it to Fodor.) According to Hurley, we should rather conceive domain-specific modules as being "horizontally modular," namely, consider the quite different notion of a module such that our various mental abilities can be described as "layers" of the mind that will both engage action and perception systems and involve certain environmental conditions.[66]

Hurley's idea, which originates in dynamical systems theory, is reformulated in a more modest version by William Bechtel, who suggests that we should think of the whole organism as an autonomous system aimed at maintaining itself, being separated from its environments yet

---

63 Esther Thelen and Linda Smith, *A Dynamic Systems Approach to the Development of Cognition and Action* (Cambridge, MA: MIT Press, 1994); J. A. Scott Kelso, *Dynamic Patterns: The Self-Organization of Brain and Behavior* (Cambridge, MA: MIT Press, 1995).
64 Giovanna Colombetti, *The Feeling Body: Affective Science Meets the Enactive Mind* (Cambridge, MA: MIT Press, 2014), 98–100.
65 Susan Hurley, "Perception and Action: Alternative Views," *Synthese* 129 (2001): 3–40.
66 Susan Hurley, *Consciousness in Action* (Cambridge, MA: Harvard University Press, 2002).

closely interacting with it.[67] Such an organism comprises component mechanisms that perform the necessary operations for maintaining themselves and are organized in such a way that they can also operate in unison. According to Bechtel, we should carve up the mind in terms of these incrementally evolved mechanisms that might interact closely with each other and their environment. Such mechanisms differ from encapsulated modules, however, because they allow for a much greater degree of crosstalk between the systems. Yet they also present a clear functional structure of the mind that can be said to be responsible for all kinds of intelligent behavior that we display.

It is eye-catching that the critique of the faculties, as it has been articulated in recent research, resembles ideas put forward by pragmatists and existentialists in the late nineteenth and early twentieth centuries. Their central idea was that the interaction between subject and world is prior to an "internal structure" or "essence" of the mind. The recent turn back to pragmatist ideas is largely due to the role that the neurosciences have started to play. Research on the location of the faculties in the brain was prominent all along in the neurosciences through the nineteenth century and large parts of the twentieth. While these studies have been central for the modularity hypothesis, the philosophical critique of modularity derives its main impulse from neuroscientific studies, which suggest that there are no faculty-specific areas in the brain. Recent critique of the modularity hypothesis is not a critique of neuroscientific approaches to the architecture of the mind per se. It is rather a critique that uses insights from neuroscience itself to criticize traditional accounts of the faculties and current versions of the modularity hypothesis alike. These new developments might make it seem as if theorizing about the modularity of mind might have been nothing but a brief interlude in the long decline of the faculties throughout modern times.

---

67 William Bechtel, "Explanation: Mechanism, Modularity, and Situated Cognition," in *The Cambridge Handbook of Situated Cognition*, edited by Robbins Ayede (Cambridge: Cambridge University Press, 2009), 155–70.

6. Conclusion

We have seen that, while theorizing about the faculties and their locations in the brain has been prominent in the neurosciences all through the nineteenth and early twentieth centuries, such theorizing was frowned on at the very same time by philosophers and psychologists. The tide turned with the rise of cognitivism and a return to the inner structures of the mind from the 1960s onward. The discussion about the modularity of the mind has its roots there. Modules, as we have pointed out, represent successor concepts of traditional faculties within current naturalistic theories about the architecture of the mind. In the work of Fodor, Carruthers, and others, modules are defined as those parts of the mind that can be individuated according to their function and, therefore, need to be understood as underlying our mental abilities. Modules do not capture folk-psychological categories but rather subpersonal mechanisms that are assumed by science. A main reason for introducing modules is the belief that behaviorist and associationist theorizing about the mind cannot explain our learning abilities on a broad scale, the way our cognitive states are content-sensitive, and how our mental abilities differ from one another.

Fodor and Carruthers are both functionalists and naturalists about the mind, yet each devises a rather different account. As we have shown, this is mainly due to the diverging naturalist paradigms operating in the background: While Fodor is arguing in the spirit of "good old-fashioned artificial intelligence,"[68] conceiving of input-systems as inference-producing encapsulated mechanisms, Carruthers applies adaptationist notions to the mental structure and therefore utilizes very different design principles.

The modularity debate is not only a debate that is situated *within* a naturalistic framework, it is also a debate *about* that very naturalistic framework. As we have shown, Fodor's moderate modularity thesis

---

68 This label was coined by John Haugeland, *Artificial Intelligence: The Very Idea* (Cambridge, MA: MIT Press, 1986), 112.

holds that the central cognitive systems are nonmodular and that cognitive science has no means of understanding the global processing of nonmodular systems. Yet since cognitive science is the only game in town when it comes to a naturalistic understanding of the mind, the prospects for naturalism look quite bleak. From a broader historical perspective, the debate initiated by Fodor is at its core a debate about our understanding of the higher mental faculties that are characteristic of the *human* mind. Are these faculties—the higher-level cognitive systems—to be explained within the naturalistic framework of cognitive science and evolutionary theory? Descartes thought that, in contrast to the lower faculties, the higher faculties of will and intellect cannot be explanatory targets within a general science of the material world. Fodor seems to agree. Evolutionary psychologists on the other hand aim to develop an account of the mind that is modular all the way up, since evolution forces us to conceive of the mind as a system that developed incrementally. Evolutionary psychologists therefore blur the distinction between lower and higher faculties. One might be skeptical about the kind of adaptationist reasoning that justifies massive modularity approaches. Yet it remains an interesting fact that evolutionary psychologists have started to doubt the unity of the higher faculties and to suggest splitting them up into specialized modules *because* they believe that the evolution of a global system such as "reason" is close to impossible in an evolutionary framework of explanation.

Current embodied approaches do not question the naturalist framework and stick with the idea that the mind developed incrementally, but they argue from a scientific perspective that assumes that there are no modules akin to task-specific locatable areas in the brain that realize mental abilities. Instead, embodied and dynamicist theories suggest that naturalist approaches to the mind fare better if they describe cognitive systems as systems that become established in the interaction between an embodied agent and a structured environment. In this way, they rekindle pragmatist ideas about the primacy of action and revive them in the framework of cognitive science. While the modularity debate has been dominated by the question about whether the faculties can be understood in a naturalist framework, it seems that most sections of cognitive science are pretty efficient nowadays in describing the mechanisms that realize cognitive abilities without assuming modules or anything faculty-like in the background.[69]

---

69 The authors would like to thank Jesse Prinz, Sven Walter, Dominik Perler, Stephan Schmid, and Thomas Jacobi for very helpful comments on earlier versions of this paper. They would also like to acknowledge financial support from the Swiss National Science Foundation (professorship grant PP00P1_139037).

Reflection
FACULTIES AND NEUROENHANCEMENT
Saskia K. Nagel

Enhancing faculties has always been part of human striving and pursuit. Influencing mental states by various means, including psychoactive substances, to achieve different perceptions, feelings, thoughts, and faculties is crucial to human cultures. A fairly recent development is the specific usage of psychopharmacological agents that have been developed for treatment purposes to enhance mental faculties. So-called neuroenhancement is the use of medical and technological means beyond their applications in classical therapy to improve faculties in healthy persons.[1] Neuroenhancement is about becoming "better at experiencing the world through all of our senses, better at assimilating and processing what we experience, better at remembering and understanding things, stronger, more competent, more of everything we want to be."[2] The most relevant way nowadays to enhance faculties is the usage of psychoactive drugs to influence physical faculties such as muscle strength, sleep or sexuality, and cognitive capacities such as attention, learning, vigilance, and memory, as well as affective states

---

[1] Erik Parens, ed., *Enhancing Human Traits: Ethical and Social Implications* (Washington, DC: Georgetown University Press, 1998).

[2] John Harris, *Enhancing Evolution: The Ethical Case for Making People Better* (Princeton: Princeton University Press, 2007).

such as emotions and moods. The usage of drugs for enhancement reaches to social and moral competences.³ It is notable, and often overlooked, that a distinction between cognitive and affective capacities, not to mention between those and physical capacities, cannot be neatly drawn, and interventions thus can never be as selective as assumed and most probably involve manifold trade-offs.

The growing trend to use enhancement interventions challenges boundaries in medicine and wider society between health and illness, between treatment and enhancement, and between normality and abnormality. As there is no a priori clear-cut dividing line between treatment and enhancement, the complex continuum with many nuances in the field of medical services renders regulatory procedures difficult. What is conceived as enhancement and how it is evaluated is strongly related to what is conceived as the nature of psychological traits, personal identity, and normality, and to how human nature and the value of it are understood, as well as to what preferences exist as to how to realize one's life project. However, while notoriously difficult to define, a distinction between treatment and enhancement is helpful for normative guidance and decision-making and necessary for dealing with questions in the health care system, for example on rationalization and prioritization.

A particularly urgent issue is the growing trend of intervening in the nervous systems of healthy children and adolescents by psychopharmacological means. Pediatric neuroenhancement is an unsettled and value-laden practice with its own developmental, ethical, social, and legal implications.⁴ While usage of amphetamines

---

3 Saskia K. Nagel, *Ethics and the Neurosciences: Ethical and Social Consequences of Neuroscientific Progress* (Paderborn: Mentis, 2010).
4 William D. Graf, Saskia K. Nagel, Leon G. Epstein, Geoffrey Miller, Ruth Nass, and Dan Larriviere, "Pediatric Neuroenhancement: Ethical, Legal, Social, and Neurodevelopmental Implications," *Neurology* 80 (2013): 1251–60.

and stimulants has a long history in particular with wide employment for military purposes, the spreading of enhancement interventions that can be witnessed in the last decade in the vulnerable group of children and adolescents needs special attention. The number of prescriptions for stimulants and psychotropic medications has substantially increased among children and adolescents over the past twenty years,[5] and there is evidence for raising off-label usage for enhancement purposes.[6] The increase in availability and consumption of the substances used for enhancement such as methylphenidate (e.g. with the brand name Ritalin) is paralleled by an extension of licit and illicit channels of diversion, that is, the channeling of regulated pharmaceuticals. Notably, the ADHD diagnosis is inexorably associated with a direct medical intervention. Thus, potential reasons for the increase in demand, production, and consumption of stimulants, in addition to the inclusion of milder ADHD diagnoses as requiring medical treatment, could be the illegal diversion of controlled medication to consumers seeking stimulants as neuroenhancements.[7]

While the prevalence of enhancement interventions is on the rise, it is still unclear what is enhanced with those interventions.[8] The popular opinion that methylphenidate enhances attention and concentration in healthy consumers does not find a solid scientific basis. Controlled studies on efficacy and—importantly—on long-term safety are rare. However, notably, the placebo effect might be particularly relevant and lead to enhancement effects, or the

---

5 International Narcotics Control Board, *Report of the International Narcotics Control Board for 2012* (Vienna: United Nations Publications, 2013).

6 Neuroenhancement should not be confused with medical interventions with potential beneficial effects on neurological disorders such as attention deficit hyperactivity disorder (ADHD).

7 William D. Graf, Geoffrey Miller, and Saskia K. Nagel, "Addressing the Problem of ADHD Medication as Neuroenhancements," *Expert Review in Neurotherapeutics* 14 (2014): 569–81.

8 Dimitris Repantis, Peter Schlattmann, Oona Laisney, and Isabella Heuser, "Modafinil and Methylphenidate for Neuroenhancement in Healthy Individuals: A Systematic Review," *Pharmacological Research* 62 (2010): 187–206.

perceptions thereof, in users. At this point in time, expectations regarding the effectiveness of stimulant drugs for enhancement purposes probably exceed their actual effects. Furthermore, dissecting which faculty is enhanced is difficult if not impossible. The expectation that stimulant medication in healthy users selectively improves concentration or the focus on a specific task, such that it would be useful as "study aid," cannot easily be tested by classical performance measurements. Cognitive functions are not isolated modules that can be targeted and tested one by one, and detailing how the plethora of aspects such as motivation, wakefulness, vigilance, memory, concentration, and mood interact in a specific test is difficult. Thus, the expected benefits of stimulants as enhancement of specific faculties still lack solid scientific foundation.

The physiological and psychological short- and long-term effects of psychotropics on the developing human brain have not yet been systematically investigated in humans, such that caution in usage is strongly advisable at the present time: one cannot judge whether the benefits outweigh the harms. Another important aspect is the role of physicians. Today, the drugs that are most widely used for pediatric neuroenhancement are prescription drugs that require contact with a physician, if they are not obtained illicitly. Traditionally, the doctor-child-parent relationship strongly relies on the fiduciary responsibility of the physician to promote and protect a child's health. Prescription practices beyond treatment of disorders should be handled with particular caution. Furthermore, the communication process about a prescription should clearly inform about risks and uncertainties and emphasize the limited evidence regarding the efficacy and safety of medications for neuroenhancement that are prescribed for healthy children and adolescents. Physicians and parents should screen for coercion at various levels, whether it is parents themselves who are coercing children to get good grades in school or be otherwise more successful, teachers and school administrators who seek to control behavior, or peers who are pressuring others, or children who seek "to follow the crowd." In the context of screening for direct or subtle coercion, it is important to rethink the practice of direct-to-consumer advertising that, while providing useful information on health issues, at the same time risks misinterpretation and a commercialization of the doctor-patient relationship, potentially eroding the professional ethos.

Besides practical questions of safety, effectiveness, risk of coercion, and the role of the fiduciary responsibility of the physician, the ethical and social problems related to pediatric neuroenhancement touch central individual and societal interests. Thus, when studying what is ethically at stake when using enhancement technologies in healthy children, one needs to discuss what is considered to be appropriate, what is valued about the natural maturation process and usual childhood behavior, and where society should intervene and draw the line when it comes to enhancement interventions. For the ethical discussion, a central question concerns the question of who decides. The particular role of the child, who is not yet autonomous in the decision-making process, needs special attention. While adults are assumed to be able to decide for themselves, and are protected by the principle of respect for autonomy, children and adolescents lack full decision-making capacities. Decision-making capacities require the abilities to receive and remember information, to engage in mutual questioning and answering, to assess relevant information, and to use information to make and justify a decision. With increasing maturity, autonomy is developing. Adolescents gradually take on more and more responsibility as their decision-making capacities mature. This is exemplified by an adolescent's capacity for assent even before informed consent matures and fully develops. It is difficult to determine a threshold for autonomy such that the relevant capabilities for autonomy can be assessed.[9] This problem is usually solved pragmatically, but still, these solutions are often unsatisfactory if cases become more complicated. Notably, for this consideration, children and adolescents possess anticipatory autonomy rights or "rights to an open future," that is, rights to reach maturity with as many open options, opportunities, and advantages as possible.[10] While one might argue that neuroenhancement extends and improves the possible futures of a child and thus should be embraced and could even be understood as being morally obligatory, the problem remains that it is the parents who restrict or otherwise shape the child's future and thereby potentially the child's development of autonomy. Moreover, close scrutiny and empirical work is desirable to study whether autonomy and the ability to develop an appropriate conception of personal responsibility for behavior is enhanced or diminished by psychopharmacological enhancement interventions.[11] Undoubtedly, parents act as stewards of their children to enable them to live their lives, while influencing their character development, setting constraints, and maintaining contingencies for an open future. Parents often have aspirations for their children to attain higher levels of performance. They shape their child's personal development through encouragement of educational activities, of music and sports.

Through advances in neurotechnologies and the presence of psychopharmacological drugs, the means and quality of

---

9 William D. Graf, Saskia K. Nagel, Leon G. Epstein, Geoffrey Miller, Ruth Nass, and Dan Larriviere, "Pediatric Neuroenhancement: Ethical, Legal, Social, and Neurodevelopmental Implications," *Neurology* 80 (2013): 1251–60.

10 Joel Feinberg, "A Child's Right to an Open Future," in *Whose Child? Parental Rights, Parental Authority and State Power*, edited by William Aiken and Hugh La Follette (Totowa, NJ: Littlefield, Adams, 1980), 124–53.

11 Saskia K. Nagel and Peter B. Reiner, "Autonomy Support to Foster Individuals' Flourishing," *American Journal of Bioethics* 13(6) (2013): 36–37; Ilina Singh, "Clinical Implications of Ethical Concepts: Moral Self-understandings in Children Taking Methylphenidate for ADHD," *Clinical Child Psychology and Psychiatry* 12(2) (2007): 167–82.

interventions during development radically change and potentially lead to increasing pressures to use enhancement means. This likely impacts the child-parent relationship, as parents might seek interventions that they believe will help their own reputations or self-images. This does not seem to be a far-fetched consideration, in view of the much-discussed "helicopter parenting," that is, parents seeing their children as projects that they want to optimize for a competitive society. The availability of means to supposedly selectively influence a child from very early on, extending to prenatal interventions, thus must be carefully accompanied by consideration of the motivations of parents and implications for children, parents, physicians, and others concerned.

Pediatric neuroenhancement has further normative dimensions, with regard to what is valued about childhood, and how the means that are used to intervene in developmental processes matter normatively. The vulnerability of children and adolescents in their maturation process cannot be overestimated. While there is no doubt that caring for children in all possible ways is a value and should be praised, and while it is safe to assume that usually parents only want their children to flourish, the ways parents seek to supposedly improve their children and the motivations why they do so deserve scrutiny. Notably, in the case of psychostimulants that children and adolescents consume, the means that are used to influence faculties have gained a special status in popular discourse in the last years. Stimulants seem to carry an aura of virtue that other drugs do not possess. As these drugs offer culturally desirable traits, such as improved productivity and heightened focus, the risks and implications, as well as more or less subtle messages that are conveyed by the usage, are not considered as much as they should be.

There is little data about dosages and duration of enhancement interventions as well as habits related to them. It is likely that a routine consumption of drugs in healthy children develops into a self-reinforcing process. If children and their surrounding adults and peers perceive drugging healthy children and adolescents as the normal way of life, the self-image of children (who are growing into adults) then might include the consumption of drugs without an underlying disease, that is, for enhancement purposes. Certainly, social attributions in families and wider social contexts play an important role for the self-images of children and for the general evaluation of drug usage in healthy children. The societal context that fosters the development of increased drug consumption by healthy children, adolescents, and adults needs close examination.

It is worthwhile to consider the climate in which the trend of pediatric neuroenhancement can flourish: With the rise of the neurosciences and their popularization, the brain is becoming a project that is to be shaped and that seems to be the key to feared or hoped-for futures. People search for advantages in their social environment, both at work and in private, and parents do the same for their children, with the best intentions. This climate encompasses all age groups and extends to antiageing efforts for the elderly. Even though at this point in time drugs are not yet perceived as being as normal as breakfast cereals, and even though there seem to be cultural differences in how the usage of enhancement means is judged, studies to investigate the influence of neuroenhancement drugs on development and on self-perception are needed to allow a timely debate.[12] Similarly, studies on public attitudes toward neuroenhancement will shed light on what people outside academia think about neuroenhancement.[13] Moreover, the debate on enhancement needs to challenge the reasoning that the neurosciences can offer neurochemical solutions to social problems. A close look at pediatric neuroenhancement requires giving up

---

12 Ilina Singh, "Not Robots: Children's Perspectives on Authenticity, Moral Agency and Stimulant Drug Treatments," *Journal of Medical Ethics* 39 (2013): 359–66.
13 Nicholas S. Fitz, Richard Nadler, Praveena Manogaran, Eugene W. J. Chong, and Peter B. Reiner, "Public Attitudes toward Cognitive Enhancement," *Neuroethics* 7 (2014): 173–88.

focusing on the brain but instead understanding the person, the child or adolescent, with a developing brain and body in his or her social context.[14] Caution should be exercised with respect to treating healthy children and adolescents with drugs and thereby potentially harming them, with respect to the risk of medicalizing social problems, and with respect to seeing children as mere brains that just need to be calibrated correctly to function.[15]

---

14 Thomas Fuchs, *Das Gehirn—ein Beziehungsorgan: Eine phänomenologisch-ökologische Konzeption* (Stuttgart: Kohlhammer Verlag, 2009); Walter Glannon, "Our Brains Are Not Us," *Bioethics* 23 (2009): 321–29.

15 I acknowledge the Sievert Stiftung für Wissenschaft und Kultur (S248/10006/2013) for its financial support.

Bibliography

LITERATURE BEFORE 1900

ʿĀmirī, Abū l-Ḥasan Muḥammad Ibn Yūsuf al-. *A Muslim Philosopher on the Soul and Its Fate.* Edited and translated by Everett K. Rowson. New Haven: American Oriental Society, 1988 (= *Amad*).

Aristotle. *Ethica Nicomachea.* Edited by I. Bywater. Oxford: Oxford University Press, 1897.

Aristotle. *De anima.* Edited by W. D. Ross. Oxford: Oxford University Press, 1956.

Aristotle. *Metaphysica.* Edited by W. Jaeger. Oxford: Oxford University Press, 1957.

Arnauld, Antoine. *Des vrayes et des fausses idées: Contre ce qu'enseigne l'Auteur de La recherche de la vérité.* Paris: Fayard, 1986.

Augustine. *City of God.* Translated by John Healy. Edinburgh: John Grant, 1909.

Descartes, René. *Oeuvres.* Edited by Charles Adam and Paul Tannery. Paris: Vrin, 1983–91.

Descartes, René. *The Philosophical Writings.* Edited and translated by John Cottingham, Robert Stoothoff, and Dugald Murdoch (vols. 1–2), and Anthony Kenny (vol. 3). Cambridge: Cambridge University Press, 1984–91.

Descartes, René. *The World and Other Writings*. Edited and translated by Stephen Gaukroger. Cambridge: Cambridge University Press, 2004.

Euripides. *Medea*. Edited by Donald J. Mastronarde. Cambridge: Cambridge University Press, 2002.

Fārābī, Abū Naṣr al-. *Falsafat Arisṭūṭālīs*. Edited by Mushin Mahdi. Beirut: Dār Majallat Shiʿr, 1961. English: *Alfarabi's Philosophy of Plato and Aristotle*. Translated by Muhsin Mahdi. Ithaca: Cornell University Press, 1962.

Fārābī, Abū Naṣr al-. *On the Perfect State* [Mabādiʾ ārāʾ ahl al-madīnat al-fāḍilah]. Edited and translated by Richard Walzer. Oxford: Oxford University Press, 1985 (= *Madīna*).

Fichte, Johann Gottlieb. *Recension des Aenesidemus oder über die Fundamente der vom Herrn Prof. Reinhold in Jena gelieferten Elementarphilosophie* (1792). In *Fichtes Werke*, edited by Immanuel Hermann Fichte, vol. 1, 1–16. Berlin: Veit und Comp., 1845/6.

Fichte, Johann Gottlieb. *Über den Begriff der Wissenschaftslehre oder der sogenannten Philosophie* (1794). In *Fichtes Werke*, edited by Immanuel Hermann Fichte, vol. 1, 27–82. Berlin: Veit und Comp., 1845/6. English: *Concerning the Concept of the Wissenschaftslehre*. In *Fichte: Early Philosophical Writings*, edited and translated by Daniel Breazeale. Ithaca: Cornell University Press, 1993.

Fichte, Johann Gottlieb. *Grundlage der gesammten Wissenschaftslehre* (1794). In *Fichtes Werke*, edited by Immanuel Hermann Fichte, vol. 1, 83–328. Berlin: Veit und Comp., 1845/6. English: *Foundations of the Entire Wissenschaftslehre*. In *The Science of Knowledge*. Edited and translated by Peter Lauchlan Heath and John Lachs. Cambridge: Cambridge University Press, 1982.

Fichte, Johann Gottlieb. *Grundriss des Eigenthümlichen der Wissenschaftslehre* (1795). In *Fichtes Werke*. Edited by Immanuel Hermann Fichte, vol. 1, 329–411. Berlin: Veit und Comp., 1845/6. English: *Outline of the Distinctive Character of the Wissenschaftslehre with Respect to the Theoretical Faculty*. In *Fichte: Early Philosophical Writings*. Edited and translated by Daniel Breazeale. Ithaca: Cornell University Press, 1993.

Fichte, Johann Gottlieb. *Versuch einer neuen Darstellung der Wissenschaftslehre* (1797). In *Fichtes Werke*. Edited by Immanuel Hermann Fichte, vol. 1, 519–34. Berlin: Veit und Comp., 1845/6. English: *An Attempt at a New Presentation of the Wissenschaftslehre (1797/8)*. Ch. 1 in *Introductions to the Wissenschaftslehre and Other Writings (1797–1800)*, edited and translated by Daniel Breazeale. Indianapolis: Hackett, 1994.

Ficino, Marsilio. *Platonic Theology*. 6 vols. English translation by Michael J. B. Allen. Latin text edited by James Hankins. Cambridge, MA: Harvard University Press, 2001–6.

Galen, Claudius of Pergamon. *Scripta Minora*. Edited by Iwan von Müller. Vol. 2. Leipzig: Teubner, 1891.

Galen, Claudius of Pergamon. *On the Natural Faculties*. With an English translation by Arthur J. Brock. Cambridge, MA: Harvard University Press, 1916.

Galen, Claudius of Pergamon. *Psychological Writings*. Edited by P. N. Singer and translated with introductions and notes by Vivian Nutton, Daniel Davies, and P. N. Singer, with the collaboration of Piero Tassinari. Cambridge: Cambridge University Press, 2014.

Gall, Franz Joseph, "Des Herrn Dr. F. J. Gall Schreiben ueber seinen geendigten Prodromus ueber die Verrichtungen des Gehirns der Menschen und der Thiere, an Herrn Jos. Fr. von Retzer." *Wielands Neuer Teutscher Merkur* 12 (1798): 311–35.

Ghazālī, Abū Ḥāmid Muḥammad al-. *Iḥyāʾ ʿulūm al-dīn*. 16 vols. Cairo: Lajnat Nashr al-Thaqāfa al-Islāmiyya, 1356–57.

Ghazālī, Abū Ḥāmid Muḥammad al-. *Al-Munqidh min al-ḍalāl*. Edited by Kamil ʿAyyād and Jamil Salība. 2nd ed. Beirut: Commission libanaise pour la traduction des chefs-d'œuvre, 1969. English: *Deliverance from Error (al-Munqidh min al-Dalal) and Five Key Texts*. Translated by Richard J. McCarthy. Louisville, KY: Fons Vitae, 1999.

Hobbes, Thomas. *Leviathan*. Edited by Richard Tuck. Cambridge: Cambridge University Press, 1991.

Homer. *Odyssey*. Edited by Thomas W. Allen. Opera III–IV. Oxford: Oxford University Press, 1919.

Homer. *Iliad*. Edited by David B. Munro and Thomas W. Allen. Opera I–II. Oxford: Oxford University Press, 1920.

Hume, David. *An Enquiry concerning Human Understanding*. Edited by Tom L. Beauchamp. Oxford: Oxford University Press, 1999.

Hume, David. *A Treatise of Human Nature*. Edited by David F. Norton and Mary J. Norton. Oxford: Oxford University Press, 2000.

Ḥunayn Ibn Isḥāq. *Risāla fī l-farq bayna l-rūḥ wa-l-nafs*. Edited by Louis Cheikho. *Al-Mashriq* 14 (1911): 94–109 (= *Farq*).

Ibn Bājja, Abū Bakr Muḥammad. *Rasāʾil Ibn Bājja l-ilāhiyya/Opera metaphysica*. Edited by Majid Fakhrī. Beirut: Dār al-nahār li-l-nashr, 1968.

Ibn Bājja, Abū Bakr Muḥammad. *Kitāb al-nafs*. Edited by M. al-Maʿṣūmī. 2nd ed. Beirut: Dār Ṣādir, 1992. English: *Ibn Bajjah's "Ilm al-Nafs."* Translated by M. S. Hasan Maʿsumi. New Delhi: Kitab Bhavan, 1992 (= *Nafs*).

Ibn Rushd, Abū l-Walīd Muḥammad Ibn Aḥmad. *Tahāfut al-tahāfut*. Edited by Maurice Bouyges. Beirut: Imprimerie Catholique, 1930. English: *The Incoherence of the Incoherence*. Translated by Simon Van Den Bergh. London: Luzac, 1954.

Ibn Rushd, Abū l-Walīd Muḥammad Ibn Aḥmad. *Talkhīṣ* [sic] *kitāb al-nafs*. Edited by Ahmad F. al-Ahwānī. Cairo: Maktaba l-nahḍa l-miṣriyya, 1950 (= *Mukhtaṣar*). English: *Epitome of "De anima."* Translated by Deborah L. Black. Unpublished translation. Available at http://individual.utoronto.ca/dlblack/translations.html.

Ibn Rushd, Abū l-Walīd Muḥammad Ibn Aḥmad. *Commentarium magnum in Aristotelis "De anima" libros*. Edited by F. Stuart Crawford. Cambridge, MA: Mediaeval Academy of America, 1953. English: *Long Commentary on the "De Anima" of Aristotle*. Translated by Richard C. Taylor. New Haven: Yale University Press, 2009.

Ibn Rushd, Abū l-Walīd Muḥammad Ibn Aḥmad. *Middle Commentary on "De anima."* Edited and translated by Alfred Ivry. Provo, UT: Brigham Young University Press, 2002 (= *Talkhīṣ*).

Ibn Sīnā, Abū ʿAlī. *Al-Ishārāt wa-l-tanbīhāt*. Edited by Jacques Forget. Leiden: Brill, 1892.

Ibn Sīnā, Abū ʿAlī. *Al-Najāt*. Cairo: Muḥyī l-Dīn al-Ṣabrī l-Kurdī, 1938. English: *Avicenna's Psychology: An English Translation of "Kitāb al-Najāt," Book II, Chapter VI*. Translated by Fazlur Rahman. London: Oxford University Press, 1952.

Ibn Sīnā, Abū ʿAlī. *Avicenna's "De Anima." Being the Psychological Part of Kitāb al-Shifāʾ*. Edited by Fazlur Rahman. Oxford: Oxford University Press, 1959. Latin: *Avicenna Latinus: Liber de anima seu sextus de naturalibus*. Edited by Simone Van Riet. 2 vols. Louvain: Peeters, 1968–72.

Ibn Sīnā, Abū ʿAlī. *The Metaphysics of the Healing*. Edited and translated by Michael E. Marmura. Provo, UT: Brigham Young University Press, 2005 (= *Ilāhiyyāt*).

Ibn Sīnā, Abū ʿAlī. *The Physics of the Healing*. Edited and translated by Jon McGinnis. Provo, UT: Brigham Young University Press, 2009.

Ibn Ṭufayl, Abū Bakr. *Ḥayy Ibn Yaqẓān*. Edited by Léon Gauthier. 2nd ed. Beirut: Imprimerie Catholique, 1936. English: *Hayy Ibn Yaqzan*. Translated by Lenn Evan Goodman. Chicago: University of Chicago Press, 2009.

John Duns Scotus. *Quaestiones in librum secundum Sententiarum*. Edited by Ludovicus Wadding. Opera omnia 13. Paris: Vivès, 1893.

Kant, Immanuel. *Gesammelte Schriften*. Edited by Königlich Preußische Akademie der Wissenschaften. 22 vols. Berlin: de Gruyter, 1902.

Kant, Immanuel. *Critique of Pure Reason*. Edited and translated by Allan Wood and Paul Guyer. Cambridge: Cambridge University Press, 1999.

Kant, Immanuel. *Critique of the Power of Judgment*. Edited by Paul Guyer. Translated by Paul Guyer and Eric Matthews. Cambridge: Cambridge University Press, 2002.

Leibniz, Gottfried Wilhelm. *On Nature Itself.* In *Philosophical Essays*, edited and translated by Roger Ariew and Daniel Garber, 155–67. Indianapolis: Hackett, 1989.

Leibniz, Gottfried Wilhelm. "Preface to the *New Essays*." In *Philosophical Essays*, edited and translated by Roger Ariew and Daniel Garber, 291–306. Indianapolis: Hackett, 1989.

Locke, John. *An Essay concerning Human Understanding.* Edited by Peter H. Nidditch. Oxford: Clarendon Press, 1975.

Malebranche, Nicolas. *Oeuvres complètes.* Edited by André Robinet. Paris: Vrin, 1958–84.

Malebranche, Nicolas. *The Search after Truth.* Translated by Thomas M. Lennon and Paul J. Olscamp. Cambridge: Cambridge University Press, 1997.

Miskawayh, Abū ʿAlī Aḥmad Ibn Muḥammad. *Tahdhīb al-akhlāq.* Edited by Qusṭanṭīn Zurayq. Beirut: American University of Beirut, 1966. English: *The Refinement of Character.* Translated by Constantine K. Zurayk. Beirut: American University of Beirut, 1968.

Petrus Hispanus. *Tractatus called afterwards Summule Logicales.* Edited by Lambert M. de Rijk. Assen: Van Gorcum, 1972.

Plato. *Opera.* Edited by J. Burnet. Oxford: Clarendon Press, 1905.

Plato. *Complete Works.* Edited by John M. Cooper. Indianapolis and Cambridge: Hackett, 1997.

Porphyry. *Isagoge.* Edited by Adolf Busse. In *Commentaria in Aristotelem Greca.* Vol. IV 4, pt. -1. Berlin: Reimer, 1887.

Rasāʾil Ikhwān al-Ṣafāʾ. 4 vols. Beirut: Dār Ṣādir, 2004.

Rāzī, Fakhr al-Dīn al-. *Kitāb al-nafs wa-l-rūḥ wa sharḥ qawā-humā.* Edited by Muḥammad Ṣaghīr Ḥasan al-Maʿṣūmī. Islamabad: Islamic Research Institute, 1968. English: *Imām Rāzī's "ʿIlm al-akhlāq."* Translated by M. S. al-Maʿṣūmī. Islamabad: Islamic Research Institute, 1969.

Ribot, Théodule. *La psychologie anglaise contemporaine.* Paris: Librairie philosophique de Ladrange, 1870.

Ribot, Théodule. *L'hérédité: Étude psychologique sur ses phénomènes, ses lois, ses causes, ses conséquences.* Paris: Librairie philosophique de Ladrange, 1873.

Ribot, Théodule. *Les maladies de la mémoire.* Paris: Félix Alcan, 1881.

Ribot, Théodule. *Les maladies de la volonté.* Paris: Félix Alcan, 1882.

Ribot, Théodule. *Les maladies de la personnalité.* Paris; Félix Alcan, 1885.

Selby-Bigge, Lewis A. *British Moralists: Selections from Writers Principally of the Eighteenth Century.* 2 vols. Oxford: Clarendon Press, 1897.

Shakespeare, William. *A Midsummer Night's Dream.* Edited by Peter Holland. Oxford: Clarendon Press, 1994.

Shakespeare, William. *Twelfth Night, or What You Will*. Edited by Roger Warren and Stanley Wells. Oxford: Clarendon Press, 1994.

Spenser, Edmund. *The Faerie Queene: Book Two*. Edited by Erik Gray. Indianapolis: Hackett, 2006.

Spinoza, Baruch de. *The Collected Works*. Edited and translated by Edward Curley. Vol. 1. Princeton: Princeton University Press, 1985.

Stobaeus. *Eclogae Physicae et Ethicae*. Edited by Kurt Wachsmuth. Leipzig: Teubner, 1884.

Suárez, Francisco. *De anima*. Edited by A. D. M. André. Opera omnia 3. Paris: Vivès, 1856.

Suárez, Francisco. *Disputationes Metaphysicae*. Edited by Carolus Berton. Opera omnia 25–26. Paris: Vivès, 1861.

Suárez, Francisco. *De anima*. Edited by Salvador Castellote. 3 vols. Madrid: X. Zubrini and Editorial Labor, 1978–91.

Thomas Aquinas. *Quaestiones disputatae*. Edited by Raymundus Spiazzi et al. Turin: Marietti, 1949.

Thomas Aquinas. *Summa theologiae*. Edited by Peter Caramello. Turin: Marietti, 1952.

Thomas Aquinas. *Sentencia libri De anima*. Editio Leonina XLV/1. Rome: Commissio Leonina and Vrin, 1984.

William of Auvergne. *De anima*. Opera omnia 2. Paris: A. Pralard, 1674. Reprint, Frankfurt: Minerva, 1963.

William of Ockham. *Opera Philosophica et Theologica*. Edited by G. Gál et al. St. Bonaventure: Franciscan Institute, 1967–88.

Wright, Thomas. *The Passions of the Minde (1601)*. Reprint, Hildesheim: Olms, 1973.

Wundt, Wilhelm. "Die Aufgaben der experimentellen Psychologie." In *Unsere Zeit III*. Leipzig: F. A. Brockhaus, 1882.

Yaḥyā Ibn 'Adī. *The Reformation of Morals*. Edited and translated by Sidney H. Griffith. Provo, UT: Brigham Young University Press, 2002.

LITERATURE AFTER 1900

Adams, Robert M. "Malebranche's Causal Concepts." In *The Divine Order, the Human Order, and the Order of Nature: Historical Perspectives*, edited by Eric Watkins, 67–104. Oxford: Oxford University Press, 2013.

Adams McCord, Marilyn. *William Ockham*. Notre Dame, IN: Notre Dame University Press, 1987.

Adamson, Peter. "The Kindian Tradition: The Structure of Philosophy in Arabic Neoplatonism." In *The Libraries of the Neoplatonists*, edited by Cristina D'Ancona, 351–70. Leiden: Brill, 2010.

Adamson, Peter, and Peter Pormann. "More Than Heat and Light: Miskawayh's *Epistle on Soul and Intellect*." *Muslim World* 102 (2012): 478–524.

Allison, Henry. "Kant's Antinomy of Teleological Judgment." In *Kant's Critique of the Power of Judgment*, edited by Paul Guyer, 219–36. Oxford: Rowman and Littlefield, 2003.

Apel, Karl-Otto. *Charles S. Peirce: From Pragmatism to Pragmaticism.* Amherst: University of Massachusetts Press, 1981.

Baier, Annette C. *A Progress of Sentiments: Reflections on Hume's "Treatise."* Cambridge, MA: Harvard University Press, 1991.

Bakker, Paul J. J. M., and Johannes M. M. H. Thijssen, eds. *Mind, Cognition and Representation: The Tradition of Commentaries on Aristotle's De anima.* Aldershot: Ashgate, 2007.

Barrett, Harold Clark. "Enzymatic Computation and Cognitive Modularity." *Mind and Language* 20 (2005): 259–87.

Barrett, Harold Clark, and Robert Kurzban. "Modularity in Cognition: Framing the Debate." *Psychological Review* 113 (2006): 628–47.

Bechtel, William. "Explanation: Mechanism, Modularity, and Situated Cognition." In *The Cambridge Handbook of Situated Cognition*, edited by Robbins Ayede, 155–70. Cambridge: Cambridge University Press, 2009.

Beere, Jonathan B. *Doing and Being: An Interpretation of Aristotle's Metaphysics Theta.* Oxford: Oxford University Press, 2009.

Beiser, Frederick C. *German Idealism: The Struggle against Subjectivism 1781–1801.* Cambridge, MA: Harvard University Press, 2002.

Bennett, M. R., and Hacker, P. M. S. *Philosophical Foundations of Neuroscience.* Oxford: Blackwell, 2003.

Biard, Joël. "Diversité des fonctions et unité de l'âme dans la psychologie péripatéticienne (XIVe–XVIe siècle)." *Vivarium* 46 (2008): 342–67.

Black, Deborah. "Intentionality in Medieval Arabic Philosophy." *Quaestio* 10 (2010): 65–81.

Borg, Emma. *Minimal Semantics.* Oxford: Oxford University Press, 2004.

Boureau, Alain. *De vagues individus: La condition humaine dans la pensée scolastique.* Paris: Les Belles Lettres, 2008.

Breland, Keller, and Marian Breland. "The Misbehavior of Organisms." *American Psychologist* 16 (1961): 681–84.

Brownlow, Frank W. *Shakespeare, Harsnett, and the Devils of Denham.* Newark: University of Delaware Press, 1993.

Bruner, Jerome. "On Perceptual Readiness." In *Beyond the Information Given*, edited by Jeremy M. Anglin, 7–42. New York: Norton, 1973.

Buckle, Stephen. *Hume's Enlightenment Tract: The Unity and Purpose of an Enquiry concerning Human Understanding.* Oxford: Clarendon Press, 2001.

Buller, David. *Adapting Minds*. Cambridge, MA: MIT Press, 2005.
Buller, David, and Valerie Hardcastle. "Evolutionary Psychology, Meet Developmental Neurobiology: Against Promiscuous Modularity." *Brain and Mind* 1 (2000): 302–25.
Burnyeat, Myles. "The Truth of Tripartition." *Proceedings of the Aristotelian Society* 106 (2006): 1–23.
Carruthers, Peter. *The Architecture of the Mind*. Oxford: Oxford University Press, 2006.
Carruthers, Peter. "The Case for Massively Modular Models of the Mind." In *Contemporary Debates in Cognitive Science*, edited by Robert Stainton, 3–21. Oxford: Blackwell, 2006.
Chomsky, Noam. "Review of Verbal Behavior." *Language* 35 (1959): 26–58.
Chomsky, Noam. *Aspects of the Theory of Syntax*. Cambridge, MA: MIT Press, 1964.
Chomsky, Noam. "Rules and Representations." *Behavioral and Brain Sciences* 3 (1980): 1–15.
Churchland, Paul. "Perceptual Plasticity and Theoretical Neutrality: A Reply to Jerry Fodor." *Philosophy of Science* 55 (1988): 167–87.
Clark, Andy. *Supersizing the Mind: Embodiment, Action, and Cognitive Extension*. Oxford: Oxford University Press, 2008.
Clarke, Edwin, and L. S. Jacyna. *Nineteenth-Century Origins of Neuroscientific Concepts*. Berkeley: University of California Press, 1987.
Colbert, Charles. *A Measure of Perfection: Phrenology and the Fine Arts in America*. Chapel Hill: University of North Carolina Press, 1997.
Colombetti, Giovanna. *The Feeling Body: Affective Science Meets the Enactive Mind*. Cambridge, MA: MIT Press, 2014.
Cooter, Roger. *The Cultural Meaning of Popular Science: Phrenology and the Organization of Consent in Nineteenth-Century Britain*. Cambridge: Cambridge University Press, 1984.
Corcilius, Klaus. *Streben und Bewegen: Aristoteles' Theorie der animalischen Ortsbewegung*. Berlin: de Gruyter, 2008.
Corcilius, Klaus, and Pavel Gregoric. "Separability vs. Difference: Parts and Capacities of the Soul in Aristotle." *Oxford Studies in Ancient Philosophy* 39 (2010): 81–119.
Corcilius, Klaus, and Dominik Perler, eds. *Partitioning the Soul: Debates from Plato to Leibniz*. Berlin: de Gruyter, 2014.
Cosmides, Leda, and John Tooby. "Cognitive Adaptations for Social Exchange." In *The Adapted Mind*, edited by Jerome Barkow, Leda Cosmides, and John Tooby, 163–228. Oxford: Oxford University Press, 1992.

Courtenay, William J. *Ockham and Ockhamism: Studies in the Dissemination and Impact of His Thought.* Leiden: Brill, 2008.
Cowie, Fiona. "Us, Them and It: Modules, Genes, Environments and Evolution." *Mind and Language* 23 (2008): 284–92.
Cross, Richard. "Ockham on Part and Whole." *Vivarium* 37 (1999): 143–67.
Cundall, Michael K. "Rethinking the Divide: Modules and Central Systems." *Philosophia* 34 (2006): 379–93.
D'Ancona, Cristina. "Degrees of Abstraction in Avicenna: How to Combine Aristotle's *De Anima* and the Enneads." In *Theories of Perception in Medieval and Early Modern Philosophy*, edited by Simo Knuuttila and Pekka Kärkkäinen, 47–71. Dordrecht: Springer, 2008.
Davies, John Dunn. *Phrenology: Fad and Science: A 19th–Century American Crusade.* New Haven: Yale University Press, 1955.
De Boer, Sander W. *The Science of the Soul: The Commentary Tradition on Aristotle's De anima, c. 1260–c. 1360.* Leuven: Leuven University Press, 2013.
Della Rocca, Michael. *Spinoza.* London: Routledge, 2008.
Des Chene, Dennis. *Life's Form: Late Aristotelian Conceptions of the Soul.* Ithaca: Cornell University Press, 2000.
Diwald, Susanne. "Die Seele und ihre geistigen Kräfte: Darstellung und philosophiegeschichtlicher Hintergrund im K. Ikhwān aṣ-Ṣafā." In *Islamic Philosophy and the Classical Tradition*, edited by Samuel M. Stern et al., 49–61. Oxford: Cassirer, 1972.
Donini, Pierluigi. "Psychology." In *The Cambridge Companion to Galen*, edited by Richard J. Hankinson, 184–209. Cambridge: Cambridge University Press, 2008.
Druart, Thérèse-Anne. "The Human Soul's Individuation and Its Survival after the Body's Death: Avicenna on the Causal Relation between Body and Soul." *Arabic Sciences and Philosophy* 10 (2000): 259–73.
Dummett, Michael. *Origins of Analytical Philosophy.* London: Duckworth, 1993.
Endress, Gerhard. "Platonizing Aristotle: The Concept of 'Spiritual' (rūḥānī) as a Keyword of the Neoplatonic Strand in Early Arabic Aristotelianism." *Studia graeco-arabica* 2 (2012): 265–79.
Everson, Stephen. *Aristotle on Perception.* Oxford: Clarendon, 1997.
Everson, Stephen. "Epicurean Psychology." In *The Cambridge History of Hellenistic Philosophy*, edited by Keimpe Algra, Jonathan Barnes, Jaap Mansfeld, and Malcolm Schofield, 542–59. Cambridge: Cambridge University Press, 2005.
Faucher, Luc, ed. *The Modularity of Emotions.* Calgary: University of Calgary Press, 2008.
Feinberg, Joel. "A Child's Right to an Open Future." In *Whose Child? Parental Rights, Parental Authority and State Power*, edited by William Aiken and Hugh La Follette, 124–53. Totowa, NJ: Littlefield, Adams, 1980.

Fitz, Nicholas S., Richard Nadler, Praveena Manogaran, Eugene W. J. Chong, and Peter B. Reiner. "Public Attitudes toward Cognitive Enhancement." *Neuroethics* 7 (2014): 173–88.
Fodor, Jerry. *The Language of Thought*. New York: Crowell, 1975.
Fodor, Jerry. *The Modularity of Mind: An Essay on Faculty Psychology*. Cambridge, MA: MIT Press, 1983.
Fodor, Jerry. "Observation Reconsidered." *Philosophy of Science* 51 (1984): 23–43.
Fodor, Jerry. "Precis of the Modularity of Mind." *Behavioral and Brain Sciences* 8 (1985): 1–42.
Fodor, Jerry. *Psychosemantics: The Problem of Meaning in the Philosophy of Mind*. Cambridge, MA: MIT Press, 1987.
Fodor, Jerry. "A Reply to Churchland's 'Perceptual Plasticity and Theoretical Neutrality.'" *Philosophy of Science* 55 (1988): 188–98.
Fodor, Jerry. "Why Should the Mind Be Modular?" In *Reflections on Chomsky*, edited by Alexander L. George, 73–95. Oxford: Blackwell, 1989.
Fodor, Jerry. *The Mind Doesn't Work That Way*. Cambridge, MA: MIT Press, 2000.
Fodor, Jerry. *LOT 2: The Language of Thought Revisited*. Oxford: Oxford University Press, 2008.
Foley, Helene P. *Female Acts in Greek Tragedy*. Princeton: Princeton University Press, 2001.
Förster, Eckart. *The 25 Years of Philosophy: A Systematical Reconstruction*. Cambridge, MA: Harvard University Press, 2012.
Fortenbaugh, William W. *Aristotle's Practical Side: On His Psychology, Ethics, Politics, and Rhetoric*. Leiden: Brill, 2006.
Fóti, Véronique. "Descartes' Intellectual and Corporeal Memories." In *Descartes' Natural Philosophy*, edited by Stephen Gaukroger, John Schuster, and John Sutton, 591–603. London: Routledge, 2000.
Franks, Paul. *All or Nothing: Systematicity, Transcendental Arguments, and Skepticism in German Idealism*. Cambridge, MA: Harvard University Press, 2005.
Fuchs, Thomas. *Das Gehirn—ein Beziehungsorgan: Eine phänomenologisch-ökologische Konzeption*. Stuttgart: Kohlhammer Verlag, 2009.
Gallistel, Charles. *The Organization of Learning*. Cambridge, MA: MIT Press, 1990.
Garber, Daniel. *Descartes' Metaphysical Physics*. Chicago: University of Chicago Press, 1992.
Garfield, Jay L., ed. *Modularity in Knowledge Representation and Natural-Language Understanding*, Cambridge, MA: MIT Press, 1987.
Garrett, Don. "Spinoza's Necessitarianism." In *God and Nature: Spinoza's Metaphysics*, edited by Yirmiyahu Yovel, 191–218. Leiden: Brill, 1991.

Garrett, Don. "Spinoza's Theory of Metaphysical Individuation." In *Individuation and Identity in Early Modern Philosophy: Descartes to Kant*, edited by Kenneth Barber and Jorge Gracia, 73–101. Albany: State University of New York Press, 1994.

Garrett, Don. *Cognition and Commitment in Hume's Philosophy*. Oxford: Oxford University Press, 1997.

Garrett, Don. "Hume's Conclusions in 'Conclusion of this Book.'" In *The Blackwell Guide to Hume's Treatise*, edited by Saul Traiger, 151–75. Oxford: Blackwell, 2008.

Gätje, Helmut. "Die inneren Sinne bei Averroes." *Zeitschrift der Deutschen Morgenländischen Gesellschaft* 115 (1965): 255–93.

Gaukroger, Stephen. *Cartesian Logic: An Essay on Descartes' Conception of Inference*. Oxford: Clarendon Press, 2002.

Geoffroy, Marc. "La tradition arabe du *Peri nou* d'Alexandre d'Aphrodise et les origines de la théorie farabienne des quatre degrés de l'intellect." In *Aristotele e Alessandro di Afrodisia nella tradizione araba*, edited by Cristina D'Ancona and Giuseppe Serra, 191–231. Padua: Il Poligrafo, 2002.

Gigerenzer, Gerd, Peter Todd, and the ABC Research Group. *Simple Heuristics That Make Us Smart*. Oxford: Oxford University Press, 1999.

Gill, Christopher. *Personality in Greek Epic, Tragedy, and Philosophy*. Oxford: Oxford University Press, 1996.

Gill, Christopher. *Naturalistic Psychology in Galen and Stoicism*. Oxford: Oxford University Press, 2010.

Ginsborg, Hannah. "Two Kinds of Mechanical Inexplicability." *Journal of the History of Philosophy* 42 (2004): 33–65.

Glannon, Walter. "Our Brains Are Not Us." *Bioethics* 23 (2009): 321–29.

Graf, William D., Geoffrey Miller, and Saskia K. Nagel. "Addressing the Problem of ADHD Medication as Neuroenhancements." *Expert Review in Neurotherapeutics* 14 (2014): 569–81.

Graf, William D., Saskia K. Nagel, Leon G. Epstein, Geoffrey Miller, Ruth Nass, and Dan Larriviere. "Pediatric Neuroenhancement: Ethical, Legal, Social, and Neurodevelopmental Implications." *Neurology* 80 (2013): 1251–60.

Greco, John, and John Turri, eds. *Virtue Epistemology: Contemporary Readings*. Cambridge, MA: MIT Press, 2012.

Gregory, Richard. *Eye and Brain: The Psychology of Seeing*. London: Weidenfeld and Nicolson, 1967.

Gregory, Richard. *The Intelligent Eye*. New York: McGraw Hill, 1970.

Gutas, Dimitri. *Avicenna and the Aristotelian Tradition*. Leiden: Brill, 1988.

Haag, Johannes. *Erfahrung und Gegenstand: Das Verhältnis von Sinnlichkeit und Verstand*. Frankfurt: Klostermann, 2007.

Haag, Johannes. "Fichte on the Consciousness of Spinoza's God." In *Spinoza and German Idealism*, edited by Eckhard Förster and Yitzhak Melamed, 100–120. Cambridge: Cambridge University Press, 2012.

Haag, Johannes. "Kant on Imagination and the Natural Sources of the Conceptual." In *Contemporary Perspectives on Early Modern Philosophy: Nature and Norms in Thought*, edited by Martin Lenz and Anik Waldow, 65–85. Dordrecht: Springer, 2013.

Hagemeier, Martin. *Zur Vorstellungskraft in der Philosophie Spinozas*. Würzburg: Königshausen und Neumann, 2012.

Hagner, Michael. *Homo Cerebralis: Der Wandel vom Seelenorgan zum Gehirn*. Frankfurt: Suhrkamp, 1992.

Hall, Robert E. "Intellect, Soul and Body in Ibn Sīnā: Systematic Synthesis and Development of the Aristotelian, Neoplatonic and Galenic Theories." In *Interpreting Avicenna: Science and Philosophy in Medieval Islam*, edited by John McGinnis and David C. Reisman, 62–86. Leiden: Brill, 2004.

Hankinson, Richard J. "Partitioning the Soul: Galen on the Anatomy of the Psychic Functions and Mental Illness." In *Partitioning the Soul: Debates from Plato to Leibniz*, edited by Klaus Corcilius and Dominik Perler, 84–106. Berlin: de Gruyter, 2014.

Hansberger, Rotraud. "Kitāb al-Ḥiss wa-l-maḥsūs: Aristotle's *Parva Naturalia* in Arabic Guise." In *Les Parva naturalia d'Aristote: Fortune antique et médiévale*, edited by Christophe Grellard and Pierre-Marie Morel, 143–62. Paris: Presses Universitaires de Paris-Sorbonne, 2010.

Harris, John. *Enhancing Evolution: The Ethical Case for Making People Better*. Princeton: Princeton University Press, 2007.

Hasse, Dag Nikolaus. *Avicenna's "De anima" in the Latin West: The Formation of a Peripatetic Philosophy of the Soul 1160–1300*. London: Warburg Institute, 2001.

Hasse, Dag Nikolaus. "Avicenna on Abstraction." In *Aspects of Avicenna*, edited by Robert Wisnovsky, 39–72. Princeton: Markus Wiener, 2001.

Hasse, Dag Nikolaus. "The Soul's Faculties." In *The Cambridge History of Medieval Philosophy*, edited by Robert Pasnau, 305–19. Cambridge: Cambridge University Press, 2010.

Hattab, Helen. *Descartes on Forms and Mechanisms*. Cambridge: Cambridge University Press, 2009.

Haugeland, John. *Artificial Intelligence: The Very Idea*. Cambridge, MA: MIT Press, 1986.

Henrich, Dieter. "Fichtes ursprüngliche Einsicht." In *Subjektivität und Metaphysik: Festschrift für Wolfgang Cramer*, edited by Dieter Henrich and Hans Wagner, 188–232. Frankfurt: Klostermann, 1966.

Henrich, Dieter. *Identität und Objektivität: Eine Untersuchung über Kants transzendentale Deduktion.* Heidelberg: C. Winter, 1976.
Hurley, Susan. "Perception and Action: Alternative Views." *Synthese* 129 (2001): 3–40.
Hurley, Susan. *Consciousness in Action.* Cambridge, MA: Harvard University Press, 2002.
International Narcotics Control Board. *Report of the International Narcotics Control Board for 2012.* Vienna: United Nations Publications, 2013.
Inwood, Brad. "Walking and Talking: Reflections on Divisions of the Soul in Stoicism." In *Partitioning the Soul: Debates from Plato to Leibniz,* edited by Klaus Corcilius and Dominik Perler, 63–83. Berlin: de Gruyter, 2014.
Ivry, Alfred. "The Ontological Entailments of Averroes' Understanding of Perception." In *Theories of Perception in Medieval and Early Modern Philosophy,* edited by Simo Knuuttila and Pekka Kärkkäinen, 73–86. Dordrecht: Springer, 2008.
James, William. *The Principles of Psychology.* New York: Dover, 1980.
Janssens, Jules. "Fakhr al-Dīn al-Rāzī on the Soul: A Critical Approach to Ibn Sīnā." *Muslim World* 102 (2012): 562–79.
Johansen, Thomas K. *The Powers of Aristotle's Soul.* Oxford: Oxford University Press, 2012.
Jolley, Nicholas. "Intellect and Illumination in Malebranche." *Journal of the History of Philosophy* 32 (1994): 209–24.
Karamanolis, George E. *The Philosophy of Early Christianity.* London: Routledge, 2014.
Kärkkäinen, Pekka. "Internal Senses." In *Encyclopedia of Medieval Philosophy,* edited by Henrik Lagerlund, 564–67. Dordrecht: Springer, 2011.
Kelso, J. A. Scott. *Dynamic Patterns: The Self-Organization of Brain and Behavior.* Cambridge, MA: MIT Press, 1995.
Kenny, Anthony. "The Homunculus Fallacy." In *The Legacy of Wittgenstein,* 125–36. Oxford: Blackwell, 1984.
King, Peter. "The Inner Cathedral: Mental Architecture in High Scholasticism." *Vivarium* 46 (2008): 253–74.
Kitcher, Philip. "The Naturalists Return." *Philosophical Review* 101 (1992): 53–114.
Knuuttila, Simo. "Aristotle's Theory of Perception and Medieval Aristotelianism." In *Theories of Perception in Medieval and Early Modern Philosophy,* edited by Simo Knuuttila and Pekka Kärkkäinen, 1–22. Dordrecht: Springer, 2008.
Kosman, Aryeh. *The Activity of Being: An Essay on Aristotle's Ontology.* Cambridge, MA: Harvard University Press, 2013.
Kukkonen, Taneli. "The Self as Enemy, the Self as Divine: A Crossroads in the Development of Islamic Anthropology." In *Ancient Philosophy of the Self,* edited by Juha Sihvola and Pauliina Remes, 205–24. Dordrecht: Springer, 2008.

Kukkonen, Taneli. "Body, Spirit, Form, Substance: Ibn Ṭufayl's Psychology." In *In the Age of Averroes: Arabic Philosophy in the Sixth/Twelfth Century*, edited by Peter Adamson, 195–214. London: Warburg Institute, 2011.

Kukkonen, Taneli. "Receptive to Reality: Al-Ghazālī on the Structure of the Soul." *Muslim World* 102 (2012): 541–61.

Kukkonen, Taneli. "Potentiality in Classical Arabic Thought." In *Handbook of Potentiality*, edited by Kristina Engelhard and Michael Quante. Dordrecht: Springer, forthcoming.

Künzle, Pius. *Das Verhältnis der Seele zu ihren Potenzen: Problemgeschichtliche Untersuchungen von Augustin bis und mit Thomas von Aquin*. Freiburg: Universitätsverlag, 1956.

Labarrière, Jean-Louis. *Langage, vie politique et mouvement des animaux: Études aristotéliciennes*. Paris: Vrin, 2004.

Lagerlund, Henrik. "John Buridan and the Problems of Dualism in the Early Fourteenth Century." *Journal of the History of Philosophy* 42 (2004): 369–87.

Langermann, Y. Tzvi. "Abū al-Faraj ibn al-Ṭayyib on Spirit and Soul." *Le Muséon* 122 (2009): 149–58.

Leijenhorst, Cees. "Attention Please! Theories of Selective Attention in Late Aristotelian and Early Modern Philosophy." In *Mind, Cognition and Representation: The Tradition of Commentaries on Aristotle's De anima*, edited by Paul J. J. M. Bakker and Johannes M. M. H. Thijssen, 205–30. Aldershot: Ashgate, 2007.

Lenz, Martin. *Mentale Sätze: Wilhelm von Ockhams Thesen zur Sprachlichkeit des Denkens*. Stuttgart: Steiner, 2003.

Lobsien, Verena Olejniczak, and Eckhard Lobsien. *Die unsichtbare Imagination: Literarisches Denken im 16. Jahrhundert*. Munich: Fink, 2003.

Long, Anthony A. "Stoic Psychology." In *The Cambridge History of Hellenistic Philosophy*, edited by Keimpe Algra, Jonathan Barnes, Jaap Mansfeld, and Malcolm Schofield, 560–84. Cambridge: Cambridge University Press, 2005.

Long, Anthony A., and David N. Sedley. *The Hellenistic Philosophers*. Cambridge: Cambridge University Press, 1987.

López-Farjeat, Luis Xavier, and Jörg Alejandro Tellkamp, eds. *Philosophical Psychology in Arabic Thought and the Latin Aristotelianism of the 13th Century*. Paris: J. Vrin, 2013.

Ludwig, Josef. *Das akausale Zusammenwirken (sympathia) der Seelenvermögen in der Erkenntnislehre des Suarez*. Munich: Karl Ludwig Verlag, 1929.

Lyons, Jack C. "Carving the Mind at Its (Not Necessarily Modular) Joints." *British Journal for the Philosophy of Science* 52 (2001): 277–302.

Machery, Eduard, and Clark Harold Barrett. "Debunking Adapting Minds." *Philosophy of Science* 73 (2006): 232–46.

Makin, Stephen. *Aristotle: Metaphysics Theta*. Translated with an introduction and commentary. Oxford: Oxford University Press, 2006.
Margolis, Eric, and Stephen Laurence, eds. *Concepts: Core Readings*. Cambridge, MA: MIT Press, 1999.
Marslen-Wilson, William, and Lorraine Komisarjevsky Tyler. "Against Modularity." In *Modularity in Knowledge Representation and Natural-Language Understanding*, edited by Jay L. Garfield, 227–35. Cambridge, MA: MIT Press, 1987.
McDowell, John. "Hegel's Idealism as a Radicalization of Kant." In *Having the World in View: Essays on Kant, Hegel, and Sellars*, 69–89. Cambridge, MA: Harvard University Press, 2009.
McDowell, John. "The Logical Form of an Intuition." In *Having the World in View: Essays on Kant, Hegel, and Sellars*, 23–43. Cambridge, MA: Harvard University Press, 2009.
McDowell, John. "Sellars on Perceptual Experience." In *Having the World in View: Essays on Kant, Hegel, and Sellars*, 3–22. Cambridge, MA: Harvard University Press, 2009.
McGinnis, Jon. "Making Abstraction Less Abstract: The Logical, Psychological, and Metaphysical Dimensions of Avicenna's Theory of Abstraction." *Proceedings of the American Catholic Philosophical Association* 80 (2006): 169–83.
McGinnis, Jon. "Avicenna's Naturalized Epistemology and Scientific Method." In *The Unity of Science in the Arabic Tradition*, edited by Shahid Rahman, Tony Street, and Hassan Tahiri, 129–52. Dordrecht: Springer, 2008.
McGinnis, Jon. *Avicenna*. Oxford: Oxford University Press, 2010.
Melamed, Yitzhak. "Spinoza's Metaphysics of Thought: Parallelisms and the Multifaceted Structure of Ideas." *Philosophy and Phenomenological Research* 200 (2011): 1–48.
Menn, Stephen. "The Origins of Aristotle's Concept of Ἐνέργεια: Ἐνέργεια and Δύναμις." *Ancient Philosophy* 14 (1994): 73–114.
Menn, Stephen. "Suárez, Nominalism, and Modes." In *Hispanic Philosophy in the Age of Discovery*, edited by Kevin White, 226–56. Washington, DC: Catholic University of America Press, 1997.
Millican, Peter. "Hume on Reason and Induction." *Hume Studies* 24 (1998): 141–59.
Morton, Samuel G. *Crania Americana: A Comparative View of the Skulls of Various Aboriginal Nations of North and South America*. Philadelphia: J. Dobson, 1939.
Nadler, Steven. "Doctrines of Explanation in Late Scholasticism and in the Mechanical Philosophy." In *The Cambridge History of Seventeenth Century Philosophy*, edited by Daniel Garber and Michael Ayers, 513–52. Cambridge: Cambridge University Press, 1998.

Nagel, Saskia K. *Ethics and the Neurosciences: Ethical and Social Consequences of Neuroscientific Progress*. Paderborn: Mentis, 2010.

Nagel, Saskia K., and Peter B. Reiner. "Autonomy Support to Foster Individuals' Flourishing." *American Journal of Bioethics* 13 (6) (2013): 36–37.

Newell, Allen, John Shaw, and Herbert Simon. "Report on a General Problem-Solving Program." *Proceedings of the International Conference on Information Processing*, 256–64. Paris: UNESCO, 1959.

Owen, David. *Hume's Reason*. Oxford: Oxford University Press, 1999.

Panaccio, Claude. *Ockham on Concepts*. Aldershot: Ashgate, 2004.

Parens, Erik, ed. *Enhancing Human Traits: Ethical and Social Implications*. Washington, DC: Georgetown University Press, 1998.

Parkinson, George Henry Radcliffe. *Spinoza's Theory of Knowledge*. Oxford: Clarendon Press, 1954.

Pasnau, Robert. *Thomas Aquinas on Human Nature*. Cambridge: Cambridge University Press, 2002.

Pasnau, Robert. "Form, Substance, and Mechanism." *Philosophical Review* 113 (2004): 31–88.

Pasnau, Robert. "The Mind-Soul Problem." In *Mind, Cognition and Representation: The Tradition of Commentaries on Aristotle's "De anima,"* edited by Paul J. J. M. Bakker and Johannes M. M. H. Thijssen, 3–19. Aldershot: Ashgate, 2007.

Pasnau, Robert, ed. *The Cambridge History of Medieval Philosophy*. Cambridge: Cambridge University Press, 2010.

Pasnau, Robert. *Metaphysical Themes 1274–1671*. Oxford: Clarendon Press, 2011.

Peppers-Bates, Susan. *Nicolas Malebranche: Freedom in an Occasionalist World*. New York: Continuum, 2009.

Perler, Dominik, ed. *Transformations of the Soul: Aristotelian Psychology 1250–1650*. Special issue, *Vivarium* 46.3. Leiden: Brill, 2008.

Perler, Dominik. "Ockham über die Seele und ihre Teile." *Recherches de théologie et philosophie médiévales* 77 (2010): 313–50.

Perler, Dominik. "The Problem of Necessitarianism (1p28–36)." In *Spinoza's Ethics: A Collective Commentary*, edited by Michael Hampe, 57–79. Leiden: Brill, 2011.

Perler, Dominik. "What Are Faculties of the Soul? Descartes and His Scholastic Background." *Proceedings of the British Academy* 189 (2013): 9–38.

Perler, Dominik. "Perception in Medieval Philosophy." In *The Oxford Handbook of the Philosophy of Perception*, edited by Mohan Matthen, 51–65. Oxford: Oxford University Press, 2015.

Perler, Dominik, and Ulrich Rudolph. *Occasionalismus: Theorien der Kausalität im arabisch-islamischen und im europäischen Denken*. Göttingen: Vandenhoeck und Ruprecht, 2000.

Pike, Nelson. "Hume's Bundle Theory of the Self: A Limited Defense." *American Philosophical Quarterly* 4 (1967): 159–65.
Pinkard, Terry. *German Philosophy 1760–1860: The Legacy of Idealism*. Cambridge: Cambridge University Press, 2009.
Pinker, Stephen. *How the Mind Works*. New York: Norton, 1997.
Pippin, Robert. *Kant's Theory of Form*. New Haven: Yale University Press, 1982.
Pitour, Thomas. *Wilhelm von Auvergnes Psychologie: Von der Rezeption des aristotelischen Hylemorphismus zur Reformulierung der Imago-Dei-Lehre Augustins*. Paderborn: Schöningh, 2011.
Plotkin, Henry. *Evolution in Mind*. London: Alan Lane, 1997.
Prinz, Jesse. "Is the Mind Really Modular?" In *Contemporary Debates in Cognitive Science*, edited by Robert Stainton, 22–36. Oxford: Blackwell, 2006.
Pyle, Andrew. *Malebranche*. London: Routledge, 2003.
Pylyshyn, Zenon. *Computation and Cognition*. Cambridge, MA: MIT Press, 1984.
Pylyshyn, Zenon. "Is Vision Continuous with Cognition? The Case for Cognitive Impenetrability of Visual Perception." *Brain and Behavioral Sciences* 22 (1999): 341–423.
Read, Rupert J., ed. *The New Hume Debate*. London: Routledge, 2000.
Repantis, Dimitris, Peter Schlattmann, Oona Laisney, and Isabella Heuser. "Modafinil and Methylphenidate for Neuroenhancement in Healthy Individuals: A Systematic Review." *Pharmacological Research* 62 (2010): 187–206.
Ridge, Michael. "Epistemology Moralized: David Hume's Practical Epistemology." *Hume Studies* 29 (2003): 165–204.
Robbins, Philip. "Minimalism and Modularity." In *Context-Sensitivity and Semantic Minimalism*, edited by Gerhard Preyer and Georg Peter, 303–19. Oxford: Oxford University Press, 2007.
Robbins, Philip. "Modularity of Mind." In *Stanford Encyclopedia of Philosophy*. http://plato.stanford.edu/archives/sum2010/entries/modularity-mind/, accessed January 2, 2015.
Rosch, Eleanor, et al. "Basic Objects in Natural Categories." *Cognitive Psychology* 8 (1976): 382–439.
Rozemond, Marleen. *Descartes's Dualism*. Cambridge, MA: Harvard University Press, 1998.
Rozemond, Marleen. "Descartes, Mind-Body Union, and Holenmerism." *Philosophical Topics* 31 (2003): 343–67.
Rozemond, Marleen. "Unity in the Multiplicity of Suárez's Soul." In *The Philosophy of Francisco Suárez*, edited by Benjamin Hill and Henrik Lagerlund, 154–72. Oxford: Oxford University Press, 2012.
Ryle, Gilbert. *The Concept of Mind*. Chicago: University of Chicago Press, 1949.

Sachs-Hombach, Klaus. "Vermögen, Vermögenspsychologie." In *Historisches Wörterbuch der Philosophie*, edited by Klaus Ritter et al., vol. 11, 727–31. Basel: Schwabe, 2001.

Samuels, Richard. "Evolutionary Psychology and the Massive Modularity Hypothesis." *British Journal for the Philosophy of Science* 49 (1998): 575–602.

Samuels, Richard. "Massively Modular Minds: Evolutionary Psychology and Cognitive Architecture." In *Evolution and the Human Mind*, edited by Peter Carruthers and Andrew Chamberlain, 13–46. Cambridge: Cambridge University Press, 2000.

Samuels, Richard. "Is the Human Mind Massively Modular?" In *Contemporary Debates in Cognitive Science*, edited by Robert Stainton, 37–56. Oxford: Blackwell, 2006.

Schmaltz, Tad. *Malebranche's Theory of the Soul: A Cartesian Interpretation*. Oxford: Oxford University Press, 1996.

Schmaltz, Tad. *Descartes on Causation*. Oxford: Oxford University Press, 2008.

Schmid, Stephan. "Spinoza on the Unity of Will and Intellect." In *Partitioning the Soul: Debates from Plato to Leibniz*, edited by Dominik Perler and Klaus Corcilius, 245–70. Berlin: de Gruyter, 2014.

Schmid, Stephan. "Suárez on Efficient Causality." In *Suárezian Causes*, edited by Jakob Leth Fink. Leiden: Brill, forthcoming.

Schuurman, Paul. *Ideas, Mental Faculties, and Method: The Logic of Ideas of Descartes and Locke and Its Reception in the Dutch Republic, 1630–1750*. Leiden: Brill, 2004.

Sebti, Meryem. "The Ontological Link between Body and Soul in Bahmanyār's *Kitāb al-Taḥṣīl*." *Muslim World* 102 (2012): 525–40.

Sellars, Wilfrid. "Empiricism and the Philosophy of Mind." *Minnesota Studies in the Philosophy of Science* 1 (1956): 253–329.

Sellars, Wilfrid. *Science, Perception and Reality*. London: Routledge and Kegan Paul, 1963.

Sellars, Wilfrid. *Science and Metaphysics: Variations on Kantian Themes*. London: Routledge, 1968.

Sellars, Wilfrid. "Kant's Transcendental Idealism." *Collections of Philosophy* 6 (1976): 165–81.

Sharples, Robert. "The Hellenistic Period: Whatever Happened to Hylomorphism?" In *Ancient Perspectives on Aristotle's "De Anima,"* edited by Gerd Van Riel and Pierre Destrée, 156–66. Leuven: Leuven University Press, 2009.

Shields, Christopher. *Order in Multiplicity: Homonymy in the Philosophy of Aristotle*. Oxford: Clarendon Press, 1999.

Shields, Christopher. "Plato's Divided Soul." In *Plato's "Republic": A Critical Guide*, edited by Mark L. McPherran, 147–70. Cambridge: Cambridge University Press, 2010.

Shields, Christopher. "Virtual Presence: Psychic Mereology in Francisco Suárez." In *Partitioning the Soul: Debates from Plato to Leibniz*, edited by Klaus Corcilius and Dominik Perler, 199–218. Berlin: de Gruyter, 2014.

Shihadeh, Ayman. "Classical Ashʿarī Anthropology: Body, Life and Spirit." *Muslim World* 102 (2012): 433–77.

Singh, Ilina. "Clinical Implications of Ethical Concepts: Moral Self-understandings in Children Taking Methylphenidate for ADHD." *Clinical Child Psychology and Psychiatry* 12 (2) (2007): 167–182.

Singh, Ilina. "Not Robots: Children's Perspectives on Authenticity, Moral Agency and Stimulant Drug Treatments." *Journal of Medical Ethics* 39 (2013): 359–66.

Snell, Bruno. *The Discovery of the Mind*. Translated by Thomas G. Rosenmeyer. New York: Harper, 1960.

Sperber, Daniel. "The Modularity of Thought and the Epidemiology of Representations." In *Mapping the Mind*, edited by Lawrence A. Hirschfeld and Susan A. Gelman, 39–67. Cambridge: Cambridge University Press, 1994.

Sperber, Daniel. *Explaining Culture: A Naturalistic Approach*. Oxford: Blackwell, 1996.

Sperber, Daniel. "In Defense of Massive Modularity." In *Language, Brain, and Cognitive Development: Essays in Honor of Jacques Mehler*, edited by Emmanuel Dupoux, 47–57. Cambridge, MA: MIT Press, 2002.

Sperber, Daniel. "Modularity and Relevance: How Can a Massively Modular Mind Be Flexible and Context Sensitive?" In *The Innate Mind: Structure and Content*, edited by Peter Carruthers, Stephen Laurence, and Stephen Stich, 53–69. Oxford: Oxford University Press, 2005.

Steinberg, Diane. "Belief, Affirmation, and the Doctrine of Conatus in Spinoza." *Southern Journal of Philosophy* 43 (2005): 147–58.

Steinberg, Diane. "Knowledge in Spinoza's *Ethics*." In *The Cambridge Companion to Spinoza's Ethics*, edited by Olli Koistinen, 140–66. Cambridge: Cambridge University Press, 2009.

Sterelny, Kim. "Language, Modularity and Evolution." In *Teleosemantics*, edited by Graham MacDonald and David Papineau. Oxford: Oxford University Press, 2003.

Sterelny, Kim. *The Evolved Apprentice: How Evolution Made Us Human*. Cambridge, MA: MIT Press, 2012.

Stich, Stephen, ed. *Innate Ideas*. Berkeley: University of California Press, 1975.

Strawson, Peter. *The Bounds of Sense: An Essay on Kant's Critique of Pure Reason*. London: Methuen, 1966.

Strohmaier, Gotthard. *Von Demokrit bis Dante*. Hildesheim: Georg Olms Verlag, 1996.
Sutton, John, and Kelly Williamson. "Embodied Remembering." In *Handbook of Embodied Cognition*, edited by Lawrence Shapiro, 315–26. New York: Routledge, 2014.
Taylor, Richard C. "The Agent Intellect as 'Form for Us' and Averroes's Critique of al-Fārābī." *Proceedings of the Society for Medieval Logic and Metaphysics* 5 (2005): 18–32.
Taylor, Richard C. "Abstraction in al-Fārābī." *Proceedings of the American Catholic Philosophical Association* 80 (2006): 151–68.
Thelen, Esther, and Linda Smith. *A Dynamic Systems Approach to the Development of Cognition and Action*. Cambridge, MA: MIT Press, 1994.
Treiger, Alexander. *Inspired Knowledge in Islamic Thought: Al-Ghazālī's Theory of Mystical Cognition and Its Avicennian Foundation*. London: Routledge, 2012.
Van der Eijk, Philip. "The Matter of Mind: Aristotle on the Biology of Psychic Processes and the Bodily Aspects of Thinking." In *Aristotelische Biologie: Intentionen, Methoden, Ergebnisse*, edited by Wolfgang Kullmann and Sabine Föllinger, 221–58. Stuttgart: F. Steiner Verlag, 1997.
Vander Waerdt, Paul A. "Aristotle's Criticism of Soul-Division." *American Journal of Philology* 108 (1987): 627–43.
Van Wyhe, John. *Phrenology and the Origins of Victorian Scientific Naturalism*. Burlington, VT: Ashgate, 2004.
Vegetti, Mario. "La medicina in Platone: IV. I 'Fedro.'" *Rivista critica di storia della filosofia* 24 (1969): 3–22.
Viljanen, Valtteri. "Field Metaphysic, Power, and Individuation in Spinoza." *Canadian Journal of Philosophy* 37 (2007): 393–418.
Viljanen, Valtteri. *Spinoza's Geometry of Power*. Cambridge: Cambridge University Press, 2011.
Von Neumann, John. *First Draft of a Report on the EDVAC*. Philadelphia: Moore School of Electrical Engineering, University of Pennsylvania, 1945.
von Staden, Heinrich. "*Dynamis:* The Hippocratics and Plato." In *Philosophy and Medicine 29: Studies in Greek Philosophy*, edited by Konstantinos J. Boudouris, 262–79. Athens: International Association for Greek Philosophy, 1999.
Warren, Howard C. *A History of the Association Psychology*. New York: Scribner, 1921.
Watson, John. "Psychology as the Behaviorist Views It." *Psychological Review* 20 (1913): 158–77.
Waxman, Wayne. *Kant's Model of the Mind*. Oxford: Oxford University Press, 1991.
Wedin, Michael V. *Mind and Imagination in Aristotle*. New Haven: Yale University Press, 1988.

Weigel, Peter. "Memory and the Unity of the Imagination in Spinoza's *Ethics*." *International Philosophical Quarterly* 49 (2009): 229–46.

Wild, Markus. "Hume on Force and Vivacity: A Teleological-Historical Interpretation." *Logical Analysis and the History of Philosophy* 14 (2011): 71–88.

Williams, Bernard. *Shame and Necessity*. Berkeley: University of California Press, 1993.

Wilson, Robert. "What Computers (Still, Still) Can't Do: Jerry Fodor on Computation and Modularity." In *New Essays in Philosophy of Language and Mind*, edited by Robert Stainton, Maite Ezcurdia, and Christopher David Viger. *Canadian Journal of Philosophy*, supp. iss. 30 (2005): 407–25.

Winkler, Kenneth P. "The New Hume." *Philosophical Review* 100 (1991): 541–79.

Winkler, Kenneth P. "Hume's Inductive Skepticism." In *The Empiricists: Critical Essays on Locke, Berkeley, and Hume*, edited by Margaret Atherton, 183–212. Lanham, MD: Rowman and Littlefield, 1999.

Winters, Barbara. "Hume on Reason." *Hume Studies* 5 (1979): 20–35.

Wolfson, Harry Austryn. "The Internal Senses in Latin, Arabic, and Hebrew Philosophic Texts." *Harvard Theological Review* 28 (1935): 69–133.

Young, Robert M. *Mind, Brain and Adaptation in the Nineteenth Century: Cerebral Localization and Its Biological Context from Gall to Ferrier*. Oxford: Clarendon Press, 1970.

Zuckert, Rachel. *Kant on Beauty and Biology: An Interpretation of the Critique of Judgment*. Cambridge: Cambridge University Press, 2007.

Index of Names

Adams, Robert M., 181n66
Adams McCord, Marilyn, 119n45, 130n66
Adamson, Peter, 78n17, 79n19, 80n22
Allison, Henry, 223n37
'Āmirī, Abū l-Ḥasan Muḥammad Ibn Yūsuf al-, 74, 86n8, 79–81, 84, 91
Apel, Karl-Otto, 259n11
Aristotle, 4–7, 9, 12, 15, 20, 31–49, 32n27, 34n36, 41n40, 42n44, 44n49, 45n50, 45n51, 46n55, 57n74, 58, 66, 67n1, 68–78, 80n21, 81n23, 82–86, 88–89, 90n34, 91–93, 95–97, 99–100, 110, 111n30, 113n35, 124n55, 131n71, 132, 136–37, 141, 150–52, 154–56, 161, 163, 175, 183, 265
Arnauld, Antoine, 177n52
Augustine, 100, 173n44, 174

Baier, Annette C., 186n72
Bakker, Paul J. J. M., 100n6
Barrett, Harold C., 283n49

Bechtel, William, 294–95, 295n67
Beere, Jonathan B., 34n29
Beiser, Frederick C., 232n47
Bennett, Max R., 104n15
Biard, Joël, 98n1
Black, Deborah, 83n26
Boureau, Alain, 100n5
Breland, Keller, 262n17
Breland, Marian, 262n17
Brownlow, Frank, 142n3
Bruner, Jerome, 280n43
Buckle, Stephen, 183n67
Buller, David, 268n29, 292n60
Burnyeat, Myles, 25n10

Carruthers, Peter, 266, 267n28, 268n29, 271, 283n51, 284–92, 296
Chomsky, Noam, 7, 7n5, 261n16, 261–62, 272, 277
Chong, Eugene W. J., 306n13

331

Churchland, Paul, 280n45
Clark, Andy, 293n62
Clarke, Edwin, 250n5
Colbert, Charles, 251n7
Colombetti, Giovanna, 294n64
Cooter, Roger, 248n1, 250
Corcilius, Klaus, 40n39, 46n53, 48n57, 50n60, 53n65, 54n68, 58n76, 130n68, 165n34
Cosmides, Leda, 267n28, 283n49, 283n50, 291
Courtenay, William J., 124n54
Cowie, Fiona, 291n58, 293n61
Cross, Richard, 123n53
Cundall, Michael K., 267n27

Davies, John D., 251
De Boer, Sander W., 124n54
Della Rocca, Michael, 164n32
Des Chene, Dennis, 108n24
Descartes, René, 6, 12, 15n14, 24n9, 45, 150n1, 151n2, 152–67, 171–75, 172n41, 173n44, 174n47, 177–78, 177n53, 181–83, 186–87, 194–96, 266, 282, 290, 297
Diwald, Susanne, 74n8
Donini, Pierluigi, 54n68
Druart, Thérèse-Anne, 91n37
Dummett, Michael, 259, 260n12

Endress, Gerhard, 80n21
Euripides, 63
Everson, Stephen, 50n60, 78n18

Fārābī, Abū Naṣr al-, 66, 72–73, 81n23, 86n30, 89, 90n34, 93–94
Faucher, Luc, 265n23
Feinberg, Joel, 304n10
Fichte, Johann Gottlieb, 200, 235–245, 235n51, 235n52, 236n53, 236n54, 236n55, 237n56, 238n57, 238n58, 240n60
Ficino, Marsilio, 142–144, 142n4, 142n5, 142n6, 143n7, 144n12
Fitz, Nicholas S., 306n13
Fodor, Jerry, 7, 7n6, 13n11, 15n14, 42n42, 266–68, 266n25, 270n31, 271–82, 272n32, 273n33, 274n34, 277n37, 278n40, 278n41, 279n42, 280n44, 280n45, 281n46, 282n48, 284–92, 288n56 294, 296–97
Foley, Helene P., 63n3, 64n4
Förster, Eckart, 210n16, 219n32, 226, 227n41, 227n42, 227n43, 229n44, 232n46, 232n47, 233n48, 234n49, 234n50, 236n55, 238n57, 238n58, 242n61, 244n65, 245n66
Fortenbaugh, William W., 75n10
Fóti, Véronique, 157n14
Franks, Paul, 211n21, 232n47, 240n60
Fuchs, Thomas, 307n14

Galen, Claudius of Pergamon, 9, 14, 54–58, 54n68, 55n70, 56n72, 57n73, 57n75, 58n76, 58n77, 75, 99n3
Gall, Franz Joseph, 13n12, 15n14, 248–50, 248n2, 254–55, 254n1, 264, 268, 274–75, 277
Garber, Daniel, 99n4, 162n30
Garrett, Don, 168n37, 172n42, 185n71, 186n72
Gätje, Helmut, 76n14
Gaukroger, Stephen, 152n3
Geoffroy, Marc, 86n30
Ghazālī, Abū Ḥāmid Muḥammad al-, 77–78, 78n16, 78n17, 87n31, 89
Gigerenzer, Gerd, 288
Gill, Christopher, 53n66, 54n68, 59n1, 64n4, 64n5
Ginsborg, Hannah, 224n38
Glannon, Walter, 307n14
Graf, William D., 300n4, 301n7, 304n9
Greco, John, 265, 264n21
Gregoric, Pavel, 48n57
Gregory, Richard, 262n18, 280n43
Gutas, Dimitri, 82n25

Haag, Johannes, 212n22, 216n26, 238n57, 238n58
Hacker, Peter M. S., 104n15
Hagemeier, Martin, 165n49
Hagner, Michael, 250n5, 255n2
Hall, Robert, 67n2
Hankinson, Richard J., 54n68, 58n76
Hansberger, Rotraud, 75n11

INDEX OF NAMES

Hardcastle, Valerie, 268n29
Harris, John, 299n2
Hasse, Dag N., 66n1, 76n14, 81n23
Hattab, Helen, 136n76
Haugeland, John, 296n68
Henrich, Dieter, 216n26, 237n56
Heuser, Isabella, 301n8
Hobbes, Thomas, 266
Homer, 20, 59, 63–64
Hume, David, 15n13, 152–55, 165, 182–97, 183n68, 184n69, 185n60, 185n61, 186n72, 188n73, 189n74, 189n75, 190n76, 190n77, 190n78, 192n79, 193n80, 193n81, 256, 258, 264, 266
Ḥunayn Ibn Isḥāq, 92
Hurley, Susan, 286, 286n53, 293n63, 294, 294n65, 294n66

Ibn Bājja, Abū Bakr Muḥammad, 69–70, 89, 91, 94
Ibn Rushd, Abū l-Walīd Muḥammad Ibn Aḥmad, 69, 86n30, 92n40
Ibn Sīnā, Abū ʿAlī, 67
Ibn Ṭufayl, Abū Bakr, 78n17, 90–91
Inwood, Brad, 50n60, 53n65
Ivry, Alfred, 90n34

Jacyna, Leon S., 250n5
James, William, 258
Janssens, Jules, 85n29, 95
Johansen, Thomas K., 41n40
John Duns Scotus, 108n24
Jolley, Nicholas, 179n59

Kant, Immanuel, 6, 197–246, 202n2, 203n3, 203n4, 205n7, 206n8, 207n10, 207n12, 207n13, 209n14, 210n17, 211n21, 213n21, 216n24, 216n23, 216n27, 217n28, 217n29, 220n33, 221n34, 223n36, 225n39, 226n40, 240n60, 255
Kärkkäinen, Pekka, 66n1, 81n23, 90n34
Kelso, J. A. Scott, 294n63
Kenny, Anthony, 104n15, 150n1
King, Peter, 98n1
Knuuttila, Simo, 66n1, 90n34

Kosman, Aryeh, 34n29, 70n6
Kukkonen, Taneli, 70n5, 78n16, 87n31, 90n35
Künzle, Pius, 100n5, 105n19, 108n24

Labarrière, Jean-Louis, 45n51
Laisney, Oona, 301n8
Lagerlund, Henrik, 66n1, 121n50, 126n61
Langermann, Y. Tzvi, 79n20
Laurence, Stephen, 17n16
Leibniz, Gottfried Wilhelm, 154, 173
Leijenhorst, Cees, 131n71
Lenz, Martin, 118n43, 212n22
Lobsien, Eckhard, 147n14
Lobsien, Verena Olejniczak, 147n14
Locke, John, 10n9, 98–99, 104, 134–36, 151n2, 152n3, 154, 256
Long, Anthony A., 50n60, 51n62, 52n64, 53n67
López-Farjeat, Luis X., 66n1
Ludwig, Josef, 133n71

Makin, Stephen, 34n29
Malebranche, Nicolas, 152–55, 162n29, 173–83, 173n44, 173n45, 174n46, 174n47, 174n49, 175n50, 175n51, 177n52, 177n54, 178n55, 178n56, 177n58, 179n60, 179n61, 180n62, 181n63, 181n64, 181n66, 186–87, 195–96, 266
Manogaran, Praveena, 306n13
Margolis, Eric, 17n16
Marušić, Jennifer, 18, 139n79
McDowell, John, 207n10, 210n17, 210n18, 211
McGinnis, Jon, 67n2, 81, 88
Melamed, Yitzhak, 170n38, 197n83, 238n57, 238n58
Menn, Stephen, 34n29, 108n25, 139n79
Miller, Geoffrey, 300n4, 301n7, 304n9
Millican, Peter, 185n71
Miskawayh, Abū ʿAlī Aḥmad Ibn Muḥammad, 74, 75n10, 78n17, 79–80
Morton, Samuel G., 251

Nadler, Richard, 306n13
Nadler, Steven, 99n4

INDEX OF NAMES

Nagel, Saskia K., 299–307, 300n3, 300n4, 301n7, 304n9, 304n11
Newell, Allen, 261, 262

Owen, David, 185n70

Panaccio, Claude, 116n40, 118n43
Parens, Erik, 299n1
Pasnau, Robert, 66n1, 100n6, 111n32, 121n50, 137n77
Peppers-Bates, Susan, 181n65
Perler, Dominik, 50n60, 53n65, 54n68, 58n76, 58n78, 100n6, 111n32, 120n46, 130n68, 133n74, 139n79, 165n34, 168n37, 197n83, 298n69
Petrus Hispanus, 109n28
Pike, Nelson, 196n82
Pinkard, Terry, 232n47
Pinker, Stephen, 267n28, 283n49
Pippin, Robert, 207n10
Pitour, Thomas, 101n9, 105n17
Plato, 20–32, 36–38, 43n46, 46–49, 53–54, 56–58, 67, 74–77, 80, 85, 92n40, 95, 142, 148
Plotkin, Henry, 267n28
Porphyry, 109n28
Prinz, Jesse, 276n37, 291n58, 292, 293, 298n69
Pyle, Andrew, 174n47
Pylyshyn, Zenon, 263n19, 266n16, 270n31

Rasā'il Ikhwān al-ṣafā', 74–75, 79, 85, 92
Rāzī, Fakhr al-Dīn al-, 85, 95–96
Read, Rupert J., 192n79
Reiner, Peter B., 304n11, 306n13
Repantis, Dimitris, 301n8
Ribot, Théodule, 256n4, 257
Ridge, Michael, 186n72
Robbins, Philip, 276n35, 277n38, 295n67
Rosch, Eleanor, 276n36
Rozemond, Marleen, 126n61, 157n13, 158n17, 197n83
Rudolph, Ulrich, 133n74
Ryle, Gilbert, 10, 10n10, 153n5, 260, 261

Sachs-Hombach, Klaus, 16n15
Samuels, Richard, 268n29, 270n30

Schlattmann, Peter, 301n8
Schmaltz, Tad, 162n30, 178n57
Schmid, Stephan, 138n78, 139n79, 165n34, 298n69
Schuurman, Paul, 152n3
Sebti, Meryem, 94n41
Selby-Bigge, Lewis A., 155n7
Sellars, Wilfrid, 201n1, 203, 204, 207, 209n15, 210n17, 210n18, 211n19, 218n31, 260n13
Shakespeare, William, 141, 143, 145, 148–149
Sharples, Robert, 49n58
Shaw, John, 261, 262
Shields, Christopher, 25n10, 111n30, 130n68
Shihadeh, Ayman, 79n20
Simon, Herbert, 261, 262
Singh, Ilina, 304n11, 306n12
Smith, Linda, 294n63
Snell, Bruno, 59n1
Spenser, Edmund, 140, 143–45
Sperber, Daniel, 268n29, 283n49, 288n55, 291
Spinoza, Baruch de, 15n13, 152–55, 163–73, 163n31, 164n32, 165n33, 167n35, 170n38, 172n41, 172n42, 191, 194–97, 245
Steinberg, Diane, 167n35, 172n39
Sterelny, Kim, 293n61
Stich, Stephen, 288n55
Stobaeus, 52
Strawson, Peter, 216n26
Strohmaier, Gotthard, 76n13
Suárez, Francisco, 98, 108n24, 124–38, 124n55, 124n56, 125n57, 125n58, 125n59, 126n60, 127n62, 127n63, 129n65, 130n68, 131n69, 131n70, 131n71, 133n72, 133n73m 133n74, 134n75, 158n164
Sutton, John, 157n14, 293n62

Taylor, Richard C., 81n23, 91n38
Tellkamp, Jörg Alejandro, 66n1
Thelen, Esther, 294n63
Thijssen, Johannes M. M. H., 100n6
Thomas Aquinas, 98, 98n1, 105–17, 107n21, 113n35, 126–28, 134, 137–38, 157
Todd, Peter, 288n57

Tooby, John, 267n28, 283n49, 283n50, 291
Treiger, Alexander, 78n16
Turri, John, 264n21

Vander Waerdt, Paul A., 46n55
Van Wyhe, John, 249n3, 249n4
Vegetti, Mario, 22n5
Viljanen, Valtteri, 172n40, 172n42, 197n83
Von Neumann, John, 263n20
Von Staden, Heinrich, 21n1, 21n2

Warren, Howard C., 256n4
Watson, John, 258n10
Wedin, Michael V., 45n51
Weigel, Peter, 165n33
Wild, Markus, 184n69

William of Auvergne, 100–05, 107, 110, 112, 116, 137
William of Ockham, 114–126, 130n66, 137
Williams, Bernard, 59n1
Williamson, Kelly, 293n62
Winkler, Kenneth P., 185n71, 192n79
Winters, Barbara, 186n72
Wolfson, Harry A., 75n12
Wright, Thomas, 141, 142n2
Wundt, Wilhelm, 258

Yaḥyā Ibn ʿAdī, 74, 75n10
Young, Robert M., 250n5

Zuckert, Rachel, 230n45

Index of Concepts

abstraction, 34, 42, 43, 81, 82, 202, 203n3, 206, 225, 238, 244
accident, 11, 79, 83, 101–103, 106, 109, 222
  accidental predication, 102
  accidental properties, 82
actuality, 32–37, 44, 69, 70, 84n28, 94, 96, 137, 143, 214, 220, 221
antiquity, 4, 9, 13, 19–22, 23n6, 28, 39, 58, 80, 143, 265, 285, 312
apperception, 208, 215, 217, 218
  transcendental, 218
appetite, 24n8, 27, 74–77, 82, 83, 88, 92n39, 168, 169, 265
Arabic philosophy, 5, 15, 58, 66–68, 70n5, 73–76, 84, 85, 91n37, 96
architecture, 254, 263, 264, 267, 268, 271, 273, 276, 278, 279, 283–287, 291, 293, 295, 296
  cognitive, 263, 264, 270
  functional, 132, 271, 272

Aristotelianism, 6, 7, 58, 71, 75–77, 82, 83, 85, 86, 88, 89, 90n34, 92, 95–97, 100, 110, 111n, 131n, 132, 136, 137, 141, 150–152, 154–156, 161, 163, 175, 183, 265
associationism, 256, 257, 264, 286, 296
autonomy, 24n9, 128, 134, 303, 304

behaviorism, 258, 259, 261, 262, 264, 286
being, 32–34, 36, 41, 47, 49, 69–71, 78, 93–95, 144, 146, 170, 193, 200, 219–225, 227, 228, 230, 231, 240, 244, 245
  human, 3–6, 8, 10, 74, 77–80, 86, 87, 90, 97–99, 101, 102–104, 107, 109, 113, 114, 117, 119, 120, 121n50, 122–124, 131, 134–136, 138, 140, 157–158, 163–164, 199, 203, 219, 259
  living, 4, 6, 8–10, 14, 15, 17, 33, 38–40, 42, 48, 68–71, 73, 93, 94, 97, 98, 104, 110, 122–124, 131, 134, 137, 138, 151, 156, 222, 223
  rational, 160, 181, 196, 199, 202, 211, 219, 220, 222, 224, 233, 245, 246, 151

INDEX OF CONCEPTS

body, 13, 21, 23n7, 24n9, 28–31, 32n27,
    34–40, 43–46, 49–54, 58, 86, 72,
    79, 92, 98–101, 107–109, 120, 121,
    124, 125, 136, 138, 141, 143, 156–160,
    163, 169, 172, 173, 175, 177n52, 178,
    255, 293, 294, 307
  bodily accidents, 79
    dead, 60, 64
  bodily constraints, 13
  bodily functions, 91, 93
  bodily instruments, 96
  living, 35–38, 44, 53, 91, 136
  bodily organ, 85, 90, 92n39, 109, 110, 112,
    120, 121, 124
    parts of the, 14, 30, 37, 40, 44, 53, 92n40,
    95, 105, 110, 136
brain, 75, 121, 247–250, 254–256, 262–264,
    269–271, 276, 277, 286, 287, 290,
    292–297, 302, 306, 307

categories, 70, 106, 112, 115, 198, 202, 203,
    205, 206, 211, 214–218, 220, 230
  schematized, 211, 215, 230
causation / cause, 9, 10, 14, 39–41, 45n50,
    54–56, 58, 99, 115, 116, 119, 123, 127, 128,
    133, 135–139, 156, 166–170, 173–175, 178,
    181, 186, 188–193, 196, 215, 222, 227, 230,
    271, 272, 296
  efficient, 14, 118, 129, 133n, 138, 139, 158
  final, 224, 225n38
  formal, 137, 138, 158
  physical, 30, 45n50
change, 30, 32–34, 49, 69, 70, 106, 136–138,
    155, 264, 273
  conceptual, 17
  corporeal, 90n34
character traits, 248, 249, 265n21
cogitation, 74n, 75, 92
  cogitative appetites, 83
  cogitative power, 85
cognitive, 21, 28, 67, 84, 156, 158, 163, 165, 167,
    169, 177, 182, 195, 196, 225, 262, 263, 266,
    267, 269–271, 273, 275, 277, 279–283,
    285, 288, 296, 302
    activities, 5, 172, 196, 231, 262, 271, 277, 294

apparatus, 81, 83
capacities, 67, 75, 154, 160, 268, 270,
    299, 300
economy, 167, 170
science, 15, 255, 261–268, 270, 271, 278,
    280–282, 289, 290, 293, 297, 298
structure, 83
system, 12, 266–268, 270, 276–278, 287, 297
theories, 9
concept, 4, 7, 8, 10, 12, 14–17, 19–21, 22n3, 23,
    24, 33, 45n50, 54, 70n5, 82, 86n35, 96,
    113, 121, 126, 193, 199–202, 204–205,
    207, 212, 214, 217, 218–233, 240, 245,
    246, 257–261, 265, 267–269, 276, 282,
    285, 288, 294, 296
  history of, 16, 17
conception, 24, 30, 34, 43, 45n50, 48n57,
    49–51, 53, 54n, 55, 58, 68, 89, 107, 114,
    118, 136–138, 152–154, 156, 158–160, 162,
    165, 171, 172, 174, 179, 181n, 185, 191–197,
    214–216, 200, 202, 206–208, 229, 230,
    237, 245, 258, 278, 284, 304

development in childhood and youth, 304
disposition, 10, 19, 20, 22n3, 23n6, 32, 33, 35,
    72, 89, 153n5, 155, 156, 161–163, 164, 171,
    177, 182, 196, 209, 214, 249, 259–261,
    265n21
  dispositional analysis, 10, 20, 25, 28, 32, 34,
    49, 56, 260
  innate, 209, 249, 257, 290
distinction, 26, 35, 46, 48, 79n20, 84, 89, 95,
    102, 137, 152, 153, 159, 167, 174, 187, 191,
    193, 194, 203, 206, 211n21, 220–223, 226,
    233, 243, 248, 264, 266, 268, 278, 280,
    282, 290, 291, 294, 297, 300
  metaphysical, 159, 177n52
  real, 108, 124, 130
  theory of, 107, 116, 128
dualism, 121n50, 158, 163, 173

embodiment, 29, 30
evolution, 283, 284, 287, 292, 297
  evolutionary anthropology, 9
  evolutionary biology, 271

INDEX OF CONCEPTS

evolutionary psychology, 283, 284, 288n55, 291–293, 297
essence, 11, 33, 34, 36, 40, 43–46, 49, 58n77, 71n7, 91, 100, 102–110, 114–116, 119, 125, 128, 130, 137, 169, 170, 173, 178, 259, 295
  essential discursivity, 233
  human, 91
  essential function, 35, 37
  essential predication, 102
  essential properties, 82
  essential unity, 92
estimation, 75, 76, 85, 265

faculties, 3–16, 19, 20, 23–32, 34, 35, 37–44, 46–58, 66–69, 71–74, 77–84, 87–90, 92–96, 102–139, 141, 143, 193–201, 203, 206, 208, 211, 212, 215, 217–219, 231–237, 241, 244, 246, 248, 249, 253, 255–257, 259, 264–265, 267–268, 271, 272, 274, 277, 282, 289, 290–291, 294–299, 302, 305
  animal, 76, 81
  Aristotelian, 75, 76
  cognitive, 5, 6, 12–14, 25, 78–80, 94, 103, 151, 159–162, 173–177, 181–183, 185n, 186, 187, 190, 192, 194, 197, 243, 245, 265, 269, 279, 281, 282, 290
  epistemic, 199, 200, 244
  hierarchy of, 4, 12, 87
  higher, 6, 12, 42, 142, 265, 266, 268, 269, 290, 291, 297
  individuation of, 11, 12, 14, 27, 278
  intellectual, 92, 248
  localization, 13
  lower, 5, 6, 12, 42, 43n45, 112, 156, 157, 265, 266, 268, 269, 271, 279, 291, 297
  mental, 10, 20, 32, 158, 159, 176–179, 195, 196, 234, 247–249, 254–261, 263–265, 268–270, 279, 286, 291, 292, 297, 299
  motive, 67, 84
  nutritive, 11, 12, 76
  ordering of, 4, 8, 12, 13
  perceptual, 3, 8, 11, 12, 85, 90, 247
  proliferation of, 12, 31, 41
  psychic, 31, 32, 58, 69, 72, 77, 93

  psychological, 90
  rational, 3–6, 8, 30, 76, 79, 92, 93, 97, 101, 102, 105, 108, 110, 112–114, 119, 127, 142
  sensory, 6, 92
  vegetative, 6, 92, 97, 103, 104, 119, 126, 127, 156, 178, 265n21
  visual, 3, 9, 12–14, 120, 121, 123, 131, 132, 134, 135, 265
fancy / fantasy, see imagination
form, 4, 8, 15n13, 22, 24, 29, 35–37, 39, 40, 43, 52, 57, 70–73, 77, 81, 82, 84, 89, 90n34, 91–93, 99, 100, 111, 112, 117–120, 121n50, 122, 123, 129, 130, 131, 133, 134, 136–139, 145, 149, 150, 151, 157, 164, 174, 189, 200, 202, 203, 205, 215, 217, 219, 225, 226, 228–230, 237, 240, 241
  a priori, 202, 203, 206, 208, 210–213
  essential, 33–38, 44
  sensible, 77, 90, 219
  sensory, 89
  substantial, 72, 82, 99, 120, 121n50, 126, 127, 129–138
function, 4, 14, 29, 35, 37, 50–53, 70, 72, 73, 80, 110, 122, 124, 125, 136, 141, 145, 159, 196, 199, 205, 208–209, 217, 229, 244, 255, 257, 271–273, 279, 281, 285, 287, 289, 292, 296
  a priori, 216
  bodily, 44, 91, 93
  cognitive, 302
  higher animal, 75, 76
  higher-order, 93
  lower-grade, 94
  psychic, 42, 52, 70
  rational, 73, 95

gender, 61, 147–149, 250
Greek philosophy, 19–22

heart, 59, 61, 62, 65, 88, 90, 121, 142, 148
hermeticism, 140, 142, 146
homunculus fallacy, 104, 122
humoral pathology, 142
hylomorphism, 34–38, 44, 71, 99, 100, 120, 136, 151, 152, 155, 157

INDEX OF CONCEPTS

identity, 35, 52, 64, 105–108, 137, 300
　female, 64
　cultural, 65
　theory, 104–106, 113, 116
imagination, 41, 75, 82, 83, 85, 93, 140–149, 151, 157, 158–160, 164–166, 170, 172, 175, 183, 184, 185n70, 186–190, 192, 195, 196, 208, 212–218, 221, 225, 226, 242, 243, 255, 258, 265, 279
　compositive, 85
intellect, 6, 10, 43, 78–80, 84–87, 90, 91, 96, 98, 103–105, 116–121, 125–129, 131, 132, 151, 153, 157, 159–165, 167, 170, 175, 179, 180, 182, 183, 186, 187, 195–197, 200, 221, 229, 240, 297, 264n21, 282, 285
　agent, 81, 85, 86, 91
　divine, 240
　practical, 87
　theoretical, 87
intellection, 67, 79, 84–86, 96
intentionality, 201, 233
internal dialogue / debate, 59, 60, 61, 64, 65
intuition, 54, 192, 204, 205, 207, 209, 210, 212, 213, 215, 217, 218, 220, 225, 226, 228, 231, 236, 240, 245
　forms of, 203, 210, 211, 219
　intellectual, 200, 219–221, 227, 228, 231, 240, 241, 244–246

judgment, 29n21, 52n63, 56, 58, 113, 118, 121, 159, 160, 165, 167, 175, 185, 204, 205, 212, 214, 219, 221n35, 223, 226, 229, 231, 234, 236, 244, 245, 290

Kantianism, 6, 200, 204, 206–208, 211n21, 219, 225, 231–233, 235–237, 244, 245

Latin philosophy, 5, 15, 58, 76

magic, *see* hermeticism
matter, 34–37, 40, 45, 51, 69, 71, 73, 80, 81, 85, 89, 90n34, 93, 99, 108, 110, 120, 121, 124–125, 130, 134, 143, 257, 176, 179, 180, 224

mechanism, 8, 99, 139, 151, 156, 174, 227, 228, 230, 231, 266, 267–274, 276, 278–280, 284, 285, 290, 291, 293, 295, 296, 298
　mechanist physics, 150, 155
melancholy, 142, 144
memory, 5, 31, 75, 143, 151, 157–159, 164, 165, 172, 175, 184, 185n70, 186, 187, 189, 192, 195, 248, 255–258, 263, 265, 275, 279, 289, 293, 299, 302
mind, 7–10, 13, 17, 19, 42n42, 59, 98, 121n50, 129, 137, 138, 141–143, 146, 149, 157–164, 166, 169–180, 182, 183, 185, 186, 190, 191, 194, 196, 197, 199, 203, 208, 209, 213, 220, 227, 242–244, 248, 254, 255, 257–274, 276–279, 281–298
　bundle theory of, 171, 194
　philosophy of, 7, 38
mind-body dualism, 121n50
modularity, 71, 93, 265, 267–271, 280, 284, 285, 291–297
　massive, 268–271, 283–285, 291
　moderate, 267–271
module, 7, 42, 265, 267–271, 274–279, 281–298, 302
　informationally encapsulated, 274
monism, 52, 164, 165, 172

naturalism, 96, 164, 269, 297
nature, 4–6, 22, 27, 29, 32, 34, 42–44, 45n50, 52, 54, 56, 57, 69–71, 73, 77, 81–83, 86, 87, 93, 105, 113, 129, 144, 152, 153, 162–164, 168, 169, 172–179, 181–183, 188, 206, 210, 221–225, 228, 230, 231, 244, 245, 249, 256, 259, 262, 263, 269, 280, 300
　causal, 32
　incorporeal, 92, 96
　intrinsic, 57
　rational, 4, 6, 29, 30
Neoplatonism, 142, 148
neuroenhancement, 299, 300–306
neuroscience, 261n14, 291, 292, 295, 296, 306
　neuroscientific advances, 304

occasionalism, 173–179, 181, 182
organology, 249

passion, 85, 112, 141, 146, 187, 191, 265
phantasy, *see* imagination
phrenology, 13n12, 247–251, 253, 254, 268
Platonism, 80
potency / potential / potentiality, 20, 32–34, 42, 68–70, 72–74, 84n28, 85, 86, 91n36, 93, 96, 106, 115, 144, 145, 148, 149, 154, 161, 162, 232, 249, 267
   potential state, 106, 115, 116
power, 8, 11, 19–24, 28, 30, 32–35, 37, 39n38, 46–49, 51, 53, 56, 72–75, 80, 85, 86, 92–94, 96, 108, 115, 122, 123, 129–132, 135, 136, 138, 140, 141, 145, 149, 154, 155, 159, 161–164, 167, 169, 171–173, 175–177, 179, 181–184, 187–189, 192–197, 248, 255–259, 261, 264, 265n21
   causal, 23n6, 28, 30, 100, 101, 103, 104, 107, 113, 155, 177, 181, 196
   cognitive, 28n17, 177, 256
   natural, 11
   of judgment, 159, 219, 221n, 223, 229, 231, 234, 244
   physiological, 20, 23n6
   rational, 46n52, 74
   principle, 4, 31, 36, 38–42, 44n49, 51, 69–72, 74, 79, 82, 88, 95, 98, 100, 101, 107–109, 113, 114, 117, 120, 131, 132, 134–138, 156, 164n32, 186, 188–190, 192, 217, 221, 223, 226, 228–230, 232, 235, 236, 238, 256, 284, 296, 303
   abstract, 136, 137
   biological, 4, 5
   Empedoclean, 72
   homonymy, 36, 111
   metaphysical, 84
   of motion and rest, 68
   of operation, 98–100, 108
   of opposites, 26, 27, 31, 47n56
   transcendent, 87
   universal, 133, 188
process, 9, 37, 73, 117, 118n43, 138, 139, 155, 156, 172, 191, 205, 210, 212–214, 218, 224, 235, 241, 242, 256, 258, 260, 263, 266–268, 270, 274, 275, 277, 278, 281–283, 285, 287–290, 297, 302, 303, 305, 306

abstractive, 67, 81, 82
causal, 14, 135
cognitive, 15, 76, 117, 119, 151, 152, 271, 280–282, 288, 294
   natural, 93
   physical, 46, 73, 174
   physiological, 9, 156
proprium, 109, 110
psychology, 7, 36n33, 47, 54, 71, 73, 75, 82, 86, 151, 249, 251, 257, 258, 259, 260, 261, 262, 278, 280, 286
   cognitive, 271
   Epicurean, 50n60
   evolutionary, 284, 288n55, 291, 292, 293
   faculty, 15, 54, 82, 96, 256, 259, 264, 267, 271, 272, 277–280, 289, 290
   philosophical, 44, 66, 76n14
   Stoic, 50n, 51
psychopharmacology, 299, 300, 301, 304

quality, 72, 90n, 106, 109, 112, 117, 126, 141, 214
   essential, 110
   necessary, 117

racism
   scientific, 253
reason, 12, 52, 53, 76, 77, 85, 140–143, 146, 147, 164n32, 167, 171, 185, 186, 190, 199–201, 219, 221, 223, 234, 243, 255, 261, 265n21, 267, 276, 277, 281, 283–286, 290, 291, 297
   disembodied, 71
   faculty of, 73, 93, 195, 196, 221n35, 226
   metaphysical, 106, 112, 135
receptivity, 84, 89, 199, 203, 205–211, 213
reductionism, 47, 50, 72, 119, 121, 154, 162, 176, 182, 192, 196
relatives, 23, 24, 54, 64
rhetoric, 141, 147

scholasticism, 6, 15, 66, 76n14, 97–101, 104, 130, 134–136, 150, 152, 154, 155, 157, 158, 160, 161n24, 162, 177, 183
self-debate, 62–64

self-preservation, 169, 171, 172
sensation, 71n7, 73, 78, 79, 82–84, 91n36, 142, 183, 184, 187, 191, 210
sense-perception, 41, 67, 80, 84, 93
senses, 53, 82, 86, 90, 142n6, 147, 160, 186–188, 190, 191, 208–211, 212n22, 213, 273, 279, 299
    inner, 67, 81n23, 143, 151
    internal, 5n2, 75, 76, 81, 85, 88
sensibility, 199, 201, 203–206, 208, 211, 212, 215, 217–222, 230, 232–234, 243
soul, 4, 6, 7, 20, 23–32, 34–49, 50–54, 56–58, 68, 69, 71, 74, 79, 81, 83, 84, 86, 90–93, 95, 96, 115–133, 135–139, 149, 152, 156–158, 160, 163, 176–178, 180, 191, 208, 212, 255, 257, 264, 265, 267, 269
    animal, 75, 80, 92, 94, 156
    division of the, 47, 96
    essence of the, 103, 105–109, 114–116, 128
    human, 25, 29, 36, 37, 43, 46, 52, 75, 86, 94, 142, 143
    hylomorphic, 36, 37, 90
    imaginative, 71
    nutritive, 72, 83
    parts of the, 25–30, 41, 42n42, 43n46, 45–49, 50, 51, 53, 56, 57, 74, 76, 100, 151, 160
    rational, 30, 57, 79–81, 86, 92, 156, 157
    sensitive, 80, 143, 147, 156
    sensory, 84
    tripartite, 23–25, 28–29, 30n22, 67, 74
    unity of the, 29, 42, 44, 157, 161n24
    vegetative, 72, 73, 75
spirit, 74–76, 79, 80n20, 90, 92n39, 140, 245
spontaneity, 199, 203, 208–210, 213, 217
Stoicism, 49–53, 75, 141
substance, 9, 49–51, 54–56, 69, 70, 86, 89, 91, 95, 96, 101, 102, 106, 109, 112, 129, 130n68, 142, 157, 158, 163, 164, 171, 173, 194, 214, 218, 220, 258, 282, 299, 301

animate, 93
human, 77
immaterial, 6, 92
living, 72
supernaturalism, 72, 130, 146, 174, 175, 180, 227, 228, 231, 269
synopsis, 208–211, 212n22
synthesis, 199, 205, 206, 208–210, 212–218, 225–231, 241, 242, 245

taxonomy, 161, 171, 181, 251, 254, 278, 279, 281
    functional, 273
teleology, 16, 42–44, 113, 114, 126, 222, 223, 224n38, 225–231, 244, 245
transcendence, 141, 174, 198, 200, 203, 206–208, 210, 214–216, 218, 221, 233, 235, 242
transcendental philosophy, 198, 199, 232, 237, 243, 244

understanding, 7, 21, 56, 65, 69, 70n5, 71, 76, 87, 88, 91n36, 98, 108, 143, 151, 158–160, 164, 173, 176, 178, 183, 185–187, 190, 199–206, 209–212, 214–215, 217–221, 222–226, 228–234, 261, 264, 270, 282, 297
    discursive, 219, 220–226, 230, 233, 234
    intuitive, 200, 220, 222, 226–231, 245, 246
unity, 29, 36, 42, 44, 91, 92, 101, 126, 128, 136, 157, 158, 161n, 165, 205, 208, 216–218, 227, 228, 231, 234, 239, 240, 242, 297
    functional, 122, 123, 158

will, 6, 10, 11, 97, 98, 101, 103, 105, 108n25, 116–119, 122, 125–129, 132, 151, 157–165, 167–170, 175, 176, 179–182, 184–187, 195, 196, 277, 279, 285, 288, 297
    agent's, 84n28
    free, 130, 168, 288
    holy, 221, 222, 227, 232

R.A. does not assum[e] to "othe" S[i,]
at b.diy affairs, but only as a desir[e]
which does not in flew the q.ip?